AF389618

The 13th EFITA International Conference

The 13th EFITA International Conference

Editors

Dionysis Bochtis
Maria Lampridi
Charisios Achillas
Dimitrios Aidonis
Dimitrios Kateris
Aristotelis Christos Tagarakis
Ilias Platis
Lefteris Benos

MDPI • Basel • Beijing • Wuhan • Barcelona • Belgrade • Manchester • Tokyo • Cluj • Tianjin

Editors

Dionysis Bochtis
Institute for Bio-Economy and
Agri-Technology (IBO),
Centre of Research and
Technology-Hellas (CERTH)
Greece

Maria Lampridi
Institute for Bio-Economy and
Agri-Technology (IBO),
Centre of Research and
Technology-Hellas (CERTH)
Greece

Charisios Achillas
Department of Supply Chain
Management, International
Hellenic University
Greece

Dimitrios Aidonis
Department of Supply Chain
Management, International
Hellenic University
Greece

Dimitrios Kateris
Institute for Bio-Economy and
Agri-Technology (IBO),
Centre of Research and
Technology-Hellas (CERTH)
Greece

Aristotelis Christos Tagarakis
Institute for Bio-Economy and
Agri-Technology (IBO),
Centre of Research and
Technology-Hellas (CERTH)
Greece

Ilias Platis
Department of Agrotechnology,
School of Agricultural Sciences,
University of Thessaly
Greece

Lefteris Benos
Institute for Bio-Economy and
Agri-Technology (IBO),
Centre of Research and
Technology-Hellas (CERTH)
Greece

Editorial Office
MDPI
St. Alban-Anlage 66
4052 Basel, Switzerland

This is a reprint of articles from the Proceedings published online in the open access journal Engineering Proceedings (ISSN: 2673-4591) (available at: https://www.mdpi.com/2673-4591/9/1).

For citation purposes, cite each article independently as indicated on the article page online and as indicated below:

LastName, A.A.; LastName, B.B.; LastName, C.C. Article Title. *Journal Name* **Year**, *Volume Number*, Page Range.

ISBN 978-3-0365-3643-9 (Hbk)
ISBN 978-3-0365-3644-6 (PDF)

Contents

About the Editors

Dionysis Bochtis works in the field of engineering for agricultural production under enhanced ICT, automation, and robotics technologies. His research/academic track record includes being appointed to several positions, such as Director of the Institute for Bio-economy and Agri-technology (IBO — CERTH); Professor (Agri-Robotics) at the University of Lincoln, U.K.; and Senior Scientist (Operations Management), Aarhus University, Denmark.

Maria Lampridi is a Mechanical Engineer focusing, through her MSc and PhD studies, on Environmental Management and Sustainability. Her main research areas include lifecycle assessment of agricultural production systems and sustainability assessment of advanced agri-technologies as parts of these systems. Currently, she is a research assistant in the Institute of Bio-economy and Agri-technology (iBO) of the Center for Research and Technology Hellas (CERTH).

Charisios Achillas is a Mechanical Engineer with a PhD in reverse logistics, working in the fields of environmental management, sustainable development and circular economy with a primary focus on reverse logistics and agri-chains. Currently, he is an Assistant Professor at the International Hellenic University, Department of Supply Chain Management, and a collaborating senior researcher at the Institute for Bio-Economy and Agri-Technology (iBO/CERTH) and the Laboratory of Heat Transfer and Environmental Engineering (Aristotle University of Thessaloniki).

Dimitrios Aidonis received his Diploma in Mechanical Engineering and a Ph.D. from the Department of Mechanical Engineering of Aristotle University of Thessaloniki. He also holds an MBA from Staffordshire University, U.K. His work deals mainly with Supply Chain, Reverse Logistics Management and Green Logistic. He is currently an Assistant Professor at the Department of Logistics of the Technological Educational Institute of Central Macedonia, Greece. In parallel, he works as a visitor Professor at the Hellenic Open University.

Dimitrios Kateris works in the fields of Agricultural Robotics, Digital Agriculture, Agricultural Engineering, Decision Support Systems (DSS), Operations Management, Information Systems, Automation Systems in Agriculture and Machine Learning. In 2015, he received his PhD in Agricultural Engineering from Aristotle University of Thessaloniki. Currently, he is an Assistant Researcher (Grade C) at the Institute for Bio-economy and Agri-technology (iBO) of the Centre for Research and Technology Hellas (CERTH).

Aristotelis Christos Tagarakis works in the fields of precision agriculture, remote sensing and sensor networks in agriculture. He received his MSc degree in "Automation in Irrigation, Farm Structures and Farm Mechanization" and his PhD degree in precision agriculture from the University of Thessaly, Greece. He is currently an Assistant Researcher (Grade C) at the Institute for Bio-economy and Agri-technology (iBO/CERTH) with expertise in precision agriculture.

Ilias Platis works in the laboratory of Technology of Aromatic-Medicinal plants and arable crops in University of Thessaly, Department of Agri-technology. He graduated (B.Sc.) from the Technological Educational Institute of Larissa, School of Agriculture, Department of Crop Production (1999). He holds an M.Sc. from the University of Thessaly in the Postgraduate Program "Integrated Management of Aromatic and Medicinal Plants" (2018).

Lefteris Benos is a physicist with a M.Sc. and a Ph.D. in Mechanical Engineering. He has expertise in modelling magnetohydrodynamic flows and the heat transfer of liquid metals and nanofluids and a strong background on finite element analysis. Since 2018, he has been working as a Postdoctoral Researcher in iBO/CERTH, focusing on the fields of Agricultural Ergonomics, Biomechanics and CFD.

Editorial

The Cutting Edge on Advances in ICT Systems in Agriculture [†]

Maria Lampridi [1], Lefteris Benos [1], Dimitrios Aidonis [2], Dimitrios Kateris [1], Aristotelis C. Tagarakis [1], Ilias Platis [3], Charisios Achillas [2] and Dionysis Bochtis [1,*]

[1] Centre of Research and Technology-Hellas (CERTH), Institute for Bio-Economy and Agri-Technology (IBO), 6th Km Charilaou-Thermi Rd., GR 57001 Thessaloniki, Greece; m.lampridi@certh.gr (M.L.); e.benos@certh.gr (L.B.); d.kateris@certh.gr (D.K.); a.tagarakis@certh.gr (A.C.T.)

[2] Department of Supply Chain Management, International Hellenic University, GR 60100 Katerini, Greece; daidonis@ihu.gr (D.A.); c.achillas@ihu.edu.gr (C.A.)

[3] Department of Agritechnology, School of Agricultural Sciences, University of Thessaly, Nea Ionia, GR 38446 Volos, Greece; i.platis@certh.gr

[*] Correspondence: d.bochtis@certh.gr

[†] Presented at the 13th EFITA International Conference, online, 25–26 May 2021.

Citation: Lampridi, M.; Benos, L.; Aidonis, D.; Kateris, D.; Tagarakis, A.C.; Platis, I.; Achillas, C.; Bochtis, D. The Cutting Edge on Advances in ICT Systems in Agriculture. *Eng. Proc.* **2021**, *9*, 46. https://doi.org/10.3390/engproc2021009046

Published: 11 March 2022

Publisher's Note: MDPI stays neutral with regard to jurisdictional claims in published maps and institutional affiliations.

Modern agriculture has to shoulder the burden of a plethora of challenges associated with demographics, climate change, and natural resources depletion [1], as well as challenges associated with a new socio-technical framework [2,3]. Therefore, there is a need to increase the effectiveness of agricultural practices and sustainability performance. Current breakthrough technologies are capable of strengthening agriculture for addressing rising needs worldwide. Information and Communication Technology (ICT) has become an integral element of Agriculture 4.0. ICT is a concise term for describing any device, software, application, or network that allows for data collection, exchange, and communication [4]. An illustrative instance of how ICT is incorporated in modern agriculture is provided in Figure 1. The procedure begins by collecting the required data from the field ("Sensors" Phase). Once the data have been gathered, a crucial process is the extraction of important information from them ("Data" Phase). Subsequently, this information is exploited during decision-making for management operations ("Decision-making" Phase). Finally, the essential actions are taken in the field ("Action" Phase) based on the decision-making, and then, this cycle of processes starts again.

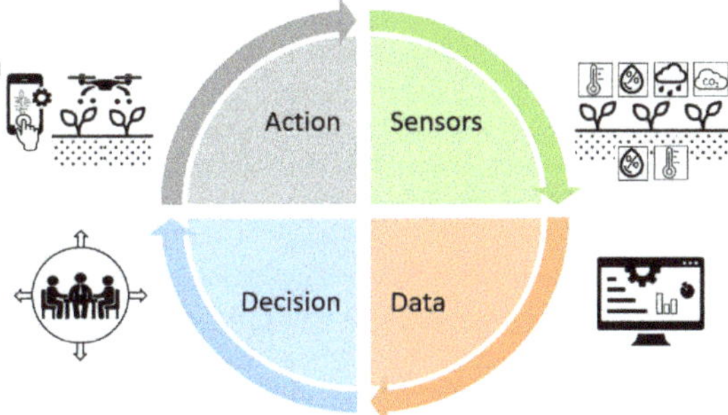

Figure 1. The four phases of modern agriculture through exploiting ICT.

The object of this editorial is to summarize the innovative approaches of applying ICT in agriculture, as presented at the 13th International Conference of the European Federation for Information Technology in Agriculture, Food and the Environment (EFITA). Overall, 45 works were presented and classified into one of the aforementioned phases depicted in Figure 1.

Eng. Proc. **2021**, *9*, 46. https://doi.org/10.3390/engproc2021009046 https://www.mdpi.com/journal/engproc

Regarding the phase associated with sensors, two review studies were presented on the use of modern technologies for monitoring the impact of wind on trees and forests [5], and the combination of the Internet of Multimedia Things, precision agriculture, and agrifood [6], indicating gaps in these fields. Furthermore, Arink et al. [7] utilized visible and near-infrared spectroscopy for a non-destructive quality assessment, while Hahn et al. [8] investigated the use of inductive and capacitive sensors against trunk dendrometer measurements towards optimizing water use. Filintas [9] presented a soil moisture depletion model for maize crops, whereas Hoxha et al. [10] developed a management decision tool for aromatic and medicinal plants based on GPS and historical inventory data. Kateris et al. [11] developed a new image-based technique for weed mapping in vineyards, considering the height of weeds at the inter-row path of vineyards, and Bataka et al. [12] used statistical methodology to compare open-source with industrial weather stations. Tasiopoulos et al. [13] proposed a methodology based on satellite images and machine learning, for accurate flood extent maps, and Aguilar and Cháves [14] focused on teaching–learning strategies at the University of Costa Rica, as a means of supporting geomatic concepts and tools. Finally, Tagarakis et al. [15] acquired kinematic data from wearable sensors during a human–robot field experiment, aiming to identify human activity signatures.

The second phase that pertains to data analysis was the subject of seven papers. Specifically, Mietzsch et al. [16] summarized the recent developments of AGROVOC, a multilingual thesaurus of the Food and Agriculture Organization of the United Nations (FAO). In [17], the quality of the digital infrastructure in agriculture was assessed, as well as issues preventing its adoption in the USA and Germany. Sáenz et al. [18] identified drought periods in Ecuador in 2001–2018, while Stratakis et al. [19] integrated ambient intelligence technologies for two precision agriculture applications. Additionally, Lallas et al. [20] proposed an ontology supporting the monitoring of the illegal wood trade, whereas Common Greenhouse Ontology was developed in [21] by considering domain experts and existing ontologies. Lastly, a system was developed for calf body weight estimation based on depth images in [22].

The phase related to decision-making concerned the majority of the presented papers. In particular, a decision support system to study the spread of *Ailanthus altissima* in particular Greek agro-ecosystems was proposed in [23], while a simple decision support system for soybean yield was presented in [24]. A survey on the Greek business contribution to the 17 Sustainable Development Goals was conducted in [25], and factors affecting ICT adoption constituted the matter of [26]. Additionally, a study by Jablanovic [27] pertained to the benefits of ICT investments in China, while Chiem et al. [28] dealt with the reasons affecting the adoption of rice contract-farming policies. Moreover, machine learning algorithms were utilized in three studies. These algorithms were used for (a) grape ripeness estimation [29], (b) predicting the daily prices of sugar and corn in Brazil [30], and (c) crop water availability mapping in the Danube Basin [31]. Based on circular economy concepts, Tagarakis et al. [32] proposed an integrated system via implementing smart farming tools, while Crovella et al. [33] highlighted a key element towards this transition, namely the collaboration between farmers and stakeholders. Silva et al. [34] proposed a data-driven framework for multi-hazard risk mapping in agriculture, whereas in [35], an automatic monitoring system for rainwater harvesting was proposed. Gigot et al. [36] implemented a methodology for wheat production, and Cicuéndez et al. [37] dealt with the identification and modeling of the gross primary production. Finally, Filintas et al. [38] investigated the effects of two irrigation and fertilization treatments on cotton yield and seed oil.

The studies related to action were those of Hahn et al. [39] and Alexandropoulos et al. [40]. In the former, a remotely controlled seawater fertilizer extraction system was developed, whereas the latter evaluated 15 farm-scale greenhouse gas-based decision support tools based on a number of criteria.

Finally, nine studies were classified as addressing cross-cutting themes. Specifically, Spykman et al. [41] and Gabriel and Gandorfer [42] focused on society's opinion of the use of digital technologies in agriculture. The requirements for adopting blockchain technology and digital technologies by small and medium farms were investigated in [43] and [44], respectively. The role of modern technologies in the implementation of sustainable agriculture was studied in [45], while recommendations for future research pertaining to Earth observation for agricultural applications were presented in [46]. A survey on the importance of interfaces and middleware in agriculture was conducted in [47]. Finally, two studies focused on the molecular and phenotypic diversity of indigenous oenological strains of *Saccharomyces cerevisiae* [48], as well as kiwifruit genotypes and cultivars evaluation for susceptibility to four strains of *Pseudomonas syringae* pv. *actinidiae* (Psa) biovar 3 [49].

In summary, the 45 studies presented at the 13th EFITA conference brought together engineers, scientists, technicians, academics, and industry people for the sake of exchanging knowledge and ideas on the state-of-the-art and future of ICT use in agriculture. As a concluding remark, ICT has the potential to be the driving force towards strengthening agriculture to meet the growing demands for food in a sustainable manner. However, finding ways to facilitate the adoption of new agri-technologies should be prioritized by focusing on farmers' education and information, while socio-economic factors that affect their assimilation in the field must be considered.

Author Contributions: Conceptualization, All; writing—original draft preparation, All; writing—review and editing, All. All authors have read and agreed to the published version of the manuscript.

Conflicts of Interest: The authors declare no conflict of interest.

References

1. Lampridi, M.G.; Sørensen, C.G.; Bochtis, D. Agricultural Sustainability: A Review of Concepts and Methods. *Sustainability* **2019**, *11*, 5120. [CrossRef]
2. Marinoudi, V.; Lampridi, M.; Kateris, D.; Pearson, S.; Sørensen, C.G.; Bochtis, D. The Future of Agricultural Jobs in View of Robotization. *Sustainability* **2021**, *13*, 12109. [CrossRef]
3. Benos, L.; Bechar, A.; Bochtis, D. Safety and ergonomics in human-robot interactive agricultural operations. *Biosyst. Eng.* **2020**, *200*, 55–72. [CrossRef]
4. Benos, L.; Tagarakis, A.C.; Dolias, G.; Berruto, R.; Kateris, D.; Bochtis, D. Machine Learning in Agriculture: A Comprehensive Updated Review. *Sensors* **2021**, *21*, 3758. [CrossRef] [PubMed]
5. Faria, J.S.R.; Silva, R.F.; Brazolin, S.; Cugnasca, C.E. Bibliometric Review on the Use of Internet of Things Technologies to Monitor the Impacts of Wind on Trees and Forests. *Eng. Proc.* **2021**, *9*, 16. [CrossRef]
6. Mourikis, A.I.; Kalamatianos, R.; Karydis, I.; Avlonitis, M. A Survey on the Use of the Internet of Multimedia Things for Precision Agriculture and the Agrifood Sector. *Eng. Proc.* **2021**, *9*, 32. [CrossRef]
7. Arink, M.; Khan, H.A.; Polder, G. Light Penetration Properties of Visible and NIR Radiation in Tomatoes Applied to Non-Destructive Quality Assessment. *Eng. Proc.* **2021**, *9*, 18. [CrossRef]
8. Hahn, F.; Espinoza, J.; Zacarías, U. Mango Leaf Monitoring with Inductive and Capacitive Sensors and Its Comparison with Trunk Dendrometer Measurements. *Eng. Proc.* **2021**, *9*, 28. [CrossRef]
9. Filintas, A. Soil Moisture Depletion Modelling Using a TDR Multi-Sensor System, GIS, Soil Analyzes, Precision Agriculture and Remote Sensing on Maize for Improved Irrigation-Fertilization Decisions. *Eng. Proc.* **2021**, *9*, 36. [CrossRef]
10. Hoxha, V.; Bombaj, F.; Ilbert, H. Construction of an Observatory, As a Management Tool Decision, Valorisation and Sustainable Preservation of the Resources of Aromatic and Medicinal Plants. *Eng. Proc.* **2021**, *9*, 23. [CrossRef]
11. Kateris, D.; Kalaitzidis, D.; Moysiadis, V.; Tagarakis, A.C.; Bochtis, D. Weed Mapping in Vineyards Using RGB-D Perception. *Eng. Proc.* **2021**, *9*, 30. [CrossRef]
12. Bataka, E.P.; Miliokas, G.; Katsoulas, N.; Nakas, C.T. A Method Comparison Study between Open Source and Industrial Weather Stations. *Eng. Proc.* **2021**, *9*, 8. [CrossRef]
13. Tasiopoulos, L.; Stefouli, M.; Voutos, Y.; Mylonas, P.; Charou, E. Machine Learning Techniques in Agricultural Flood Assessment and Monitoring Using Earth Observation and Hydromorphological Analysis. *Eng. Proc.* **2021**, *9*, 40. [CrossRef]
14. Aguilar, J.; Cháves, M. A Review of a Teaching– Learning Strategy Change to Strengthen Geomatic Concepts and Tools in the Biosystems Engineering Academic Studies at the Universidad de Costa Rica. *Eng. Proc.* **2021**, *9*, 44. [CrossRef]
15. Tagarakis, A.C.; Benos, L.; Aivazidou, E.; Anagnostis, A.; Kateris, D.; Bochtis, D. Wearable Sensors for Identifying Activity Signatures in Human-Robot Collaborative Agricultural Environments. *Eng. Proc.* **2021**, *9*, 5. [CrossRef]
16. Mietzsch, E.; Martini, D.; Kolshus, K.; Turbati, A.; Subirats, I. How Agricultural Digital Innovation Can Benefit from Semantics: The Case of the AGROVOC Multilingual Thesaurus. *Eng. Proc.* **2021**, *9*, 17. [CrossRef]
17. Bernhardt, H.; Schumacher, L.; Zhou, J.; Treiber, M.; Shannon, K. Digital Agriculture Infrastructure in the USA and Germany. *Eng. Proc.* **2021**, *9*, 1. [CrossRef]

18. Sáenz, C.; Litago, J.; Wiese, K.; Recuero, L.; Cicuéndez, V.; Palacios-Orueta, A. Drought Periods Identification in Ecuador between 2001 and 2018 Using SPEI and MODIS Data. *Eng. Proc.* **2021**, *9*, 24. [CrossRef]
19. Stratakis, C.; Stivaktakis, N.; Bouloukakis, M.; Leonidis, A.; Doxastaki, M.; Kapnas, G.; Evdaimon, T.; Korozi, M.; Kalligiannakis, E.; Stephanidis, C. Integrating Ambient Intelligence Technologies for Empowering Agriculture. *Eng. Proc.* **2021**, *9*, 41. [CrossRef]
20. Lallas, E.; Karageorgos, A.; Ntalos, G. An Ontology Based Approach for Regulatory Compliance of EU Reg. No 995/2010 in Greece. *Eng. Proc.* **2021**, *9*, 38. [CrossRef]
21. Bakker, R.; van Drie, R.; Bouter, C.; van Leeuwen, S.; van Rooijen, L.; Top, J. The Common Greenhouse Ontology: An Ontology Describing Components, Properties, and Measurements inside the Greenhouse. *Eng. Proc.* **2021**, *9*, 27. [CrossRef]
22. Yamamoto, Y.; Ohkawa, T.; Ohta, C.; Oyama, K.; Nishide, R. Depth Image Selection Based on Posture for Calf Body Weight Estimation. *Eng. Proc.* **2021**, *9*, 20. [CrossRef]
23. Voutos, Y.; Godsil, N.; Sotiropoulou, A.; Mylonas, P.; Bouchagier, P.; Exarchos, T.; Martinis, A.; Kabassi, K. Capturing and evaluating the effects of the expansive species Ailanthus altissima on agro-ecosystems at the Ionian Islands. *Eng. Proc.* **2021**, *9*, 19. [CrossRef]
24. Nitta, A.; Chonan, Y.; Hayashi, S.; Nakamura, T.; Tsuji, H.; Murakami, N.; Nishide, R.; Ohkawa, T.; Ozawa, S. An Easily Installed Method of the Estimation of Soybean Yield Based on Meteorological Environments with Regression Analysis. *Eng. Proc.* **2021**, *9*, 26. [CrossRef]
25. Chrysos-Anestis, A.; Achillas, C.; Folinas, D.; Aidonis, D.; Anestis, M.C. Sensitivity of Greek Organisations in Sustainability Issues. *Eng. Proc.* **2021**, *9*, 4. [CrossRef]
26. Romanelli, T.L.; Muñoz-Arriola, F.; Colaço, A.F. Conceptual Framework to Integrate Economic Drivers of Decision Making for Technology Adoption in Agriculture. *Eng. Proc.* **2021**, *9*, 43. [CrossRef]
27. Jablanovic, V. Investment in Information and Communication Technology in Agriculture and Soybean Production Stability: The Case of China. *Eng. Proc.* **2021**, *9*, 34. [CrossRef]
28. Tuyen, M.C.; Sirisupluxana, P.; Bunyasiri, I.; Hung, P.X. Rice Contract Farming in Vietnam: Insights from a Qualitative Study. *Eng. Proc.* **2021**, *9*, 6. [CrossRef]
29. Vrochidou, E.; Bazinas, C.; Papakostas, G.A.; Pachidis, T.; Kaburlasos, V.G. A Review of the State-of-Art, Limitations, and Perspectives of Machine Vision for Grape Ripening Estimation. *Eng. Proc.* **2021**, *9*, 2. [CrossRef]
30. Silva, R.F.; Barreira, B.L.; Cugnasca, C.E. Prediction of Corn and Sugar Prices Using Machine Learning, Econometrics, and Ensemble Models. *Eng. Proc.* **2021**, *9*, 31. [CrossRef]
31. Migdall, S.; Dotzler, S.; Gleisberg, E.; Appel, F.; Muerth, M.; Bach, H.; Weikmann, G.; Paris, C.; Marinelli, D.; Bruzzone, L. Crop Water Availability Mapping in the Danube Basin Based on Deep Learning, Hydrological and Crop Growth Modelling. *Eng. Proc.* **2021**, *9*, 42. [CrossRef]
32. Tagarakis, A.C.; Dordas, C.; Lampridi, M.; Kateris, D.; Bochtis, D. A Smart Farming System for Circular Agriculture. *Eng. Proc.* **2021**, *9*, 10. [CrossRef]
33. Crovella, T.; Paiano, A.; Lagioia, G.; Cilardi, A.M.; Trotta, L. Modelling Digital Circular Economy framework in the Agricultural Sector. An Application in Southern Italy. *Eng. Proc.* **2021**, *9*, 15. [CrossRef]
34. Silva, R.F.; Fava, M.C.; Saraiva, A.M.; Mendiondo, E.M.; Cugnasca, C.E.; Delbem, A.C.B. A Theoretical Framework for Multi-Hazard Risk Mapping on Agricultural Areas Considering Artificial Intelligence, IoT, and Climate Change Scenarios. *Eng. Proc.* **2021**, *9*, 39. [CrossRef]
35. Brenes, R.; Torres, A.; Aguilar, J.; Aguilar, R. Awareness Raising and Capacity Building through a Scalable Automatic Water Harvest Monitoring System to Improve Water Resource Management in Monteverde Community, Costa Rica. *Eng. Proc.* **2021**, *9*, 45. [CrossRef]
36. Gigot, C.; Hamernig, D.; Deytieux, V.; Diallo, I.; Deudon, O.; Gourdain, E.; Aubertot, J.-N.; Robin, M.-H.; Bancal, M.-O.; Huber, L.; et al. Developing a Method to Simulate and Evaluate Effects of Adaptation Strategies to Climate Change on Wheat Crop Production: A Challenging Multi-Criteria Analysis. *Eng. Proc.* **2021**, *9*, 21. [CrossRef]
37. Cicuéndez, V.; Litago, J.; Sánchez-Girón, V.; Recuero, L.; Sáenz, C.; Palacios-Orueta, A. Identification and Modeling Carbon and Energy Fluxes from Eddy Covariance Time Series Measurements in Rice and Rainfed Crops. *Eng. Proc.* **2021**, *9*, 9. [CrossRef]
38. Filintas, A.; Nteskou, A.; Katsoulidi, P.; Paraskebioti, A.; Parasidou, M. Rainfed and Supplemental Irrigation Modelling 2D GIS Moisture Rootzone Mapping on Yield and Seed Oil of Cotton (Gossypium hirsutum) Using Precision Agriculture and Remote Sensing. *Eng. Proc.* **2021**, *9*, 37. [CrossRef]
39. Hahn, F.; González, C.J.; Delfín, C.M. Production of Fertilizer from Seawater with a Remote Control System. *Eng. Proc.* **2021**, *9*, 29. [CrossRef]
40. Alexandropoulos, E.; Anestis, V.; Bartzanas, T. Farm-Scale Greenhouse Gas Emissions' Decision Support Systems. *Eng. Proc.* **2021**, *9*, 22. [CrossRef]
41. Spykman, O.; Emberger-Klein, A.; Gabriel, A.; Gandorfer, M. Society's View on Autonomous Agriculture: Does Digitalization Lead to Alienation? *Eng. Proc.* **2021**, *9*, 12. [CrossRef]
42. Gabriel, A.; Gandorfer, M. Have City Dwellers Lost Touch with Modern Agriculture? In Quest of Differences between Urban and Rural Population. *Eng. Proc.* **2021**, *9*, 25. [CrossRef]
43. Aranda, R.S.; Silva, R.F.; Cugnasca, C.E. Requirements Identification for a Blockchain-Based Traceability Model for Animal-Based Medicines. *Eng. Proc.* **2021**, *9*, 11. [CrossRef]

44. Sodano, V. Tailored Digitization for Rural Development. *Eng. Proc.* **2021**, *9*, 14. [CrossRef]
45. Popescu, L.; Safta, A.S. Analysis of the Concept of Feasibility in Sustainable Agricultural Systems. *Eng. Proc.* **2021**, *9*, 13. [CrossRef]
46. Charvat, K.; Safar, V.; Kubickova, H.; Horakova, S.; Mildorf, T. Strategic Research Agenda for Utilisation of Earth Observation in Agriculture. *Eng. Proc.* **2021**, *9*, 35. [CrossRef]
47. Treiber, M.; Bernhardt, H. NEVONEX—The Importance of Middleware and Interfaces for the Digital Transformation of Agriculture. *Eng. Proc.* **2021**, *9*, 3. [CrossRef]
48. Karampatea, A.; Tsakiris, A.; Kourkoutas, Y.; Skavdis, G. Molecular and Phenotypic Diversity of Indigenous Oenological Strains of Saccharomyces cerevisiae Isolated in Greece. *Eng. Proc.* **2021**, *9*, 7. [CrossRef]
49. Thomidis, T.; Goumas, D.E.; Zotos, A.; Triantafyllidis, V.; Kokotos, E. Susceptibility of Twenty-Three Kiwifruit Cultivars to Pseudomonas syringae pv. actinidiae. *Eng. Proc.* **2021**, *9*, 33. [CrossRef]

Proceeding Paper

Bibliometric Review on the Use of Internet of Things Technologies to Monitor the Impacts of Wind on Trees and Forests [†]

José S. R. Faria [1,*], Roberto F. Silva [2], Sérgio Brazolin [3] and Carlos E. Cugnasca [1]

[1] Polytechnic School, University of São Paulo (USP), São Paulo 05508-010, Brazil; carlos.cugnasca@usp.br
[2] Institute of Mathematics and Computer Sciences (ICMC), University of São Paulo (USP), São Carlos 13566-590, Brazil; roberto.fray.silva@gmail.com
[3] Infrastructure and Environment Center, Institute for Technological Research (IPT), São Paulo 05508-901, Brazil; brazolin@ipt.br
* Correspondence: josesindefaria@usp.br
† Presented at the 13th EFITA International Conference, online, 25–26 May 2021.

Abstract: The presence of trees brings several health benefits to urban populations. However, wind damage is an important cause of falling trees, causing considerable damages. This study involved a bibliometric review on the use of Internet of Things technologies for monitoring trees. A research protocol was designed and implemented, involving a thorough search of the Scopus database. After applying the exclusion criteria and content filters, the abstracts and titles of the resulting 313 documents were analyzed. Two analyses were performed; (i) an analysis of the evolution of the area based on the study metadata; (ii) a cluster analysis of the words present in the abstracts and titles of the identified documents. The first analysis showed: (i) the current growth of this area of research; (ii) that the most important fields of study were agricultural, biological, environmental, and terrestrial and planetary sciences; (iii) that the most relevant journal was *Ecology and Forest Management*. The second analysis resulted in the identification of three clusters: (i) wind impact; (ii) variables and experiments; (iii) forest management. The main gap observed was that few studies have used IoT technologies as tools for preventive or corrective actions related to wind and storm impacts on trees and forests.

Keywords: bibliometric review; Internet of Things; monitoring; tree; windthrow

Citation: Faria, J.S.R.; Silva, R.F.; Brazolin, S.; Cugnasca, C.E. Bibliometric Review on the Use of Internet of Things Technologies to Monitor the Impacts of Wind on Trees and Forests. *Eng. Proc.* **2021**, *9*, 16. https://doi.org/10.3390/engproc2021009016

Academic Editors: Charisios Achillas and Lefteris Benos

Published: 24 November 2021

Publisher's Note: MDPI stays neutral with regard to jurisdictional claims in published maps and institutional affiliations.

1. Introduction

There are several advantages to the afforestation of cities. Some of those advantages are: (i) reductions of more than 90% in terms of sun incidence, consequently reducing temperature and light exposure [1]; (ii) absorption of carbon dioxide [2]; (iii) filtration of particulate matter (which may cause lung diseases) [2]; (iv) oxygen release [2]. These aspects help to improve a city's air quality. Inadequate maintenance of urban trees can result in trees falling, damaging property and people [3], both in rural and urban areas [4]. In urban areas, fallen trees can bring the urban infrastructure to a standstill, limiting the access and use of buildings, streets, pipelines, and public transportation, among others, and can expose lives and properties to several risks [3,5]. The monitoring of trees using Internet of Things (IoT) technologies can assist in analyzing and predicting the risk of falling trees in urban environments, making it possible to collect and store useful data in real time using wireless sensor networks [6]. The concept of IoT can be described as the combination of sensors, actuators, and network devices in a unique relationship and environment. These objects are considered intelligent and can store and transmit essential information to improve decision-making [7]. IoT technologies allow the connection between intelligent objects and people, collecting and transmitting information anywhere, anytime, and through any medium [8]. This study aims to conduct a bibliometric review on the use of IoT technologies to monitor the impacts of wind on trees and forests. The main studies, keywords, journals, and events will be identified. We will also identify and describe the main keyword clusters and how they are associated in the relevant knowledge domains.

2. Methodology

This article uses the bibliometric review methodology [9], following the study by Silva [10]. The methodology used was composed of six steps: (1) Keyword identification, based on an in-depth literature review and the study by [11]. The keywords used were: trees, forest, wind, windthrow, windstorms, drag, IoT, accelerometer, and anemometer. (2) Database selection and research strategy, using the Scopus database [12]. (3) Definition of the research protocol and exclusion criteria. The following filters were used: language (English), subject areas (agriculture, among others), and period (2010 to 2020). (4) Evaluation of the resulting works, based on the titles, keywords, and abstracts. (5) Quantitative analysis and identification of the most important works, events, journals, and the evolution of the use of IoT for the monitoring of trees and forests. (6) Clustering and data visualization, identifying the main keywords in each cluster. For the cluster analysis, we used the VosViewer software [13].

3. Results and Discussions

Before the application of the exclusion criteria and filters, the keywords search resulted in 1031 documents. Then, the content filters were applied, resulting in 440 documents. The exclusion criteria were then applied to eliminate non-relevant papers, resulting in 313 documents. These were then analyzed. Figure 1 illustrates the growth in the number of published studies throughout the years in this final dataset.

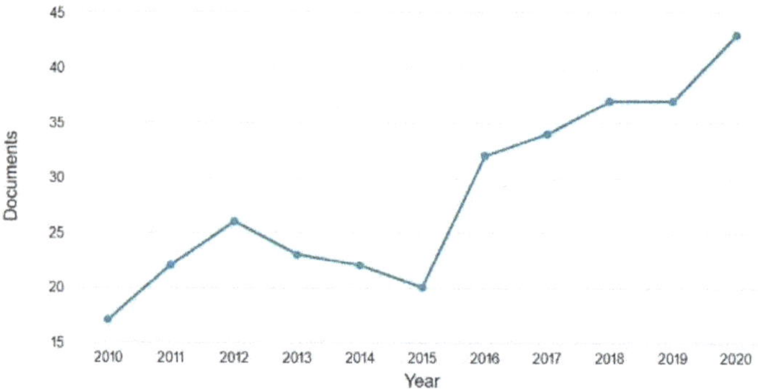

Figure 1. Number of published studies per year in the final dataset.

It can be observed in Figure 1 that the number of relevant published studies has increased considerably since 2015. This demonstrates an increasing interest by the academic community in studying the impacts of wind on trees and forests. The institutions, foundations, and countries with the most cited documents were from developed nations, such as the United States and countries in the European Union. Two areas of study prevailed in the surveyed documents, namely agricultural and biological sciences and environmental science, both linked to the rural and urban environments. Based on the analysis of the main sources of the studies in the final dataset, it is important to note that: (i) journals presented a considerably higher number of citations; (ii) the most relevant journal was *Forest Ecology and Management*; (iii) the most relevant conference was the 6th Intl Conf. on Internet of Things: Systems, Mgmt., and Security (IOTSMS). These observations are essential for researchers, pointing out relevant journals and conferences that should be considered for identifying or publishing research in the researched domain. A cluster analysis was then conducted. The following hyperparameters were considered: (i) the clustering was conducted on the titles and abstracts of the relevant works; (ii) only words that appeared more than fifteen times in the text corpus were considered. These hyperparameters were selected based on a series of experiments with different values and configurations. The final cluster analysis identified 37 keywords, divided into three clusters. Figure 2 illustrates the clusters identified and their related keywords. In the three clusters, the researchers were looking for ways to identify and reduce the damage caused by wind in the trees and forests, focusing on the rural environment.

Figure 2. Visualization of the results of the cluster analysis in the final dataset.

The first cluster was called "wind impact". This cluster represented (i) the impacts of wind and storms on forests and trees and (ii) how the wind, in its various forms, leads to tree rupture. This cluster's main keywords are shown in Figure 2. The majority of the studies were related to the rural environment. The second cluster was called "variables and experiments". This cluster represented several experiments carried out in the field and laboratories to estimate the impacts of wind and storms on trees and forests. Additionally, it also considered the modeling of tree species and their behavior when impacted by wind and storms. Its main keywords are shown in Figure 2. The third cluster was called "Forest management". It was related to the different strategies and methodologies used for managing the damage caused by wind and storms in forests. Therefore, its focus was on dealing with the events after they have happened, mitigating their impacts. Additionally, a number of papers dealt with risk analysis, using this approach to evaluate tree rupture problems. The main keywords are shown in Figure 2.

It is vital to note that we found few studies in cities and urban environments. However, the results obtained for rural environments and the models used for monitoring the impacts using IoT technologies could be adapted to the urban environments. Nevertheless, this would demand an assessment of the different models and technology. The study by [14] is one of the most relevant ones related to the use of the IoT paradigm and its technologies for monitoring trees and forests, focusing on the detection of forest fires. However, it can also be adapted to monitor the impacts of wind and storms on forests. The main limitations observed in the present study were: (i) the lack of studies considering trees and forests typical of the southern hemisphere; (ii) the lack of studies using IoT technologies to monitor the impacts of wind and storms on trees and forests.

4. Conclusions and Future Works

There are numerous benefits to having trees and forests in urban and rural areas. However, the impacts of inadequate maintenance may lead to trees falling, causing enormous asset and personal damages. In order to tackle this issue, the main studies in the literature that used IoT technology to monitor trees in urban and rural areas were identified in this study. It was observed that the number of relevant studies has been increasing considerably since 2015, showing the relevance of the topic. The cluster analysis showed that three main clusters exist, related to: (i) wind impacts; (ii) experimentation in laboratories and fields; (ii) forest management, considering risk mitigation and analysis. These were analyzed, as well as their main keywords. It was observed that most studies have focused on developed countries' species and rural environments. Additionally, only a few studies have used the IoT paradigm and the related technologies to deal with this problem. Future studies should: (i) consider the impacts of wind and storms on trees and forests typical of the southern hemisphere; (ii) adapt the current monitoring methodologies and models for the urban environment; (iii) explore the use of IoT technologies to monitor the impacts of wind and storms on trees and forests in real time.

Author Contributions: Conceptualization, J.S.R.F. and R.F.S.; methodology, J.S.R.F. and R.F.S.; investigation, J.S.R.F.; writing—original draft preparation, J.S.R.F.; writing—review and editing, R.F.S. and C.E.C.; supervision, C.E.C. and S.B. All authors have read and agreed to the published version of the manuscript.

Funding: This study was supported by FAPESP grant number 2020/01673-2, Itaú Unibanco S.A., through the Itaú Scholarship Program, at the Centro de Ciência de Dados (C2D), Universidade de São Paulo, Brazil, and Instituto de Pesquisas Tecnológicas do Estado de São Paulo (IPT).

Institutional Review Board Statement: Not applicable.

Conflicts of Interest: The authors declare no conflict of interest.

References

1. Buckeridge, M. Árvores urbanas em São Paulo: Planejamento, economia e água. *Estud. Av.* **2015**, *29*, 85–101. [CrossRef]
2. Livesley, S.J.; McPherson, E.G.; Calfapietra, C. The Urban Forest and Ecosystem Services: Impacts on Urban Water, Heat, and Pollution Cycles at the Tree, Street, and City Scale. *J. Environ. Qual.* **2016**, *45*, 119–124. [CrossRef] [PubMed]
3. Brazolin, S.; Tomazello Filho, M.; Yojo, T.; de Oliveira Neto, M.A.; Albuquerque, Á.R.; Sette Júnior, C.R. Propriedades físico-mecânicas do lenho deteriorado por fungos apodrecedores de árvores de Tipuana tipu. *CERNE* **2015**, *20*, 183–190. [CrossRef]
4. Danquechin Dorval, A.; Meredieu, C.; Danjon, F. Anchorage failure of young trees in sandy soils is prevented by a rigid central part of the root system with various designs. *Ann. Bot.* **2016**, *118*, 747–762. [CrossRef] [PubMed]
5. Hou, G.; Chen, S. Probabilistic modeling of disrupted infrastructures due to fallen trees subjected to extreme wind in urban community. *Nat. Hazards* **2020**, *102*, 1323–1350. [CrossRef]
6. de Carvalho Junior, A.D.; Hariki, V.S.; Goldman, A. Sensing Trees in Smart Cities with Open-Design Hardware. In Proceedings of the IEEE 17th International Symposium on Network Computing and Applications (NCA), Boston, MA, USA, 31 October–2 November 2018; pp. 1–4.
7. Gubbi, J.; Buyya, R.; Marusic, S.; Palaniswami, M. Internet of Things (IoT): A vision, architectural elements, and future directions. *Futur. Gener. Comput. Syst.* **2013**, *29*, 1645–1660. [CrossRef]
8. Shammar, E.A.; Zahary, A.T. The Internet of Things (IoT): A survey of techniques, operating systems, and trends. *Libr. Hi Tech* **2019**, *38*, 5–66. [CrossRef]
9. Valérie, D.; Pierre, A.G. Bibliometric idicators: Quality masurements of sientific publication. *Radiology* **2010**, *255*, 342–351.
10. Silva, T.C.; Araujo, E.C.G.; Lins, T.R.D.S.; Reis, C.A.; Sanquetta, C.R.; Rocha, M.P.D. Non-Timber Forest Products in Brazil: A Bibliometric and a State of the Art Review. *Sustainability* **2020**, *12*, 7151. [CrossRef]
11. Koizumi, A.; Motoyama, J.; Sawata, K.; Sasaki, Y.; Hirai, T. Evaluation of drag coefficients of poplar-tree crowns by a field test method. *J. Wood Sci.* **2010**, *56*, 189–193. [CrossRef]
12. Baas, J.; Schotten, M.; Plume, A. Scopus as a curated, high-quality bibliometric data source for academic research in quantitative science studies. *Quant. Sci. Stud.* **2020**, *1*, 377–386. [CrossRef]
13. VOSviewer. Centre for Science and Technology Studies. Leiden University. Available online: https://www.vosviewer.com/ (accessed on 16 January 2020).
14. Vega-Rodriguez, R.; Sendra, S.; Lloret, J.; Romero-Diaz, P.; Garcia-Navas, J.L. Low Cost LoRa based Network for Forest Fire Detection. In Proceedings of the Sixth International Conference on Internet of Things: Systems, Management and Security (IOTSMS), Granada, Spain, 19–22 October 2019; pp. 177–184.

Proceeding Paper

A Survey on the Use of the Internet of Multimedia Things for Precision Agriculture and the Agrifood Sector [†]

Alvertos Ioannis Mourikis *, Romanos Kalamatianos, Ioannis Karydis and Markos Avlonitis

Department of Informatics, Ionian University, 49132 Corfu, Greece; rkalam@ionio.gr (R.K.); karydis@ionio.gr (I.K.); avlon@ionio.gr (M.A.)
* Correspondence: c20mour@ionio.gr; Tel.: +30-2661-087-264
† Presented at the 13th EFITA International Conference, online, 25–26 May 2021.

Abstract: The Internet of Things (IoT) has already penetrated an ever-increasing array of daily aspects of life. IoTs bridge the analog and digital worlds in an unprecedented manner and degree by providing in situ sensing. Adding to the IoT the capability to collect interrelated multi-modal sensing, the use of the Internet of Multimedia Things (IoMTs) has recently been exhibited to significantly enhance the role of Information and Communication Technologies (ICT) in numerous applications, and most importantly in agrifood systems. In this work, we review key recent works in the conjunction of the three domains of IoMT, agrifood and precision agriculture and present open research directions.

Keywords: Internet of Multimedia Things; precision agriculture; agrifood sector; open research directions

Citation: Mourikis, A.I.; Kalamatianos, R.; Karydis, I.; Avlonitis, M. A Survey on the Use of the Internet of Multimedia Things for Precision Agriculture and the Agrifood Sector. *Eng. Proc.* **2021**, *9*, 32. https://doi.org/10.3390/engproc 2021009032

Academic Editors: Dimitrios Aidonis and Aristotelis Christos Tagarakis

Published: 3 December 2021

Publisher's Note: MDPI stays neutral with regard to jurisdictional claims in published maps and institutional affiliations.

1. Introduction

A combination of a plethora of factors, including (but not limited to) the man-made climate change crisis [1,2], the world's positive population growth, the ever-increasing penetration and ubiquitousness of ICT in almost all aspects of life, as well as the requirement for qualitative and quantitative nutrition of contemporary humans, all lend themselves to the amalgamation of contributing domains of precision agriculture/farming in agrifood systems. Agrifood systems, in their broadest sense, refer to crops, livestock, forestry, fisheries, and aquaculture [3], among domains producing food. Given this wide breadth of agrifood systems' contribution to food, their importance to humans cannot be overstated. Nevertheless, these systems are under strain and their sustainability is a multi-faceted challenge [4–6].

The ICTs for the collection of information and knowledge mining as well as the customised application of methods in the field, with, among other technologies, artificial intelligence, drones, and the various versions of Internet of Things (IoT) have been dubbed the latest, digital, (r)evolution of agriculture [7–10]. The central tenet of this evolution is the application of the appropriate method to the proper location at the right time and of the suitable intensity, i.e., the notion of precision farming [11]. Accordingly, in this work, we are motivated by the recent developments in ICT to address existing and new challenges of the agrifood domain. This work's contribution is in collecting, processing, homogenizing, and drawing a "bird's-eye view" presentation from research in multimedia IoT (IoMT), agrifood, and precision farming/cultivation, as well as their intersection. The rest of the work is organized as follows: Section 2 presents definitions and working examples of the three domains of IoMT, agrifood, and precision farming that act as pillars to our work. At the same time, Section 2 presents some of the key influential and recent works for each of the three domains of IoMT, agrifood, and precision agriculture, both as distinct directions as well as in conjunction. The work is concluded in Section 3, where open research directions that future research on the domain may draw upon are detailed.

2. Definitions and Working Examples

2.1. Agrifood

Agriculture has had a fair share of evolutionary periods the last approximately 10,000 years, each of which had, in turn, a profound effect on humanity [7,10,12]: the first agrarian revolution, approx. 10,000 BC, was driven by the transition from nomadic life to settlements, and thus organised agriculture with the aid and efficiency of tools such as ploughs; the second (17th–19th centuries) originated the practice of crop rotation, the initiation of technical methodologies to support human labour, as well as the reordering of farmland management the third (approx. 1910–2000) stemmed from extended use of complex mechanisation, the industrialisation of farming, as well as the use of fertilisation and genetic engineering on crops. These changes (key, among several others) allowed the evolution of the population from approx. 4 million in 8000 BC to 400 million in 1700s [13] to 6.1 billion in 2000.

Most of these advancements had a significant effect on more than crops' cultivation, reaching domains such as livestock, forestry, fisheries, aquaculture, etc. Thus, the term agriculture, in the context of this work, and as per the common definition "the science, art, or practice of cultivating the soil, producing crops, and raising livestock and in varying degrees the preparation and marketing of the resulting products" [14], is utilised in a broad manner, with focus on the production of nutritional outputs for humans from flora and fauna, thus resulting in the term "agrifood". Agrifood systems address the needs of a very significant part of humans' nutrition, which is colloquially referred to as "food from farm to fork" and the effects of unsuccessful such systems have grave consequences leading to "resources' pressure and human suffering" [3].

2.2. Internet of (Multimedia) Things

The group of technologies within the Internet of Things (IoT) categorisation provide an intermediary between the physical and virtual worlds and a layer of interconnected, distributed and in situ sensing and processing of information [15]. Under the umbrella of IoT, a wide diversity of variations as to the "thing"'s capabilities, characteristics and scenarios of application have been proposed [16]. Herein, we focus on the variety of IoTs that enhance common IoT sensors, focusing solely on scalar measurements, found in most agrifood scenarios by providing both scalar and interrelated multi-modal information, based on which a wealth of more advanced agrifood decision scenarios can successfully be met, i.e., the Internet of Multimedia Things (IoMT). IoMTs offer unique advantages in the digitisation of agrifood processes in all aspects of the "from cultivator to consumer" chain.

Visual stimuli information, such as images and/or videos, has been shown to have a wide breadth of applications in agriculture, especially when used in conjunction with computer vision methods [17]: identification of cultivation's products as per size, shape, colour, holistic condition and potential output/yield [18], over-accumulation or deficiency of nutrients [19], invasive species and pests [20], as well as estimation of developmental parameters of the cultivation. Similarly, auditory cue sensors are more common than expected in various sensor sets used in agriculture [21] for purposes such as invasive species and pest identification [22], rainfall estimation [23], soil characteristic evaluation [24], *volume measurement* of agricultural products [25], and many more.

Moreover, IoMTs can support ICT services that will promote high-quality products by: (a) providing consumers with details about each product and producer (origin, certifications, farming practices, milling and packaging details, historical facts, producer's page, contextual information, review from other consumers, etc.); (b) providing producers with suggestions and support from experts to improve the quality of their products and to be guided through a sustainable cultivation process (reducing water and pesticide usage, having access to local meteorological information, etc.); (c) providing consumers with specific information (understandable scientific details about each product, nutraceutical analysis, etc.) and suggestions from experts; and (d) creating a social-level web of interactions where all interested parties can comment, review, interact and state their opinion about the products, generating a culture of informed consumption, digital promotion and a "crowd-based acceptance certification".

2.3. Precision Farming

Precision (or smart) agriculture/farming/cultivation (or "Agriculture 4.0"), in this context almost all synonymous, has been dubbed the fourth agrarian (r)evolution. A healthy discussion as to this (r)evolution's necessity and its envisaged positive and/or negative consequences is already underway [26]. One of the most dramatic results of this evolution is expected through the integration of all stakeholders of agrifood, even at the conceptual level, thanks to the ICT advancements, as described in Section 2.2.

Artificial intelligence represents a pillar of precision farming [27] in order to achieve processes oriented at soil sciences and soil-specific cultivation management, the management of geoinformation, calculation of fertilisation rate, customisation of cultivation solutions based on site-specific characteristics and the use/definition of cultivation management zones, extended use of high-precision location information of assets and cultivation, yield mapping, methods for variable rate of herbicides/pesticides, high specificity adaptable irrigation rate, near real-time remote, high precision and various sensing measurements, automatic cultivation asset navigation and work-load mechanisation through drones/robotics, and proximal sensing of cultivation and its environment [11].

3. Conclusions

The fourth agrarian (r)evolution is upon us, and a number of interesting open research directions exist: the philosophical and ethical quandaries posed by extensive dependencies on technological advancements and AI; the inclusivity of precision agriculture's methods to all stakeholders (from policy-makers, to producers, to consumers); the mitigation strategies for the unexpected and/or unwanted (side-)effects of the evolution; the balancing of the role of technological innovations vs. innovations in information analysis and decision support in the evolution; the environmental parameters of the evolution (e.g., the footprint of the life-cycle of all parts of the evolution), and many others. As with all new and exciting ventures, the fourth agrarian (r)evolution needs to evolve to its best self in order to reap the benefits and address in the best possible way the nutrition of humankind.

Author Contributions: All authors have contributed equally to the processes of conceptualisation, methodology, validation, investigation, resources, writing—original draft preparation and review and editing. All authors have read and agreed to the published version of the manuscript.

Funding: This research received no external funding.

Acknowledgments: The financial support of the European Union and Greece (Partnership Agreement for the Development Framework 2014–2020) under the Regional Operational Programme Ionian Islands 2014–2020, for the project "Olive Observer" is gratefully acknowledged.

Conflicts of Interest: The authors declare no conflict of interest. The funders had no role in the design of the study; in the collection, analyses, or interpretation of data; in the writing of the manuscript, or in the decision to publish the results.

References

1. Figueres, C.; Rivett-Carnac, T. *The Future We Choose: Surviving the Climate Crisis*; Vintage: New York, NY, USA, 2020.
2. Crist, E. Beyond the Climate Crisis: A Critique of Climate Change Discourse. *Telos* **2007**, *141*, 29–55.
3. Herrero, M.; Jones, P.; Thornton, P.; Chen, M.; Gerland, P.; Gilbert, M. Chapter 33-Agrifood Systems. In *Sustainable Food and Agriculture*; Campanhola, C., Pandey, S., Eds.; Academic Press: Cambridge, MA, USA, 2019; pp. 305–330. [CrossRef]
4. Barrett, H.; Rose, D.C. Perceptions of the Fourth Agricultural Revolution: What's In, What's Out, and What Consequences are Anticipated? *Sociol. Rural.* **2021**. [CrossRef]
5. Devaux, A.; Goffart, J.-P.; Kromann, P.; Andrade-Piedra, J.; Polar, V.; Hareau, G. The Potato of the Future: Opportunities and Challenges in Sustainable Agri-food Systems. *Potato Res.* **2021**. [CrossRef]
6. El Bilali, H.; Strassner, C.; Ben Hassen, T. Sustainable Agri-Food Systems: Environment, Economy, Society, and Polic. *Sustainability* **2021**, *13*, 6260. [CrossRef]

7. Lombardo, S.; Sarri, D.; Corvo, L.; Vieri, M. Approaching to the Fourth Agricultural Revolution: Analysis of Needs for the Profitable Introduction of Smart Farming in Rural Areas. HAICTA 2017. pp. 521–532. Available online: https://flore.unifi.it/retrieve/handle/2158/1112565/296930/360.pdf (accessed on 22 April 2021).
8. Zhai, Z.; Martínez, J.F.; Beltran, V.; Martínez, N.L. Decision support systems for agriculture 4.0: Survey and challenges. *Comput. Electron. Agric.* **2020**, *170*, 105256. [CrossRef]
9. De Clercq, M.; Vats, A.; Biel, A. Agriculture 4.0: The Future of Farming Technology. 2018; pp. 11–13. Available online: https://skyfarms.io/wp-content/uploads/2020/08/84-OliverWyman-World-Government-Report-Agriculture-4.0.pdf (accessed on 22 April 2021).
10. Rose, D.C.; Chilvers, J. Agriculture 4.0: Broadening Responsible Innovation in an Era of Smart Farming. *Front. Sustain. Food Syst.* **2018**. Available online: https://doi.org/10.3389/fsufs.2018.00087 (accessed on 23 April 2021). [CrossRef]
11. Mulla, D.; Khosla, R. Historical evolution and recent advances in precision farming. In *(Επιμ.), Soil-Specific Farming*; Lal, R., Stewart, B., Eds.; Taylor Francis: Abingdon, UK, 2016; pp. 1–35. Available online: https://doi.org/10.1201/b18759 (accessed on 23 April 2021).
12. Faris, J.D. Wheat Domestication: Key to Agricultural Revolutions Past and Future. In *Genomics of Plant Genetic Resources: Volume 1. Managing, Sequencing and Mining Genetic Resources*; Springer: Dordrecht, The Netherlands, 2014; pp. 439–464. [CrossRef]
13. Lal, R.; Reicosky, D.; Hanson, J.D. Evolution of the plow over 10,000 years and the rationale for no-till farming. *Soil Tillage Res.* **2007**, *93*, 1–12. [CrossRef]
14. Merriam-Webster. Agriculture. Retrieved from Dictionary by Merriam-Webster. 25 July 2021. Available online: https://www.merriam-webster.com/dictionary/agriculture (accessed on 19 April 2021).
15. Kopetz, H. Internet of things. In *Real-Time Systems*; Springer: Boston, MA, USA, 2011; pp. 307–323.
16. Kassab, W.; Darabkh, K.A. A–Z survey of Internet of Things: Architectures, protocols, applications, recent advances, future directions and recommendations. *J. Netw. Comput. Appl.* **2020**, *163*, 102663. [CrossRef]
17. Patrício, D.I.; Rieder, R. Computer vision and artificial intelligence in precision agriculture for grain crops: A systematic review. *Comput. Electron. Agric.* **2018**, *153*, 69–81. [CrossRef]
18. Mazloumzadeh, S.; Shamsi, M.; Nezamabadi-pour, H. Fuzzy logic to classify date palm trees based on some physical properties related to precision agriculture. *Precis. Agric.* **2010**, *11*, 258–273. [CrossRef]
19. Wei, F.; Yan, Z.; Tian, Y.; Cao, W.; Xia, Y.; Li, Y. Monitoring leaf nitrogen accumulation in wheat with hyper-spectral remote sensing. *Acta Ecol. Sin.* **2008**, *28*, 23–32. [CrossRef]
20. Kalamatianos, R.; Karydis, I.; Doukakis, D.; Avlonitis, M. DIRT: The Dacus Image Recognition Toolkit. *J. Imaging* **2018**, *4*, 129. [CrossRef]
21. Shafi, U.; Mumtaz, R.; García-Nieto, J.; Hassan, S.A.; Zaidi, S.; Iqbal, N. Precision Agriculture Techniques and Practices: From Considerations to Applications. *Sensors* **2019**, *19*, 3796. [CrossRef] [PubMed]
22. Potamitis, I.; Rigakis, I.; Fysarakis, K. Insect Biometrics: Optoacoustic Signal Processing and Its Applications to Remote Monitoring of McPhail Type Traps. *PLoS ONE* **2015**, *10*, e0140474. [CrossRef] [PubMed]
23. Avanzato, R.; Beritelli, F. An Innovative Acoustic Rain Gauge Based on Convolutional Neural Networks. *Information* **2020**, *11*, 183. [CrossRef]
24. Rillig, M.C.; Bonneval, K.; Lehmann, J. Sounds of Soil: A New World of Interactions under Our Feet? *Soil Syst.* **2019**, *3*, 45. [CrossRef]
25. Nishizu, T.; Ikeda, Y. Volume Measuring System by Acoustic Method for Agricultural Products, 1: Precision and Accuracy of Volume Measuring System by Applying Helmholtz Resonance Phenomena. *J. Jpn. Soc. Agric. Mach.* **1995**, *57*, 47–54. Available online: https://agris.fao.org/agris-search/search.do?recordID=JP9603020 (accessed on 15 April 2021).
26. Barrett, C.B.; Benton, T.G.; Cooper, K.A.; Fanzo, J.; Gandhi, R.; Herrero, M.; James, S.; Kahn, M.; Mason-D'Croz, D.; Mathys, A.; et al. Bundling innovations to transform agri-food systems. *Nat. Sustain.* **2020**, *3*, 974–976. [CrossRef]
27. Khanna, A.; Kaur, S. Evolution of Internet of Things (IoT) and its significant impact in the field of Precision Agriculture. *Comput. Electron. Agric.* **2019**, *157*, 218–231. [CrossRef]

Proceeding Paper

Light Penetration Properties of Visible and NIR Radiation in Tomatoes Applied to Non-Destructive Quality Assessment †

Merel Arink [1], Haris Ahmad Khan [1] and Gerrit Polder [2,*]

1   Farm Technology Group, Wageningen University & Research, P.O. Box 16,
     6700 AA Wageningen, The Netherlands; merel.arink@gmail.com (M.A.); haris.khan@wur.nl (H.A.K.)
2   Greenhouse Horticulture Group, Wageningen University & Research, P.O. Box 644,
     6700 AP Wageningen, The Netherlands
*   Correspondence: gerrit.polder@wur.nl
†   Presented at the 13th EFITA International Conference, online, 25–26 May 2021.

Abstract: Tomato is an important food product for which the development of non-destructive quality assessment methods is of great interest. Using visible and near-infrared (NIR) spectroscopy, the sugar content, acidity and even taste can be estimated through the use of chemometric methods (e.g., partial least squares regression). In the case of reflection spectra, which are the common modality for imaging spectroscopy, the question arises regarding how much of the interior of the tomato contributes to the measured spectra. An experiment was performed with tomatoes of four different types: beef tomato, classic round tomato, cocktail tomato, and snack tomato. The tomatoes were sliced at different thicknesses and imaged on a 98% reflective white background and a 4% reflective black background. Spectral images were acquired with VNIR (400–1000 nm) and NIR (900–1700 nm) imaging spectrographs. The difference between the spectra with a white and black background was used to determine the relationship between the wavelength and the light penetration depth. The results show that at wavelengths between 600 and 1100 nm, light penetrates the tomatoes up to a distance of 20 mm. The relation more or less follows the law of Lambert–Beer. This relation was the same for all four types of tomatoes. These results help the interpretation of chemometric models based on reflection (imaging) spectroscopy.

Keywords: light penetration; NIR radiation; tomato; quality assessment; non-destructive

Citation: Arink, M.; Khan, H.A.; Polder, G. Light Penetration Properties of Visible and NIR Radiation in Tomatoes Applied to Non-Destructive Quality Assessment. *Eng. Proc.* **2021**, *9*, 18. https://doi.org/10.3390/engproc2021009018

Academic Editors: Charisios Achillas and Lefteris Benos

Published: 24 November 2021

Publisher's Note: MDPI stays neutral with regard to jurisdictional claims in published maps and institutional affiliations.

1. Introduction

The ripeness of a tomato is very important. A tomato that is ripened before picking has a better flavor and overall quality compared to tomatoes ripened after picking. Lycopene is a carotenoid pigment related to ripening. It is responsible for the red color of the tomato [1]. Sugars in tomatoes are closely related to fruit quality and yield, but also in fruit set, ripening, composition, and growth [2]. Sugars, mainly fructose and glucose, account for about 65% of the soluble solids of ripe tomatoes [3]. The soluble solids value expressed in Brix is determined by the variety, method of cultivation, and environmental conditions. A refractometer can be used to measure the percentage of solids in juice. This method is destructive, labor intensive, and expensive. As an alternate, visible and NIR spectroscopy can be used as a non-destructive quality measurement method for features such as sugar content, acidity, and taste. Imaging spectroscopy, also known as hyperspectral imaging [4], is a special mode of spectroscopy that generates a spatial map of spectral variation [5]. Each vector in the array represents the reflection spectrum at the specific pixel location.

The interaction between light and the tomato is a complicated phenomenon; it involves both absorption and scattering [6]. Photons scatter multiple times before being absorbed or exited from the fruit. The absorption of light is primarily related to chemical components such as sugar [7]. The scattering of light is influenced by physical and structural features, such as particle size, density, and cellular structure. Since absorption and scattering are intertwined, it is challenging to relate the spectra to the concentration of the compounds. The Lambert–Beer law ascribes the absorption of light with the properties of the material it passes through. The Lambert–Beer law states that the intensity of the electromagnetic wave decreases exponentially from the surface of a sample, for homogeneous media:

$$I/I_0 = e^{-\mu d} \tag{1}$$

where I is the transmitted light, I_0 is the incident light, μ is the attenuation coefficient, and d is the thickness of the sample. There is a growing interest in using imaging spectroscopy in reflection mode to estimate the quality and attributes of food products. The question arises regarding how much of the interior of a fruit contributes to the measured spectra. In this paper, we answer this question by investigating light penetration in four different types of tomatoes. Knowing the relation between light absorption and penetration depth provides insight into the contribution of internal compounds to the measured spectra.

2. Materials and Methods

Four types of tomatoes (beef tomato, classic round tomato, cocktail tomato, and snack tomato) were imaged with an imaging spectrograph, sliced at different thicknesses on a black and a white surface. When light penetrates through the slice onto the surface, the white surface reflects the light resulting in a higher measured reflectance compared to the light-absorbing black surface. When the tomato slice has a thickness for which all light is absorbed before reaching the background, there will be no difference between the measured reflectance on both surface colors. This idea is visualized in Figure 1.

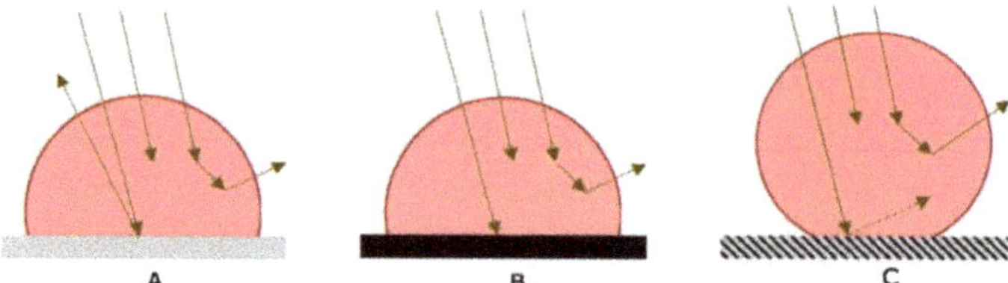

Figure 1. Visualization if (**A**,**B**) the transmitted light influences the reflected light for the different colored backgrounds, and if not (**C**). The striped surface for C indicates a similar result for both background colors.

A white Teflon (PTFE) plate that reflects 98% of the incoming light between 300 and 1900 nm was used as a white background. A black nylon polyurethane-coated fabric that reflects 4% of light between the wavelengths 400 to 2400 nm was used as a black background. Two spectral line scan cameras, Specim FX10 and Specim FX17, were used for the spectral measurements. The spectral range for the Specim FX10 camera is 400 to 1000 nm, with a resolution of 5.5 nm. The spatial resolution is 1024 pixels. The exposure time was 6 ms. The spectral range for the Specim FX17 camera is 900 to 1700 nm, with a spectral resolution of 8 nm and the spatial resolution is 640 pixels (Specim, 2020b). The exposure time was 3 ms. Scan speed was adjusted to obtain square pixels. Two 150 W Dolan Jenner PL900 tungsten halogen lamps with glass fiber optic line arrays of 0.5 mm × 152.4 mm and a quartz rod lens formed a line-shaped beam of light on the samples. A 1000 W halogen tube was used for illumination in the infrared spectrum; the tube operated at 75% of the nominal voltage (170 V), which resulted in an output of 570 W. An exponential model was fitted on the data of the first sample and validated on the second sample of each tomato type for the wavelengths 500, 700, 900, 1100, 1300, and 1500 nm.

3. Results

Between the thickness of 0 to 20 mm and wavelengths 600 to 1100 nm, a definite difference in reflectance between the black and white background was observed. For the other combinations of thicknesses and wavelengths, there was not a clear reflectance difference between the two different backgrounds (Figure 2). The reflectance values showed similar values for the different types of tomatoes. Exponential models were fitted for the six wavelengths. Figure 3 shows the data points of both samples and the fitted model on the first sample for the classic tomato at 500 and 900 nm. From this figure, the exponential relationship can clearly be seen for 900 nm. For wavelengths between 700 and 1300 nm and for the beef, classic, and cocktail tomatoes, the models showed comparable results. At 500 nm, for each type of tomato, the relation is more linear.

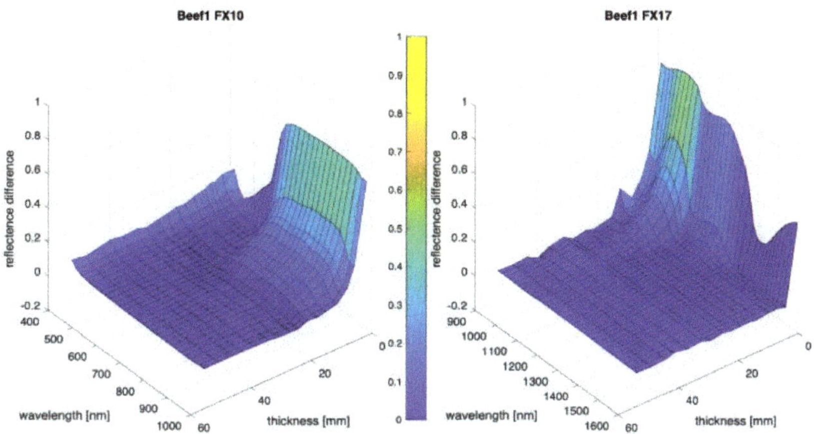

Figure 2. Three-dimensional plots of thickness, wavelength, and reflectance difference for beef tomato sample 1.

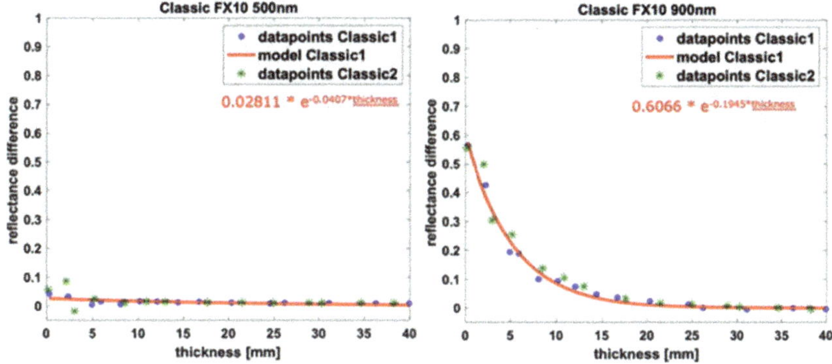

Figure 3. The plot of the model for the classic tomato and data points of the first and second samples for 500 and 900 nm.

Since the models were based on the data of the first samples (training data), the R2 for this data should be close to one. This was the case for most models, but not for classic 500 nm, cocktail 500 nm, snack 500 nm, and 1500 nm. The R2 on the second sample (validation data) was below 0.75 for the beef 500 nm and 1500 nm, classic and cocktail 1500 nm, and all of the other snack models.

4. Discussion

When looking at the 3D surface plots of the thickness, wavelengths, and reflectance difference for the different types and samples of the tomatoes, only combinations of a thickness between 0 to 20 mm and a wavelength between 600 to 1100 nm showed an evident increase in the signal, proving that for these wavelengths the interior of the tomato contributes to the reflection spectra. For the other combinations of thickness and wavelength, there was no evident surface increase visible. This suggests that for all wavelengths, an interior deeper than 20 mm does not contribute to the measured spectra. For the wavelengths 400 to 600 and 1100 to 1700, the interior seemed to barely affect the result. The models for the snack tomato all performed badly, most likely because of the very few data points available. The models for the wavelength 500 nm did also not reach the minimum R2 of 0.75. This might have been a result of the uncertainty of the sensor at the beginning of the range due to the lack of irradiance of the light source in this wavelength. From the literature, it is known that temperature has a major impact on optical properties in the NIR spectrum of the sample. In this research, the temperature was not taken into account; future research is required to evaluate the effect of temperature on the measured spectra.

5. Conclusions

The interior of a tomato contributes to the result of reflection imaging spectroscopy for a maximum tomato thickness of 20 mm when the wavelength of the measurements lies between 600 to 1100 nm. An exponential relation was visible between the decreasing thickness of the tomato and the increasing reflectance. This is in accordance with the Lambert–Beer law, although this law is only defined for homogeneous material. Outside the wavelength range 600 to 1100 nm or when the tomato thickness was larger than 20 mm, the interior did not significantly influence the result of reflection imaging spectroscopy. The tomato type also did not influence the light penetration.

Author Contributions: Conceptualization, G.P. and H.A.K.; methodology, G.P. and M.A.; software, M.A.; validation, M.A., H.A.K., and G.P.; formal analysis, M.A.; investigation, M.A.; resources, G.P.; data curation, M.A., H.A.K., and G.P.; writing—original draft preparation, M.A.; writing—review and editing, H.A.K., G.P.; supervision, H.A.K., G.P. All authors have read and agreed to the published version of the manuscript.

Funding: Funding was provided by the Dutch KB (knowledge base) program 37 "Healthy & Safe Food Systems" of Wageningen UR and Foundation TKI Horticulture Propagation Materials within the Fresh on Demand project.

Institutional Review Board Statement: Not applicable.

Informed Consent Statement: Not applicable.

Data Availability Statement: The data presented in this study are available on request from the corresponding author.

Conflicts of Interest: The authors declare no conflict of interest.

References

1. Llorente, B.; D'Andrea, L.; Ruiz-Sola, M.A.; Botterweg, E.; Pulido, P.; Andilla, J.; Loza-Alvarez, P.; Rodriguez-Concepcion, M. Tomato fruit carotenoid biosynthesis is adjusted to actual ripening progression by a light-dependent mechanism. *Plant J.* **2016**, *85*, 107–119. [CrossRef] [PubMed]
2. Kanayama, Y. Sugar metabolism and fruit development in the tomato. *Hortic. J.* **2017**, *84*, 417–425. [CrossRef]
3. Helyes, L.; Pék, Z.; Lugasi, A. Function of the variety technological traits and growing conditions on fruit components of tomato (Lycopersicon lycopersicum L. Karsten). *Acta Aliment.* **2008**, *37*, 427–436. [CrossRef]
4. Polder, G.; Gowen, A. The hype in spectral imaging. *J. Spectr. Imaging* **2020**, *9*, a4. [CrossRef]
5. Ravikanth, L.; Jayas, D.S.; White, N.D.; Fields, P.G.; Sun, D.W. Extraction of spectral information from hyperspectral data and application of hyperspectral imaging for food and agricultural products. *Food Bioprocess Technol.* **2017**, *10*, 1–33. [CrossRef]
6. Qin, J.; Lu, R. Measurement of the optical properties of fruits and vegetables using spatially resolved hyperspectral diffuse reflectance imaging technique. *Postharvest Biol. Technol.* **2008**, *49*, 355–365. [CrossRef]
7. Tuchin, V. Tissue optics and photonics: Light-tissue interaction II. *J. Biomed. Photonics Eng.* **2016**, *2*, 030201. [CrossRef]

Proceeding Paper

Mango Leaf Monitoring with Inductive and Capacitive Sensors and Its Comparison with Trunk Dendrometer Measurements [†]

Federico Hahn [1,*], Juan Espinoza [2] and Ulises Zacarías [2]

[1] Irrigation Department, Universidad Autonoma Chapingo, Texcoco 56230, Mexico

[2] IAUIA, Universidad Autonoma Chapingo, Texcoco 56230, Mexico; juanesphdz@gmail.com (J.E.); uzah13@gmail.com (U.Z.)

* Correspondence: fhahn@correo.chapingo.mx; Tel.: +52-553-242-3033

† Presented at the 13th EFITA International Conference, online, 25–26 May 2021.

Abstract: Mango is one of the main fruits grown in Mexico that are exported worldwide, but the trees consume a lot of water, and irrigation scheduling should be implemented to optimize water use. Dendrometers were installed in fruit trees to optimize water usage during 2019 and 2020. A capacitor with Teflon clamps pressurized the leaf, and its dielectric changed with leaf water content. Additionally, Hall sensors were installed in leaves to study the effect of water during mango production. It was found that capacitance tend to be more sensitive than magnetic field monitoring. Higher changes were noted during midday with warm weather. Thresholds from the capacitance and Hall effect sensors can provide signals for irrigation scheduling.

Keywords: dendrometer; mango twig; capacitor sensing; magnetic sensing; irrigation scheduling

Citation: Hahn, F.; Espinoza, J.; Zacarías, U. Mango Leaf Monitoring with Inductive and Capacitive Sensors and Its Comparison with Trunk Dendrometer Measurements. *Eng. Proc.* **2021**, *9*, 28. https://doi.org/10.3390/engproc2021009028

Academic Editors: Dimitrios Kateris and Maria Lampridi

Published: 2 December 2021

Publisher's Note: MDPI stays neutral with regard to jurisdictional claims in published maps and institutional affiliations.

1. Introduction

The amount of water available for irrigation is limited and mango trees require a large quantity during flowering and fruit growth. Pests and diseases in fruit trees will be present after a water shortage [1]. Different irrigation scheduling methods can be used, being ideal if the plant water status is known [2]. Plant water status monitoring relies on the leaf or stem water status or stomatal conductance. Sap flow, trunk diameter and leaf turgor pressure are some of the variables studied in fruit and forest trees [2]. The near infrared region (NIR) spectral reflectance of leaves is positively correlated to leaf thickness. Leaf tissues with variable water content present different NIR reflectance signatures at 760 nm, 970 nm and 1450 nm [3]. This technique is becoming useful in cameras transported by unmanned aerial vehicles (UAV) [4]. In trees, stem daily variations (SDV) change from tenths to hundreds of microns depending on stem diameter. Dendrometers measure trunk variations using linear variable differential transformers (LVDT), linear resistors and strain gauges. Stem diameter increases or shrinks daily due to soil water availability and atmospheric demand. For scheduling irrigation, maximum daily shrinkage (MDS) and stem growth rate (SGR) are the best indicators. For young trees, SGR is a better indicator than MDS, as the latter could be more affected by growth than by water stress [5]. Both stem and branch indicators are successfully used to control irrigation but are still expensive for growers [6].

Leaf turgor has a significant effect on stomatal behavior and can be used to study leaf water content. A magnetic leaf patch-clamp pressure probe (ZIM) is used to obtain leaf turgor measurements [7]. This probe does not work well when leaves suffer from dehydration, as vapor accumulates in the leaf tissues and the pressure chip does not sense properly [4]. Capacitive sensors are used in precision agriculture to estimate leaf moisture content as their dielectric constant change [8].

Eng. Proc. **2021**, *9*, 28. https://doi.org/10.3390/engproc2021009028   https://www.mdpi.com/journal/engproc

Three sensors were developed to monitor mango tree water status in the tree trunk and tree leaves. In a twig of a mango tree, the sensors were evaluated during a season. Capacitance leaf meters are easy to install, and their use and operation was analyzed.

2. Equipment Design

The monitoring equipment makes use of dendrometers to measure trunk variations, leaf capacitance sensors and a sensor to measure magnetic field changes through the day. The dendrometer measures the axial variations (thickness) of the trunk or branch during the day. An ESP32-WROOM-32 microcontroller board (Espressif, Shanghai, China) controls the dendrometer by using a linear resistor (model 9605R1.7KL2.0, BEISensors, Thousand Oaks, CA, USA) to provide an analog output voltage according to the trunk diameter. A current source avoids output voltage fluctuations when battery voltage discharges. The TTGO-ESP32-WROOM-32 card has a Bluetooth module for wireless transmission of data to a cell phone.

A simple capacitance meter uses 2-cooper parallel plates with the mango leaf between them being the dielectric constant; the plates were pushed from the outside by two Teflon washers. The capacitor plates are 15 mm long and 4 mm wide. A 555 timer was configured as a monostable multivibrator (Figure 1). The pin 22 of the ESP32 applies a trigger pulse to the input terminal (2) of the 555. A pulse at output (3) changes its state from low (0) to high (VCC). The capacitor charges exponentially and when it reaches 2/3 VCC, the output returns to low (0), representing the end of the pulse. The pulse duration (T) corresponds to the value determined by:

$$T = 1.1\ RC, \tag{1}$$

Figure 1. Complete electronic circuit showing the dendrometer linear resistance, the temperature sensor, the Hall effect sensor and the capacitance meter.

The output (3) of the 555 is connected to pins 25 and 26 of the ESP32 which are programmed as interrupts 0 and 1. ESP32-pin 25 linked to interrupt 0 detects the rising edge of the output pulse and instantly will stop any activity taking place. In this moment, the ESP32 will act as a counter. When the 555 output pulse ends, interrupt 1 (ESP32-pin 26) will detect its falling edge. The counter will stop, and its value will represent the RC charging period given by Equation (1). If the resistance is known, the capacitor can be calculated. These pulses will be read ten times and the average capacitance will be saved in the mass storage device (MSD). Every 10 min the capacitance will be studied.

Leaf temperature was monitored throughout the day with a remote infrared temperature sensor (MLX90614, Melexis, Nashua, NH, USA) fixed 5 cm over the leaf. Communication takes place through the I2C bus, making it easy to read, and several sensors can be connected. It was connected to pins 36 and 33 of ESP32 (Figure 1).

The Hall effect module (mod KY-035, Keyes, Hamilton, ON, Canada) is a magnetic field sensor, that provides an analog voltage output according to the field strength. The sensor has a plate where an electric current circulates. When a magnetic field interacts, the electrons are displaced generating a potential difference. A high gain amplifier generates an output voltage proportional to the intensity of the magnetic field. The Hall effect sensor is attached to one side of a clamp and a 200-gauss magnet is hold by the other side (Figure 1). The leaf remains between the magnet and the sensor and is not pressed by them. The distance between the Hall sensor and the magnet was of 5 mm for magnetic field monitoring.

3. Field Measurements

Field experiment was conducted during the 2019-2021 period in a high-density commercial mango farm at Loma Bonita ($17°25'47''$ N, $-101°11'19''$ W, 17 m ASL), in the state of Guerrero, Mexico. The trial site is characterized by basaltic clay-loam red soil, being its average bulk density of 1.1 g cm^{-3}. A 0.15 m deep circular furrow was dug to retain water beneath the trees canopy.

Measurements were taken with the dendrometer during the entire season in 7 cm diameter branches of 20-year-old trees. Ten trees older than 30 years were selected in 2021 and cut down with a chainsaw and watered to begin the rejuvenation process. The grafts were prepared, shaved, and attached to the vegetative shoots before being tightly covered with plastic to prevent moisture from entering (Figure 2).

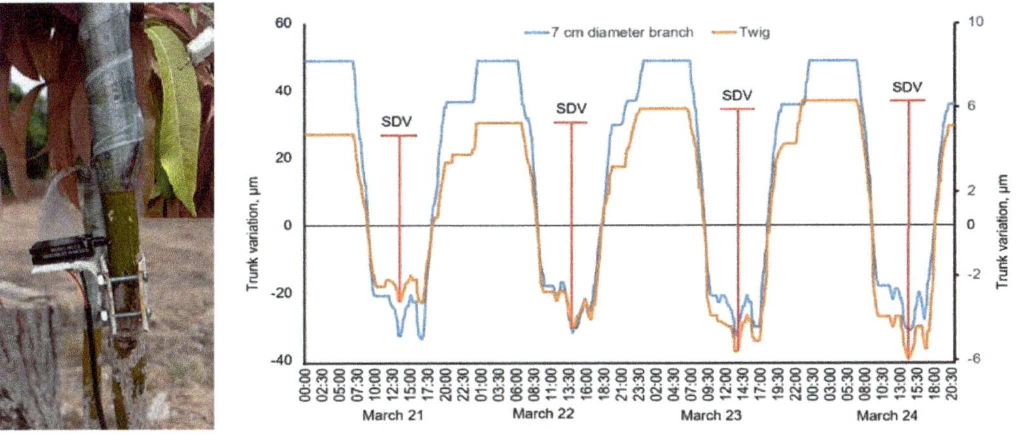

Figure 2. Capacitor sensor and dendrometer installed in a twig of a mango tree and dendrometer measurements.

Dendrometer installation requires the dissection of a tree branch in order to fix the linear resistance (Figure 2). The encapsulated capacitive sensor was placed close to the 555 timer on the upper leaves. The sensor that measures the magnetic variations was also placed on a leaf at the top of the canopy. If the magnet and the Hall sensor were placed in the same clamp to press the leaf, noise in the measurements was appreciated when amplification was 12 and sampling period was 20 s. A simple EMA (exponential moving average) filter reduces sampling noise. The sampling time was carried out every minute and the filter α used a constant of 0.2 to optimize the measurements.

4. Results and Discussion

Dendrometer stem daily variations (SDV) are constant for the 7 cm diameter branch (Figure 2). This blue line varies from a maximum value of 49 μm at midnight to -30 μm at midday. In the twig, continuous SDV growth is noted over the four days, both noted at midnight and midday. SDV was 7.5 on 21 March 2021 and increased daily to achieve 12.2 on 24 March 2021. Vegetative twigs irrigated daily showed an SDV increase of 62.6%.

The capacitance sensor was first tested by introducing a dry coffee paper filter and then watered. Capacitance values were 5.61 and 4.51 pF, respectively, meanwhile air capacitance is 4.44 pF. Leaf capacitance monitoring during a cloudy and windless day is shown in Figure 3a. An increase in capacitance starts at 13:30 and peaks at 15:30 as shown with the red square, showing a similar behavior as encountered by Afzal [8]. A good relationship during the day is observed between temperature and capacitance. On the third day, water was applied, and capacitance decreased. The capacitance of wet leaves was higher than dry leaves (Table 1). EMA low pass filter used an $\alpha = 0.2$, a sampling time of 20 s, and a Hall sensor resolution of 10. Experiments showed that the Hall effect sensor measurements are related to plant water status and its temperature. Under controlled conditions, measurements were significantly better, obtaining a quasi-constant line when the plant was not watered. Hall sensor measurements decreased when temperature increased (Figure 3b).

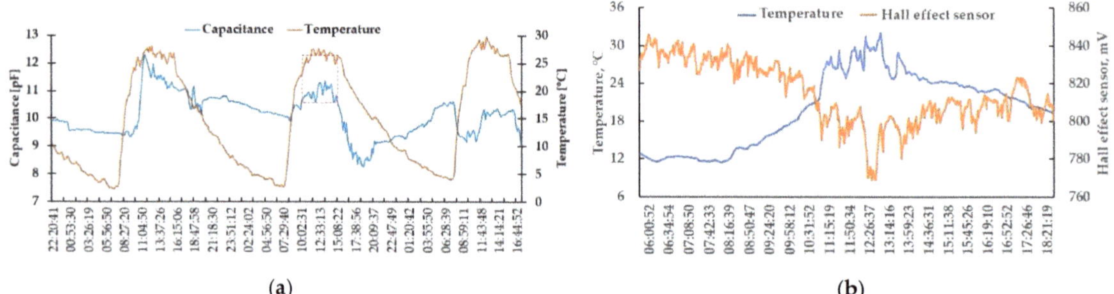

| | (a) | | (b) |

Figure 3. (**a**) Capacitance and temperature measured in a tree leaf for three days.; (**b**) magnetic field variation expressed in voltage during one day in a mango twig.

Table 1. Leaf thickness, capacitance and voltage provided by the Hall effect sensor for dry and wet samples.

	Thickness, μm	Capacitance, pF	Magnetic Field, mV
Wet twig leaf	204 ± 4.3	6.7 ± 2.6	830 ± 2.1
Dry tree leaf	226 ± 7.8	9.7 ± 1.8	780 ± 1.9
Wet tree leaf	242 ± 11.2	6.3 ± 2.6	840 ± 1.9

Author Contributions: Conceptualization, F.H., J.E. and U.Z.; writing—original draft preparation, F.H., J.E. and U.Z.; writing—review and editing, U.Z.; supervision, F.H. All authors have read and agreed to the published version of the manuscript.

Funding: This research was funded by the University Autonoma Chapingo, grant number 20014-DTT-65.

Data Availability Statement: Data were obtained by farmers and can be obtained from Dr. Hahn.

Conflicts of Interest: The authors declare no conflict of interest.

References

1. Osroosh, Y.; Peters, R.T.; Campbell, C.S.; Zhang, Q. Comparison of Irrigation Automation Algorithms for Drip-Irrigated Apple Trees. *Comput. Electron. Agric.* **2016**, *128*, 87–99. [CrossRef]
2. Fernández, J.E. Plant-based sensing to monitor water stress: Applicability to commercial orchards. *Agric. Water Manag.* **2014**, *142*, 99–109. [CrossRef]
3. Nicolai, B.M.; Beullens, K.; Bobelyn, E.; Peirs, A.; Saeys, W.; Theron, K.I.; Lammertyn, J. Non-destructive measurement of fruit and vegetable quality by means of NIR spectroscopy: A review. *Postharvest Biol. Technol.* **2007**, *46*, 99–118. [CrossRef]
4. Fernández, J.E. Plant-Based Methods for Irrigation Scheduling of Woody Crops. *Horticulturae* **2017**, *3*, 35. [CrossRef]
5. Nortes, P.A.; Pérez-Pastor, A.; Egea, G.; Conejero, W.; Domingo, R. Comparison of changes in stem diameter and water potential values for detecting water stress in young almond trees. *Agric. Water Manag.* **2005**, *77*, 296–307. [CrossRef]

6. Lu, P. Mango water relations and irrigation scheduling. In Proceedings of the 1st International Conference on Mango and Date Palm: Culture and Export, Faisalabad, Pakistan, 20–23 June 2005.
7. Zimmermann, D.; Reuss, R.; Westhoff, M.; Gessner, P.; Bauer, W.; Bamberg, E.; Bentrup, F.W.; Zimmermann, U. A novel, non-invasive, online monitoring, versatile and easy plant-based probe for measuring leaf water status. *J. Exp. Bot.* **2008**, *59*, 3157–3167. [CrossRef] [PubMed]
8. Afzal, A.; Duiker, S.W.; Watson, J.E.; Luthe, D. Leaf thickness and electrical capacitance as measures of plant water status. *Am. Soc. Agric. Biol. Eng.* **2017**, *60*, 1063–1074. [CrossRef]

engineering proceedings

MDPI

Proceeding Paper

Soil Moisture Depletion Modelling Using a TDR Multi-Sensor System, GIS, Soil Analyzes, Precision Agriculture and Remote Sensing on Maize for Improved Irrigation-Fertilization Decisions [†]

Agathos Filintas

Department of Agricultural Technologists, Campus Gaiopolis, University of Thessaly, 41500 Larisa, Greece; filintas@uth.gr

† Presented at the 13th EFITA International Conference, Online, 25–26 May 2021.

Abstract: The effects of three drip irrigation (IR1: Farmer's, IR2:Full (100%ET$_c$), IR3:Deficit (80%ET$_c$) irrigation), and two fertilization (Ft1, Ft2) treatments were studied on maize yield and biomass by applying new agro-technologies (TDR—sensors for soil moisture (SM) measurements, Precision Agriculture, Remote Sensing—NDVI (Sentinel-2 satellite sensor), soil-hydraulic analyses and Geostatistical models, SM-rootzone modelling-2D-GIS mapping). A daily soil moisture depletion (SMDp) model was developed. The two-way-ANOVA statistical analysis results revealed that irrigation (IR3 = best) and fertilization treatments (Ft1 = best) significantly affect yield and biomass. Deficit irrigation and proper fertilization based on new agro-technologies for improved management decisions can result in substantial improvement on yield (+116.10%) and biomass (+119.71%) with less net water use (−7.49%) and reduced drainage water losses (−41.02%).

Keywords: geostatistical modelling on maize's yield; 2-D TDR-GIS soil moisture mapping; drip irrigation modeling; precision agriculture; remote sensing and NDVI; soil and hydraulic analyses

Citation: Filintas, A. Soil Moisture Depletion Modelling Using a TDR Multi-Sensor System, GIS, Soil Analyzes, Precision Agriculture and Remote Sensing on Maize for Improved Irrigation-Fertilization Decisions. *Eng. Proc.* **2021**, *9*, 36. https://doi.org/10.3390/engproc 2021009036

Academic Editors: Dimitrios Kateris and Maria Lampridi

Published: 28 December 2021

Publisher's Note: MDPI stays neutral with regard to jurisdictional claims in published maps and institutional affiliations.

1. Introduction

Maize is a major irrigated crop worldwide that requires large quantities of water seasonally (400–800 m^3) [1]. Worldwide, the maize cultivated area and yield are 194.18 million ha and 1118.56 million metric tons (2019/2020) [2]. The agricultural sector accounts for 70% of global [1,3,4] and for 59% of Europe's [5] freshwater withdrawals. At present, many countries worldwide are experiencing a scarcity of fresh water [1,3,4] for potable and irrigation use. Global water demand is projected to increase by 55% between 2000 and 2050 and the climate change will have adverse impact on world water resources and food production with high degree of regional variability and scarcity [6], with the irrigation water amount being the main factor limiting crop production [1,4,7].

With the ever-increasing competition for finite water resources worldwide and the steadily rising demand for agricultural commodities, the call to improve the efficiency and productivity of water use for crop production, to ensure future food and crop products security and address the uncertainties associated with climate change, has never been more urgent [8]. The global challenge for the coming decades will be increasing crop production with less water using precision agriculture as a management strategy that helps farmers to improve crop production and water efficiency. The aim of the study was to determine the effects of three drip irrigation (IR1: Farmer's, IR2: Full [100%ET$_c$], IR3: Deficit [80%ET$_c$] irrigation), and two fertilization (Ft1, Ft2) treatments on maize yield and biomass by applying new agro-technologies.

Eng. Proc. **2021**, *9*, 36. https://doi.org/10.3390/engproc2021009036

2. Materials and Methods

2.1. Experimental Plot Design, Soil Sampling and Laboratory Soil and Hydraulic Analysis

The 2.90 ha field had a factorial split plot design with main factor the three drip irrigation treatments: (a) IR1: Farmer's, (b) IR2: Full (100%ET_c) and c) IR3: Deficit (80%ET_c) irrigation. The sub factor was two fertilization treatments: Ft1: N-P-K = 333.63-9.16-34.86 Kg ha^{-1}, and Ft2: N-P-K = 270.03-9.16-34.86 Kg ha^{-1}). A GPS receiver was used to identify locations of soil samples that were collected at depth 0–40 cm and analyzed at the laboratory. The soil's pH was measured in a 1:2 soil/water extract with a glass electrode and a pH meter. Soil organic matter was analyzed by chemical oxidation with 1 mol L^{-1} $K_2Cr_2O_7$ and titration of the remaining reagent with 0.5 mol L^{-1} $FeSO_4$ [9]. Cation-exchange capacity (CEC) was analyzed by (i) saturation of cation exchange sites with Na by "equilibration" of the soil with pH 8.2, 60% ethanol solution of 0.4N NaOAc-0.1N NaCI; and (ii) extraction with 0.5N $MgNO_3$. Total Na and CI were determined in the extracted solution [9]. The soil's nitrate and ammonium nitrogen were extracted with 0.5 mol L^{-1} $CaCl_2$ and estimated by distillation in the presence of MgO and Devarda's alloy, respectively. Available phosphorus P (Olsen method) was extracted with 0.5 mol L^{-1} $NaHCO_3$ and measured by spectroscopy [9]. Exchangeable potassium K forms were extracted with 1 mol L^{-1} CH_3COONH_4 and measured with a flame Photometer. Field Capacity (FC) and wilting point (WP) were measured with the porous ceramic plate method with 1/3 Atm for FC and 15 Atm for WP [1]. Maize (*Zea mays* L., var. *1921 Hybrid*) was seeded at the end of March and harvested at the end of September.

2.2. Soil Moisture Measurements, Digital 2-D-GIS Moisture Maps Utilizing GIS, Precision Agriculture and Geostatistics

The TDR (time domain reflectometry) method was used to measure soil moisture because it gives accurate results within an error limit of $\pm1\%$ [1,7,10,11]. A TDR instrument and multisensory probes were used [1,7,12], placed at 0–15, 15–30, 30–45, 45–60 and 60–75 cm depths for measuring volumetric water content ($\theta vi, \dots, \theta vn$) ($i = 1, 2, \dots, n$ and $n = 5$) of the root zone. Data were imported daily in a GIS geodatabase utilizing Precision Agriculture (PA) and geostatistics [1,7,12] in order to model and produce soil moisture 2-D maps of maize's root zone profile.

2.3. Remote Sensing Crop's NDVI, Evapotranspiration and Net Irrigation Requirement

Climatic data were obtained from a nearby meteorological station. The effective rainfall was calculated according to USDA-SCS (1970) [13]. The NDVI Vegetation Index [1] was calculated every week using remote sensing (RS) data (Sentinel-2 Satellite sensor) for studying spatial crop development and coefficients. The reference evapotranspiration was computed based on the F.A.O. Penman–Monteith method [1,7,8] and the net irrigation requirement (NIR) was calculated using a soil–water–crop–atmosphere model [1,7]. The crop evapotranspiration (ET_c) and actual evapotranspiration (ET_a) were computed using crop coefficients obtained from remote sensing—NDVI [1,7,8]. The moisture rootzone depletion was calculated using a daily soil moisture depletion (SMDp) model (1) [1,7,8]:

$$Dr,(i) = Dr,(i-1)_{(TDR)} - Pe(i) - NIR(i) - GW(i) + ET_c(i) - DP(i) \tag{1}$$

where: $Dr,(i)$ = root zone depletion at the end of day i (mm), $Dr,(i-1)_{(TDR)}$ = TDR sensors measured rootzone soil-water content at the end of day $i-1$ (mm), Pe = effective rainfall (mm), $NIR(i)$ = net irrigation requirement (mm), $GW(i)$ = groundwater contribution (capillary rise) from water table on day i (mm), $ET_c(i)$ = evapotranspiration on day i (mm), $DP(i)$ = drainage water loss out of the root zone by deep percolation on day i (mm).

3. Results and Discussion

3.1. Study Area, Soil-Hydraulic Analysis and 2-D Moisture Maps Utilizing GIS, Precision Agriculture and Geostatistics

The experiment was carried out in a farm field located at Viotia Prefecture in Central Greece. The study area is characterized by a typical Mediterranean climate [1,4,7] with a cold winter, hot summer and low precipitation in spring and summer. Results found from sensors and analyses were used as input variables to delineate soil moisture profile digital 2D-GIS maps of the maize's root zone. SM is the major factor for crops' enhanced growth and production [1,7]. Spatial analysis revealed an excellent moisture distribution.

Results of laboratory soil and hydraulic analysis revealed that the field's soil was suitable for maize's growth [1,8,9,13] and it was characterized as Sandy Clay (SC) [1,9,13]. The soil organic matter was 2.12% ($\pm$0.17), bulk specific gravity of the soil was 1.44 g cm^{-3} ($\pm$0.04), plant available water was 0.0975 cm cm^{-1} ($\pm$0.02), pH at (1:2) soil/water extract was 8.01 ($\pm$0.23) and the cation-exchange capacity of soil was 21.47 cmol kg^{-1} ($\pm$1.12) (a sufficient level). The N-NO$_3$ was found 11.05 mg kg^{-1} ($\pm$2.23) (a marginal level) and the N-NH$_4$ was found 3.57 mg kg^{-1} ($\pm$1.04) (a low level). The phosphorus P-Olsen was found 21.71 mg kg^{-1} ($\pm$2.21) (a sufficient level), and the potassium K-exchangeable was found at high concentration levels (497 mg kg^{-1} ($\pm$20.34)).

3.2. Daily Soil Moisture Depletion (SMDp) Model and NDVI Vegetation Index

The NDVI (Normalized Difference Vegetation Index) [1] was calculated using RS data (Sentinel-2 satellite sensor) for monitoring spatial crop development and coefficients for the ET$_c$ and SMDp model (Figure 1). The NDVI vegetation index quantifies crops' vegetation by measuring the difference between near-infrared (which vegetation strongly reflects) and red light (which vegetation absorbs). The daily TDR moisture measurements and 2D-GIS SM mapping, the accurate estimated ET$_c$ and monitoring of system inflows and estimated surface outflow and NDVI vegetation index mapping, showed a reliable daily Soil Moisture Depletion model [1,7] for the four stages of maize crop growth.

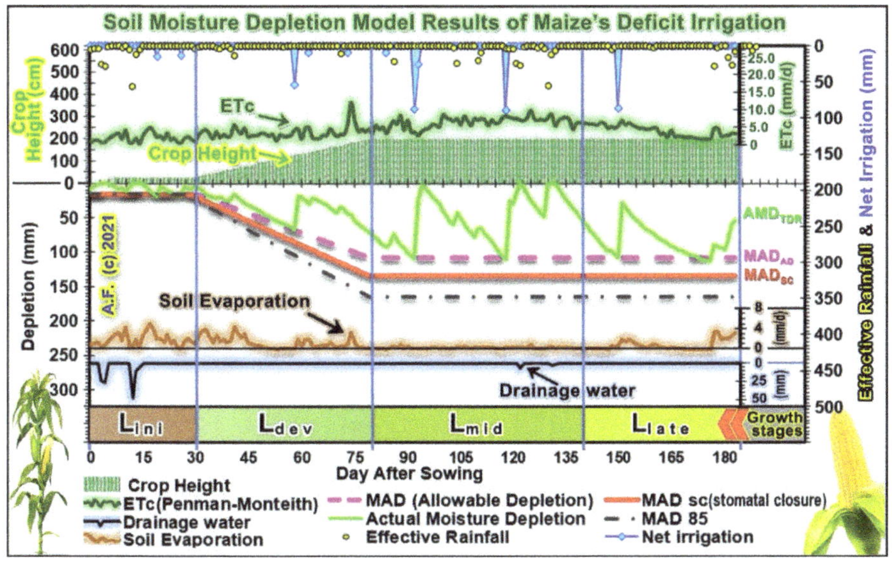

Figure 1. Treatment IR3 results of the daily soil moisture depletion model for the four stages of maize crop growth, also depicting MAD$_{AD}$ (Management Allowed Depletion) and MADsc (MAD at stomatal closure).

3.3. Statistical Analysis, Maize's Yield and Biomass Results

The two-way-ANOVA statistical analysis ($p = 0.05$) using IBM-SPSS (v.26) [1,7,14,15] revealed that the irrigation treatments (IR3: Deficit (80%ET_c) irrigation (best)) and the fertilization treatments (Ft1: (best)) significantly affect maize's yield and biomass. Whenever a prolonged shortage of soil moisture occurs during the most sensitive growth stages (flowering (L_{mid} stage) and grain filling (L_{late} stage)) (Figure 1), this usually has as a result that the maize's crop growth is reduced because of less or excess water supply by the farmers and the yield is consequently lower [1,7,8]. Deficit irrigation and proper fertilization based on new agro-technologies for improved management decisions can result in substantial improvement in yield (+116.10%) and biomass (+119.71%) with less net water use (−7.49%) and reduced drainage losses (−41.02%).

4. Conclusions

Prolonged shortage of soil moisture often occurs during the most sensitive growth stages (flowering (L_{mid} stage) and grain filling (L_{late} stage)) of many crops. In practice, the usual outcome is that the maize's crop growth is reduced because of less (water stress) or excess (saturation) water supply by farmers who cannot estimate the correct soil moisture depletion, the MAD_{AD} (Management Allowed Depletion) and the net water requirements of the crop, so yield is consequently lower. Proper fertilization and deficit irrigation based on new agro-technologies for improved management decisions can result in substantial improvement on yield (+116.10%) and biomass (+119.71%) with less net water use (−7.49%) and reduced drainage water losses (−41.02%) and sustainable management.

Funding: This research received no external funding.

Institutional Review Board Statement: Not applicable.

Informed Consent Statement: Not applicable.

Data Availability Statement: Data are available on reasonable request to the corresponding author.

Acknowledgments: The technical support provided by the local farmers of Viotia region in Central Greece is gratefully acknowledged.

Conflicts of Interest: The author declares no conflict of interest.

References

1. Filintas, A. Land Use Evaluation and Environmental Management of Biowastes, for Irrigation with Processed Wastewaters and Application of Bio-Sludge with Agricultural Machinery, for Improvement-Fertilization of Soils and Crops, with the Use of GIS-Remote Sensing, Precision Agriculture and Multicriteria Analysis. Ph.D. Thesis, University of the Aegean, Mitilini, Greece, 2011.
2. USDA. *World Agricultural Production*, 2021th ed.; United States Department of Agriculture: New York, NY, USA, 2021; p. 43.
3. FAO. *Coping with Water Scarcity: An Action Framework for Agriculture and Food Security*; FAO: Rome, Italy, 2012; p. 100.
4. Stamatis, G.; Parpodis, K.; Filintas, A.; Zagana, E. Groundwater quality, nitrate pollution and irrigation environmental management in the Neogene sediments of an agricultural region in central Thessaly (Greece). *Environ. Earth Sci.* **2011**, *64*, 1081–1105. [CrossRef]
5. EEA. *Use of Freshwater Resources in Europe, CSI 018*; European Environment Agency (EEA): Copenhagen, Denmark, 2019.
6. Islam, S.M.F.; Karim, Z. World's Demand for Food and Water: The Consequences of Climate Change. In *Desalination-Challenges and Opportunities*; Farahani, M.H.D.A., Vatanpour, V., Taheri, A.H., Eds.; IntechOpen: London, UK, 2019; Chapter 4; pp. 1–27. [CrossRef]
7. Filintas, A.; Wogiatzi, E.; Gougoulias, N. Rainfed cultivation with supplemental irrigation modelling on seed yield and oil of *Coriandrum sativum* L. using Precision Agriculture and GIS moisture mapping. *Water Supply* **2021**, *21*, 2569–2582. [CrossRef]
8. Allen, R.; Pereira, L.; Raes, D.; Smith, M. *Crop Evapotranspiration. Drainage & Irrigation Paper N°56*; FAO: Rome, Italy, 1998.
9. Page, A.L.; Miller, R.H.; Keeney, D.R. *Methods of Soil Analysis Part 2: Chemical and Microbiological Properties*; Agronomy, ASA and SSSA: Madison, WI, USA, 1982; p. 1159.
10. Dioudis, P.; Filintas, A.; Koutseris, E. GPS and GIS based N-mapping of agricultural fields' spatial variability as a tool for non-polluting fertilization by drip irrigation. *Int. J. Sustain. Dev. Plan.* **2009**, *4*, 210–225. [CrossRef]
11. Dioudis, P.; Filintas, A.; Papadopoulos, A. Corn yield response to irrigation interval and the resultant savings in water and other overheads. *Irrig. Drain. J. Int. Comm. Irrig. Drain.* **2009**, *58*, 96–104. [CrossRef]

12. Filintas, A.; Dioudis, P.; Prochaska, C. GIS modeling of the impact of drip irrigation, of water quality and of soil's available water capacity on *Zea mays* L, biomass yield and its biofuel potential. *Desalination Water Treat.* **2010**, *13*, 303–319. [CrossRef]
13. USDA-SCS. *Irrigation Water Requirements*; Technical R. No. 21; USDA Soil Conservation Service: Washington, DC, USA, 1970.
14. Norusis, M.J. *IBM SPSS Statistics 19 Advanced Statistical Procedures Companion*; Pearson: London, UK, 2011.
15. Hatzigiannakis, E.; Filintas, A.; Ilias, A.; Panagopoulos, A.; Arampatzis, G.; Hatzispiroglou, I. Hydrological and rating curve modelling of Pinios River water flows in Central Greece, for environmental and agricultural water resources management. *Desalination Water Treat.* **2016**, *57*, 11639–11659. [CrossRef]

Proceeding Paper

Construction of an Observatory, As a Management Tool Decision, Valorisation and Sustainable Preservation of the Resources of Aromatic and Medicinal Plants [†]

Valter Hoxha [1,*], Florjan Bombaj [1,2] and Hélène Ilbert [3]

[1] Department of Economics, Mediterranean University of Albania, 1023 Tirana, Albania; florjan.bombaj@supagro.fr

[2] CIRAD, INRAE, Institut Agro, UMR Innovation, Université Montpellier, 34060 Montpellier, France

[3] CIRAD, CIHEAM-IAMM, INRAE, Institut Agro, UMR MOISA, Université Montpellier, 34060 Montpellier, France; ilbert@iamm.fr

* Correspondence: valter.hoxha@umsh.edu.al; Tel.: +355-68-82-36-474

† Presented at the 13th EFITA International Conference, online, 25–26 May 2021.

Abstract: Today, the actors of the aromatic and medicinal plants (MAPs) sector are facing several problems related to the management, exploitation, marketing and valorisation of these resources. The objective of this presentation is to build a MAPs monitoring model based on two very important sources of information: Global Positioning System (GPS) tracks for the plant gatherer Linden (Tilia argentea), and historical inventory data of the year 1988. The results show that the experimental model of the database enables the storage, processing and cross-referencing of historical data with the GPS geographic information provided by gatherers.

Keywords: observatory; medicinal and aromatic plants; human sensor (GPS)

Citation: Hoxha, V.; Bombaj, F.; Ilbert, H. Construction of an Observatory, As a Management Tool Decision, Valorisation and Sustainable Preservation of the Resources of Aromatic and Medicinal Plants. *Eng. Proc.* **2021**, *9*, 23. https://doi.org10.3390/engproc2021009023

Academic Editors: Dimitrios Kateris and Maria Lampridi

Published: 30 November 2021

Publisher's Note: MDPI stays neutral with regard to jurisdictional claims in published maps and institutional affiliations.

1. Introduction

Albania is a country rich in biodiversity of medicinal and aromatic plants (MAPs). They are a major economic, social and environmental resource for the country, especially for hilly and the mountainous areas. Each year, the MAPs chain employs between 70,000 and 100,000 people [1,2], for an average value of USD 25 million [3]. At the global level, according to the COMTRADE database, the annual growth rate of this market is 9% [4].

To meet this growing global demand, competition between industries and between the same firms of the same industry is very strong. For them to ensure the raw material, and thus to know the potential of habitats and to check its origin, is information of the utmost importance [5,6].

In Albania, currently the key issue is that there is no real time and space information about the origins of the collected quantity, geographic distribution and the potential of their habitats [2,7]. Therefore, the final objective of this research is to build a model observatory for MAPs that will serve as a decision support tool for all actors in the sector. Our hypothesis is: through the information we obtain from gatherers and human sensors (Global Positioning System—GPS), we can build an observatory of MAPs, aiming to sustainably manage these resources.

2. Materials and Methods

The methodology used to build and test an experimental model of the operation of an observatory goes through several stages. The first stage was the modelling of a database that would serve as an experimental model for monitoring the MAPs sector. For this, we used the MERISE methodology [8].

The second stage was the collection of useful information that served as input for model testing, by producing as output two important pieces of information that were used for decision-making by MAPs branch actors: mapping of the used habitat and statistical analysis of collection time. For this, we used two sources of information. The first comes from the plant collector and their activity, while the second source comes from the inventories carried out by the forest service agencies. The data collected from the plant collector are qualitative and quantitative, and have to do with the plant name, its geographical location, the quantity collected by the users of these plants, as well as geographical information related to the tracking of the collector's activity using a GPS instrument. To collect this information, we used the "5W1H" method.

Meanwhile, to recover the geographical information, in the district of Tepelena, in the village of Dragot, we equipped the plant collector with GPS in order to provide us with all the GPS tracks that accompany active plant collection during the year.

3. Results and Discussion

In our case, this database of MAPs observatory is made up of four large associative tables (ta) and six tables (t) which, by merging them, form the associative tables. The "central" table is the t_observations table. This table contains the quantitative or qualitative values of all the observations. Each row of the table corresponds to a unique observation.

3.1. Conceptual Model Testing

Most of the time, the use of the OMAPs database consists of making specific requests from the v_observations global view. For us, two types of data use are tested:

1. The use of data based on the constitution of a habitat exploited from the aggregation of individual harvesting areas;
2. Comparison of historical data and reconstructed exploited habitats.

3.1.1. Use of GPS Data from Picking Areas

Picking areas are the basic elements that will help to restore the exploited habitat of a plant. To aggregate only the picking areas of the Tepelena district, the above SQL query realises a spatial intersection between the picking areas and the Tepelena district.

The result of this query is displayed in the form of a map and a data table (attribute data) in Figure 1. This cartographic display can also be represented as a bar histogram to better visualise the quantity of plants extracted from the exploited habitat (red bar in Figure 1).

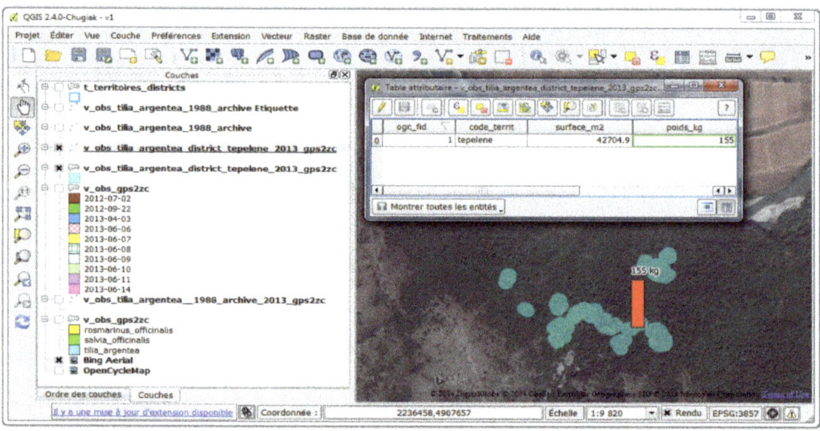

Figure 1. Display in cartographic and histogram form of the piker's GPS data.

3.1.2. Cross-Referencing of Historical Data with GPS Data

The final test is to show that it is possible to aggregate data from individual picking areas, in order to make them directly comparable to certain data from archives provided at the district level. To "cross" these two types of data, they must be at the same scale, or even be attached to the same geographic reference entity. The constructed SQL query makes it possible to compare, in the same representation, the 1988 data from the archives (available only at the district level) and the 2013 data based on the modelling of the harvesting areas also aggregated at the district level (Figure 2).

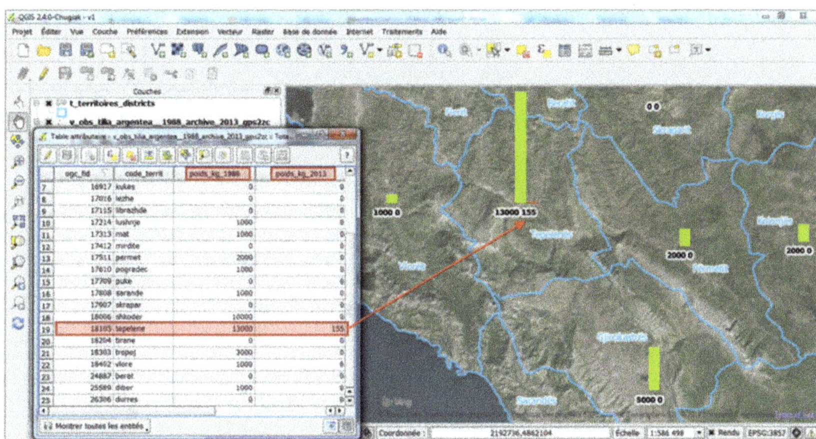

Figure 2. Comparison between the inventory for the year 1988 and the habitat reconstructed in 2013.

It is important to note that the comparison carried out is not representative because it would be necessary to have all the modelled collection zones of all the collectors at the district level for the year 2013, whereas only some areas are available in the Tepelena district.

4. Conclusions

The experimental model of this observatory shows that it is concretely possible to harmonise heterogeneous data and to visualise the results in a synthetic and systematic way from a single base. In addition, the first tests made it possible to test the structuring of the OMAPs database.

In Albania, as probably in other regions of the world, the exploitation and preservation of natural resources could be objectified via a MAPs observatory, making it possible to manage structured repositories, despite the initial heterogeneity of the data.

Author Contributions: Conceptualisation, V.H. and F.B.; methodology, V.H. and H.I.; software, V.H.; validation, H.I., V.H. and F.B.; investigation, V.H and H.I.; writing—original draft preparation, V.H.; writing—review and editing, H.I. and F.B.; visualisation, V.H.; supervision, F.B. All authors have read and agreed to the published version of the manuscript.

Funding: This research received no external funding.

Data Availability Statement: Data sharing not applicable.

Acknowledgments: The technical and human support provided by CIHEAM-IAM of Montpellier is gratefully acknowledged.

Conflicts of Interest: The authors declare no conflict of interest.

References

1. Naka, K.; Musabelliu, B. Social and Economic Relevance of NTFPs in Albania. In *Albanian National Forest Inventory (ANFI) Special Study*; USAID Project; Agricutlural University of Tirana: Tirana, Albania, 2003; p. 60.
2. USAID. *Albania Medicinal and Aromatic Plants: Value Chain Assesment*; Technical report; USAID—Albania Agriculture Competitiveness (AAC) Program: Tirana, Albania, 2010; p. 85.
3. Lekoçaj, J.; Hoxha, V. Information System for Medicinal and Aromatic Plants Value Chain: Case Study of Albania. *Bus. Syst. Rese.* **2019**, *10*, 31–41. [CrossRef]
4. Ilbert, H.; Hoxha, V. Marché mondial des plantes: Analyse des échanges et de la position de l'Albanie et de l'Algérie. In *Le Marché des Plantes Aromatiques et Médicinales: Analyse des Tendances du Marché Mondial et des Stratégies Economiques en Albanie et en Algérie*; Ilbert, H., Hoxha, V., Sahi, L., Courivaud, A., Chailan, C., Eds.; Options Méditerranéennes: Série B; Etudes et Recherches, CIHEAM-IAMM: Montpellier, France, 2016; pp. 17–44.
5. Schippmann, U.; Leaman, D.; Cunningham, A.B. A comparison of cultivation and wild collection of medicinal and aromatic plants under sustainability aspects. In *Medicinal and Aromatic Plants*; Bogers, R.J., Craker, L.E., Lange, D., Eds.; Springer: Dodlerk, The Netherlands, 2006; pp. 75–95.
6. Leaman, D.J. Soulager la pression. *Planète Conserv.* **2009**, *39*, 8–10.
7. Hoxha, V. Quelles méthodes pour la gestion durable de la ressource des plantes aromatiques et médicinales? Analyse des inventaires historiques en Albanie et modélisation des habitats à partir des traces GPS des cueilleurs en vue de la construction d'un observatoire. Thèse de doctorat, Université Paul Valéry, UM3, Montpellier, France, 2014; p. 263.
8. Morley, C. *Management d'un Projet Système d'Information. Principes, Techniques, Mise en Œuvre et Outils, 6ème éd.*; Paris: Dunod, France, 2008; 458p.

Proceeding Paper

Weed Mapping in Vineyards Using RGB-D Perception [†]

Dimitrios Kateris [1,*], Damianos Kalaitzidis [1], Vasileios Moysiadis [1], Aristotelis C. Tagarakis [1] and Dionysis Bochtis [1,2]

[1] Institute for Bio-Economy and Agri-Technology (iBO), Centre of Research and Technology-Hellas (CERTH), 6th km Charilaou-Thermi Rd, 57001 Thessaloniki, Greece; d.kalaitzidis@certh.gr (D.K.); v.moisiadis@certh.gr (V.M.); a.tagarakis@certh.gr (A.C.T.); d.bochtis@certh.gr (D.B.)
[2] farmB Digital Agriculture P.C., Doiranis 17, 54639 Thessaloniki, Greece
* Correspondence: d.kateris@certh.gr
† Presented at the 13th EFITA International Conference, online, 25–26 May 2021.

Abstract: Weed management is one of the major challenges in viticulture, as long as weeds can cause significant yield losses and severe competition to the cultivations. In this direction, the development of an automated procedure for weed monitoring will provide useful data for understanding their management practices. In this work, a new image-based technique was developed in order to provide maps based on weeds' height at the inter-row path of the vineyards. The developed algorithms were tested in many datasets from vineyards with different levels of weed development. The results show that the proposed technique gives promising results in various field conditions.

Keywords: vineyard; weeds; RGB-D camera; point cloud; image processing; UGV

Citation: Kateris, D.; Kalaitzidis, D.; Moysiadis, V.; Tagarakis, A.C.; Bochtis, D. Weed Mapping in Vineyards Using RGB-D Perception. *Eng. Proc.* **2021**, *9*, 30. https://doi.org/10.3390/engproc2021009030

Academic Editors: Charisios Achillas and Dimitios Aidonis

Published: 2 December 2021

Publisher's Note: MDPI stays neutral with regard to jurisdictional claims in published maps and institutional affiliations.

1. Introduction

Successful grape production entails the implementation of management practices that control weeds which maintain grapevine performance and conserve soil quality. Integrated weed management, coupled with the use of autonomous robotic platforms (UGVs and UAVs), allows effective weed management, as a beneficial methodology for the environment [1].

It is a fact that robotics technology is used more frequently as years go by. Crop production is one of the fields robotics, providing alternative ways to weed management by mitigating its biotic and abiotic stresses. For instance, automatic weeding robots nowadays have replaced chemical herbicides to a large extent [2]. Weed spots are detected in cultivation with the help of UGV image acquisition and specific algorithms [3,4], in order for an autonomous robotic system to control weeds by herbicide spraying or through mechanical procedures [5].

In this direction, this work proposes a new image-based technique with the implementation of an RGB-D camera mounted on an autonomous navigated robotic platform in order to depict the dispersion of the weeds at the inter-row path of the vineyards. The objective is to enable the weed mapping at inter-row path reducing labor cost and time. The real-time data from an RGB-D camera are filtered with a color and shape filter consecutively in order to reduce the "useless" information. Subsequently, the proposed technique was tested in many datasets in different vineyards with different levels of weed development.

2. Materials and Methods

The experimental procedure took place in a vineyard located at Ktima Gerovasiliou (Epanomi, Thessaloniki, Greece, 40°27′3″, 22°55′32″) during the summer period of 2020 and 2021.

2.1. UGV Platform and Sensors

An autonomous robotic platform (Thorvald, Saga Robotics, Oslo, Norway) was used as a robotic platform for data acquisition. The robotic platform moved autonomously in a straight line, parallel to the vine row and in the center of the inter-row path. An RGB-D stereo camera was placed on a height-adjustable bar at the front side of the robotic platform. Furthermore, a Lidar sensor (Velodyne VLP 16) was used to scan the space around the vehicle and create a two-dimensional map containing the obstacles detected in space. Finally, a GNSS receiver (Spectra SP80 RTK) was used in order to inform the navigation system about the position of the vehicle in the field with latitude and longitude coordinates. All the sensors were placed on the vehicle in such a way without dead spots in the data that were received from them. Figure 1 shows the Thorvald robotic platform and the mounted sensors for platform navigation and weed mapping in the vineyard.

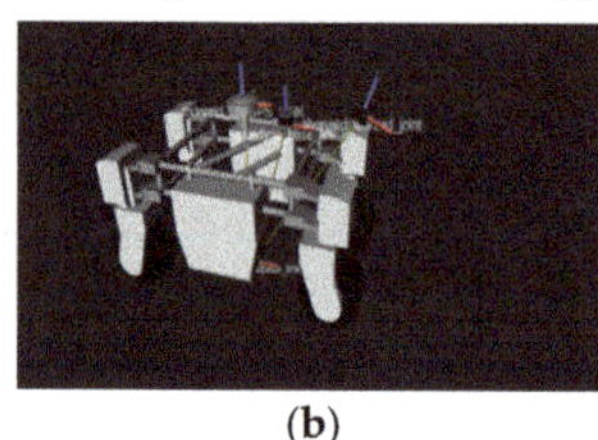

(a)　　　　　**(b)**

Figure 1. The robotic platform (**a**) in the field and (**b**) the exact positions of the sensors on the robotic platform.

2.2. Data Acquisition

The robotic platform moved at 1.5 km h^{-1}, producing insignificant vibrations at this speed. The ZED 2 stereo camera was mounted at the frond side of the platform and was used exclusively for weed mapping in the vineyard. A recording system was set up for data acquisition (Jetson TX2). This sampling system was equipped with the necessary instrumentation for data collection and remote operation mode. The sampling frequency was 15 Hz, ensuring high-quality data. The ZED 2 supplied RGB images (1280 × 720 pixels) together with depth information. The ZED 2 camera provided an image both in the form of photography (Figure 2a) and in the three-dimensional form with the creation of a point cloud (Figure 2b).

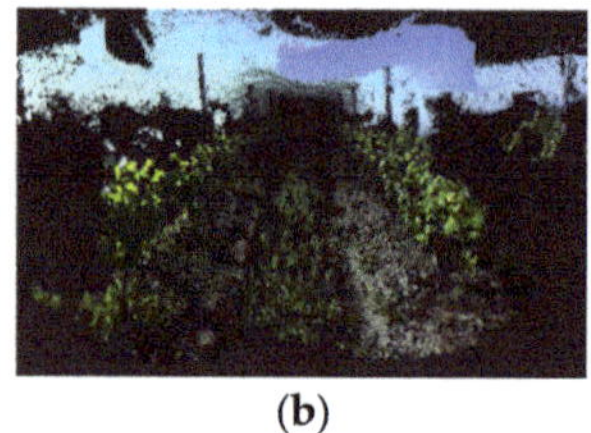

(a)　　　　　**(b)**

Figure 2. Images supplied by the ZED 2 stereo camera of the vineyard inter-row path. (**a**) from the RGB channel and (**b**) the point cloud.

2.3. Methodology

In order to detect the weeds in the inter-row path, the PCL_ROS library was used to locate the weeds in the data (point cloud) that were collected from the vineyard. This library comprises tools and filters that allow us to identify the weeds in the inter-row path and remove the "useless" information such as the vines and the soil. In this work, a simple technique was used to detect the weeds in the vineyard. This technique was based on the fact that the weeds are located at the inter-row of the vineyard and their height is over the field ground, also they have a green hue. Considering this assumption, two filters were developed in order to retain the information that meets these conditions. The entire process is presented in Figure 3.

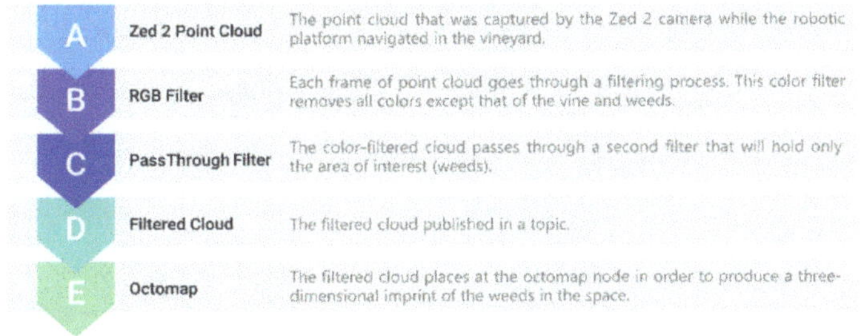

Figure 3. Data-processing flow chart.

Initially, the color filter removes all colors except that of the vine and weeds. To remove the remaining colors, a second filter was created in order to remove all shades except those of green. The RGB filter values were selected after a number of tests took place at different time periods in the vineyard. The results of the first filter filtration are presented in Figure 4. As it is clearly shown, the removal of "useless" colors in weed detection and the retention of the information necessary for detection are observed. To be more specific, the removal of the blue color is observed in Figure 4, due to the noise of the camera as well as the removal of the soil that exists in the corridor. After this filtering, the remaining information includes the weeds and the vineyard.

Figure 4. The point cloud after the color filter.

In the second phase, the filtered point cloud was also filtered in terms of its dimensions. All the areas that were not at the inter-row path were removed from the point cloud. The results of the application of the second filter are presented in Figure 5. It is obvious that the vineyard rows were removed and the only information remaining in the point cloud is from the inter-row path.

Figure 5. The point cloud after the shape filter.

After the filtering process of the point cloud, the remaining information concerns the weeds in the inter-row path of the vineyard. Initially, all points of the point cloud were taken at distances concerning the coordinate system of the frame ZED_LINK coordinates. The ROS tf_tranform library was used for the process of transforming the point cloud from the ZED_LINK frame to the world frame. After the end of this transformation process, the point cloud points obtained with the camera were converted to point cloud utm coordinates. This point cloud could then geotag and was ready to present the height distribution map of the inter-row path weeds in the vineyard (Figure 6).

Figure 6. Height distribution map of the inter-row path weeds in the vineyard.

3. Conclusions

To conclude, this work was based on an image-based technique using an RGB-D camera with the aim to show the dispersion of the weeds at the inter-row path of a vineyard. While pre-processing, the RGB-D camera's real-time data were transformed into different color spaces for noise reduction. The above algorithms and techniques were tested in data from real vineyards, on a UGV platform with promising results, where weeds are not at the same development phase.

Author Contributions: Conceptualization, D.K. (Dimitrios Kateris); methodology, D.K. (Dimitrios Kateris), D.K. (Damianos Kalaitzidis); software, D.K. (Damianos Kalaitzidis); validation, D.K. (Damianos Kalaitzidis) and V.M.; formal analysis, D.K. (Dimitrios Kateris) and D.K. (Damianos Kalaitzidis); investigation, V.M.; resources, D.K. (Dimitrios Kateris); data curation, D.K. (Damianos Kalaitzidis); writing— original draft preparation, D.K. (Dimitrios Kateris); writing—review and editing, A.C.T. and D.B.; visualization, D.K. (Damianos Kalaitzidis); supervision, D.B.; project administration, D.K. (Dimitrios Kateris); funding acquisition, D.B. All authors have read and agreed to the published version of the manuscript.

Acknowledgments: This work was supported by the HORIZON 2020 Project "BACCHUS: Mobile Robotic Platforms for Active Inspection and Harvesting in Agricultural Areas" (project code: 871704) financed by the European Union, under the call H2020-ICT-2018-2020.

Conflicts of Interest: The authors declare no conflict of interest.

References

1. Jiménez-Brenes, F.M.; López-Granados, F.; Torres-Sánchez, J.; Peña, J.M.; Ramírez, P.; Castillejo-González, I.L.; de Castro, A.I. Automatic UAV-based detection of Cynodon dactylon for site-specific vineyard management. *PLoS ONE* **2019**, *14*, e0218132. [CrossRef] [PubMed]
2. Wu, Z.; Chen, Y.; Zhao, B.; Kang, X.; Ding, Y. Review of weed detection methods based on computer vision. *Sensors* **2021**, *21*, 3647. [CrossRef] [PubMed]
3. Benos, L.; Tagarakis, A.C.; Dolias, G.; Berruto, R.; Kateris, D.; Bochtis, D. Machine Learning in Agriculture: A Comprehensive Updated Review. *Sensors* **2021**, *21*, 3758. [CrossRef] [PubMed]
4. Anagnostis, A.; Tagarakis, A.C.; Asiminari, G.; Papageorgiou, E.; Kateris, D.; Moshou, D.; Bochtis, D. A deep learning approach for anthracnose infected trees classification in walnut orchards. *Comput. Electron. Agric.* **2021**, *182*, 105998. [CrossRef]
5. Moreno, H.; Rueda-Ayala, V.; Ribeiro, A.; Bengochea-Guevara, J.; Lopez, J.; Peteinatos, G.; Valero, C.; Andújar, D. Evaluation of vineyard cropping systems using on-board rgb-depth perception. *Sensors* **2020**, *20*, 6912. [CrossRef] [PubMed]

Proceeding Paper

A Method Comparison Study between Open Source and Industrial Weather Stations [†]

Evmorfia P. Bataka [1,*], Georgios Miliokas [2], Nikolaos Katsoulas [2] and Christos T. Nakas [1,3]

1   Laboratory of Biometry, Department of Agriculture, Crop Production and Rural Environment, University of Thessaly, Fytokou St., 384 46 Volos, Greece; cnakas@uth.gr
2   Laboratory of Agricultural Constructions–Greenhouses, Department of Agriculture, Crop Production and Rural Environment, University of Thessaly, Fytokou St., 384 46 Volos, Greece; giwrgos.pg@gmail.com (G.M.); nkatsoul@uth.gr (N.K.)
3   University Institute of Clinical Chemistry, Inselspital, Bern University Hospital, University of Bern, 3010 Bern, Switzerland
*   Correspondence: bataka@uth.gr
†   Presented at the 13th EFITA International Conference, online, 25–26 May 2021.

Abstract: Open-source devices are widespread and have been available to everyone over the past decade. The low cost of such devices boosts the creation of instruments for various applications such as smart farming, environmental monitoring, animal behavior monitoring, human health monitoring, etc. This research aims to use statistical methods to assess agreement and similarity in order to compare an open-source weather station that was constructed and programmed from scratch with an industrial weather station. The experiment took place in the experimental Greenhouses of the University of Thessaly, Velestino, Greece, for 7 consecutive days. The topology of the experiment consisted of 30 open-source weather stations and three industrials, creating three clusters with a ratio of 10 open-source to 1 industrial. The results revealed low to high agreement across the measurement range, with high variability, possibly due to factors that were not considered in the statistical model.

Keywords: agreement; similarity; open-source hardware; open-source software

Citation: Bataka, E.P.; Miliokas, G.; Katsoulas, N.; Nakas, C.T. A Method Comparison Study between Open Source and Industrial Weather Stations. *Eng. Proc.* **2021**, *9*, 8. https://doi.org/10.3390/engproc2021009008

Academic Editors: Charisios Achillas and Lefteris Benos

Published: 19 November 2021

1. Introduction

The open-source philosophy [1] is applied in numerous areas of our society. The need for inexpensive and trustworthy devices in scientific and industrial applications fomented the hardware and software open-source community [2]. The reliability of such devices is questioned since non-specialized individuals can produce devices expected to have certain specifications theoretically but fail to meet them in the field. Thus, a formal method should be applied, and the outcomes can be proof of the device's effectiveness. The new device can be validated and assess its interchangeability compared to already established devices using appropriate statistical methods. Statistical methods to assess agreement and similarity are used extensively in medicine, applied chemistry [3,4] and other fields to compare approved and widely used devices or methods with experimental ones that might be either less expensive, less intrusive or even both [5]. The need for the application of this method in agriculture is vital since a great number of devices for environmental monitoring emerged during the past years [6,7]. Various comparison methods are used that are inappropriate and reveal misleading results, according to Altman et al. [8]. We constructed an inexpensive and user-friendly device for environmental monitoring and assessed its efficiency in field conditions. Then, we adapted existing statistical methods to assess agreement and similarity between our device and existing certified industrial ones to produce indices and graphs that are easily interpreted by individuals with no need for advanced knowledge in statistics.

2. Materials and Methods

2.1. Open-Source Weather Station

The open-source weather station constructed for this experiment uses a microcontroller and components listed in Table S1 (Supplementary Materials). It measures the air temperature every five minutes instantly. The device has the ability to store data in a private server provided by the Adafruit company, Adafruit IO [9], via WiFi. Figures S1 and S2 (Supplementary Materials) display the electronic parts and final product, respectively. The cost of the device is displayed in Table S2, which is around EUR 54 (2020 market prices). We programmed the microprocessor using Arduino IDE [10]. The libraries used are listed in Table S3 (Supplementary Materials). In Figure S3 (Supplementary Materials), the flowchart presents the sequence of the weather station's functions.

2.2. Industrial Weather Station

The industrial weather station is a Thygro SDI-1 (Figure S4, Supplementary Materials) [11] and is equipped with air temperature, humidity and solar radiation sensor, solar panels and radiation shield. The temperature accuracy is ±0.2 °C, and the resolution is ±0.015 °C. The total cost is EUR 3000, while the service cost for remote monitoring is additional.

2.3. Experimental Design

The experiment took place in the experimental greenhouses, University of Thessaly, Greece. The topology consists of three industrial weather stations and 30 open-source, surrounding each industrial as shown in Figure S5. This configuration reduces the cost of the experiment substantially. The cultivation was tomato, and one of the industrial weather stations was located in chamber 1 and the rest in chamber 2. The duration of the experiment was 7 days.

2.4. Method Comparison Strategy

A method comparison study sets two important goals that we adapted in our research: According to Choudharry [12], the primary goal is to quantify the extent of agreement between two measurement methods and determine whether they have sufficient agreement so that we can use them 'interchangeably'. The secondary goal is to compare important characteristics of the measurement methods such as biases, precision and sensitivities to find the sources of their disagreement. The latter comparison is set as the evaluation of similarity.

In our setting, the new method is the open-source weather station and the reference method is the industrial station. Indices and graphical illustrations aid the researcher in easily understanding and interpret the results with no need for advanced knowledge in statistics.

The official method comparison technique is the assessment of agreement and similarity using ad hoc statistical analysis [3,4]. Common practices used to compare two different methods/devices can be misleading or inappropriate. A short exposition is given in Table S4 (Supplementary Materials).

There are many indices and graphs used to interpret agreement between two methods or devices [13]. Each index has advantages and disadvantages, and a combination can be used for better understanding the results. The same applies for the graphical representations.

Choudharry et al. [12] proposed binormal and mixed effect models that produce indices by extracting the model's parameters, useful graphs and indices for agreement, such as Concordance Correlation Coefficient (CCC), Total Deviation Index (TDI) and similarity indices, such as precision ratio, fixed bias and sensitivity. We followed this approach and also produced a similarity index, called fixed bias, which is the mean difference of the paired measurements of the devices.

The model used the method as the factor, which has two levels, the industrial as method 1 and the open source as method 2. Each pair of stations, which is the subject in our case, is the random factor. The model we adapted and the index formulas can be found in Table S5 (Supplementary Materials).

Our data are longitudinal since they are produced in 5-minute intervals. However, to avoid model complexity since the instances are too many, we modeled each occasion (5 min interval) separately for every day. Table S6 (Supplementary Materials) displays the number of occasions per day.

3. Results

Visual Representation of the Results

The graphs produced for the visual representation of the data are the CCC vs. Occasion, the TDI vs. Occasion and the Fixed Bias vs. Occasion. Every graph has a common x-axis that corresponds to the occasions per day, and the points are color-coded for every pair based on pre-defined temperature categories. Only Day 1 is displayed below, and the rest can be found in the Supplementary Materials (Figures S6–S11).

For Day 1, the CCC vs. Occasions plot (Figure 1a) reveals low to high agreement for the whole temperature range due to the high variance of the index. The index is not consistent for the different temperature categories. The TDI vs. Occasions plot (Figure 1b) estimated that 90% of the absolute value of differences lies between 2.3 °C and 4.7 °C for temperatures lower than 27 °C and between 1.25 °C and around 6 °C for more than 27 °C. Temperatures more than 29 °C report higher variance than the others. The Fixed Bias vs. Occasions plot (Figure 1c) estimates −2.1 °C to 0.4 °C fixed bias for temperatures between 23 °C to 27 °C, −2.2 °C to around 0.75 °C fixed bias for temperatures between 27 °C and 29 °C and around −1.25 °C to 2.25 °C for the rest of the categories. The trend and the corresponding standard errors were estimated using a LOESS smoother.

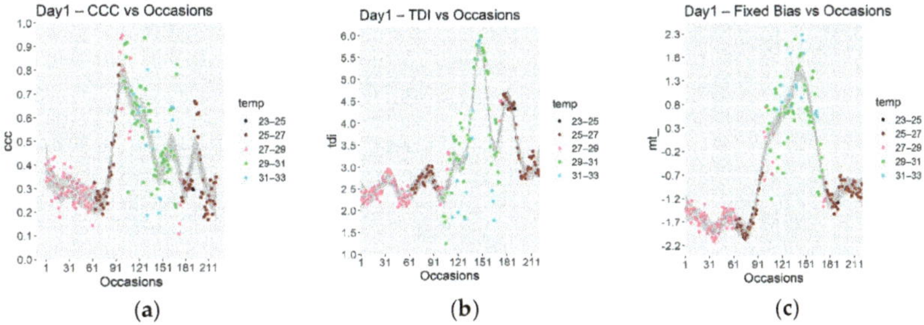

Figure 1. (a) CCC vs. Occasions plot for Day 1 for every temperature range category. (b) TDI vs. Occasions plot for Day 1 for every temperature range category. (c) Fixed bias vs. Occasions for Day 1 for every temperature range category.

Overall, from occasions 91 to 181, it seems that the variability is higher each day. This is probably due to the air humidity difference as shown in Figure S11. Thus, humidity is a possible factor that could explain the high variability on the indices.

4. Discussion

Our comparison study uses statistical methods to successfully assess agreement and similarity for open-source devices. We combine three useful indices and interpret them as complements. The analysis of the data revealed high to low agreement with increased variability for higher temperatures. The open-source weather station performance might be acceptable for specific applications, and the low production cost might encourage researchers to produce them. After examining our preliminary data, we conclude that there is a need for more research to reduce the variance of the indices across the temperatures. Specifically, adding other factors, such as air humidity, vegetation density or the location of each pair in the greenhouse, will certainly improve the model since there may be many confounding variables. Finally, including pooled data from the whole duration of the experiment and not only per day will reveal trends and will help retrieve hidden factors that affect the agreement.

Supplementary Materials: The following are available online at https://www.mdpi.com/article/10.3390/engproc2021009008/s1, Figure S1: Finalized PCB, Figure S2: The final product, Figure S3: Flowchart, Figure S4: Industrial Weather Station, Figure S5: The topology of the experiment, Figures S6–S11: Day 2 to Day 7 plots, Figure S12: Humidity for the greenhouse chambers, Table S1: Components List, Table S2: Cost, Table S3: Libraries, Table S4: Inadequate or misleading practices. Table S5: Model adapted for the analysis, Table S6: Number of occasions per day.

Author Contributions: Conceptualization, E.P.B. and C.T.N.; methodology, E.P.B. and C.T.N.; software, E.P.B. and C.T.N.; validation, E.P.B. and C.T.N.; formal analysis, E.P.B. and C.T.N.; investigation, E.P.B. and C.T.N.; resources, N.K.; data curation, E.P.B. and G.M.; writing—original draft preparation, E.P.B. and C.T.N.; writing—review and editing, E.P.B., C.T.N., G.M. and N.K.; visualization, E.P.B.; supervision, C.T.N.; project administration, C.T.N.; funding acquisition. All authors have read and agreed to the published version of the manuscript.

Funding: This research received no external funding.

Institutional Review Board Statement: Not applicable.

Informed Consent Statement: Not applicable.

Data Availability Statement: Data are contained within the article or Supplementary Material.

Conflicts of Interest: The authors declare no conflict of interest.

References

1. Introduction to the Open-Source Philosophy and the Benefits of Sharing. Available online: https://www.academia.edu/44461841/Introduction_to_the_Open_Source_Philosophy_and_The_Benefits_of_Sharing (accessed on 20 March 2021).
2. Community and Collaboration Open-Source Initiative. Available online: https://opensource.org/community (accessed on 19 April 2021).
3. Tractenberg, R.E.; Jonklaas, J.; Soldin, S.J. Agreement of immunoassay and tandem mass spectrometry in the analysis of cortisol and free t4: Interpretation and implications for clinicians. *Int. J. Anal. Chem.* **2010**, *2010*, 234808. [CrossRef] [PubMed]
4. Zaki, R.; Bulgiba, A.; Ismail, N.A. Testing the agreement of medical instruments: Overestimation of bias in the Bland-Altman analysis. *Prev. Med.* **2013**, *57*, 80–82. [CrossRef] [PubMed]
5. Bataka, E.P.; Rodovitis, V.G.; Zarpas, K.D.; Papadopoulos, N.T.; Nakas, C.T. FruiTemp: Design, Implementation and Analysis for an Open-Source Temperature Logger Applied to Fruit Fly Host Experimentation. *Appl. Sci.* **2021**, *11*, 6003. [CrossRef]
6. Fisher, D.K.; Fletcher, R.S.; Anapalli, S.S. Evolving Open-Source Technologies Offer Options for Remote Sensing and Monitoring in Agriculture. *AIT* **2010**, *10*, 1–10. [CrossRef]
7. Martinez, J.; Egea, G.; Aguera, J.; Perez-Ruiz, M. A cost-effective canopy temperature measurement system for precision agriculture: A case study on sugar beet. *Precis. Agric.* **2017**, *18*, 95–110. [CrossRef]
8. Altman, D.G.; Bland, J.M. Measurement in medicine: The analysis of method comparison studies. *J. R. Stat. Soc.* **1983**, *32*, 307–317. [CrossRef]
9. Adafruit IO Basics. Available online: https://learn.adafruit.com/series/adafruit-io-basics (accessed on 1 February 2021).
10. Arduino IDE. Available online: https://www.arduino.cc/en/software. (accessed on 25 July 2001).
11. Thygro SDI-12 Temperature and Humidity Sensor. Available online: https://symmetron.gr/uploads/ThygroSDI-12Brochure_EN.pdf (accessed on 24 May 2021).
12. Choudhary, P.K.; Nagaraja, H.N. *Measuring Agreement*, 1st ed.; Wiley: Hoboken, NJ, USA, 2017; pp. 1–251.
13. Lin, L.; Hedayat, A.S.; Wu, W. *Statistical Tools for Measuring Agreement*, 1st ed.; Springer: New York, NY, USA, 2012; pp. 1–160.

Proceeding Paper

Machine Learning Techniques in Agricultural Flood Assessment and Monitoring Using Earth Observation and Hydromorphological Analysis [†]

Lampros Tasiopoulos [1,2], Marianthi Stefouli [3], Yorghos Voutos [4], Phivos Mylonas [4] and Eleni Charou [1,*]

1   NSCR "Demokritos", 15341 Athens, Greece; labrostasiopoulos@gmail.com
2   Ernst & Young, 15125 Athens, Greece
3   Hellenic Survey of Geology & Mineral Exploration, 13672 Athens, Greece; stefouli@igme.gr
4   Department of Informatics, Ionian University, 49100 Corfu, Greece; c16vout@ionio.gr (Y.V.);
    fmylonas@ionio.gr (P.M.)
*   Correspondence: exarou@iit.demokritos.gr
†   Presented at the 13th EFITA International Conference, Online, 25–26 May 2021.

Abstract: Climate change could exacerbate floods on agricultural plains by increasing the frequency of extreme and adverse meteorological events. Flood extent maps could be a valuable source of information for agricultural land decision makers, risk management and emergency planning. We propose a method that combines various types of data and processing techniques in order to achieve accurate flood extent maps. The application aims to find the percentage of agricultural land that is covered by the floods through an automatic map estimation methodology based on the freely available Sentinel-2 (S2) satellite images and machine learning techniques.

Keywords: flood assessment; remote sensing; data processing; machine learning

Citation: Tasiopoulos, L.; Stefouli, M.; Voutos, Y.; Mylonas, P.; Charou, E. Machine Learning Techniques in Agricultural Flood Assessment and Monitoring Using Earth Observation and Hydromorphological Analysis. *Eng. Proc.* **2022**, *9*, 40. https://doi.org/10.3390/engproc2021009040

Academic Editors: Dimitrios Kateris and Maria Lampridi

Published: 31 December 2021

Publisher's Note: MDPI stays neutral with regard to jurisdictional claims in published maps and institutional affiliations.

1. Introduction

Floods constitute one of the greatest threats to the property and safety of human communities. Flash floods are events occurring on a small spatial scale within a short time, under conditions of rapid production of surface runoff [1]. Climate change could provoke more frequent, intense and adverse meteorological phenomena resulting in flash floods [2]. In this work, we used remote sensing (RS) data that offer a synoptic and repeated view over the areas of study [3]. More specifically, we exploited image data from the Sentinel-2 (S2) mission of the European Space Agency (ESA) that offers advanced spatial, spectral resolution and revisit frequencies.

Two major flood events occurred during summer 2020 in Greece, Evia floods and Ianos Cyclone floods. Thus, the objective of the study was to propose an efficient methodology to combine multitemporal/multisource data and processing techniques in order to extract information on flooded areas and produce more accurate maps.

1.1. Pilot Areas

Three sites were selected as pilot study areas: the Evia Politika area, Kefalonia and Thessalia plains in Greece (Figures S1, S2a, S2b and Table 1).

Table 1. Study site region, location and surface water conditions.

Region	Location	Surface Water Conditions	Extent (km^2)
Evia	Politika	Flooded areas due to severe rainfall event—3 fatalities	84
Kefalonia	Municipality Pilareon	Flooding event—natural hazards	59
Thessalia	Enipeas Pinios rivers	Flooded cotton fields	76

1.2. Data

The data used are Sentinel-2 [4] multi-temporal satellite data, orthophotos DEM (Digital Elevation Model) and ancillary land cover/use maps. The acquisition dates of satellite images for the respective areas are:

- Evia: 29 July 2020, 3 August 2020, 13 August 2020, 28 August 2020;
- Kefalonia: 05 September 2020, 20 September 2020;
- Thessalia: 31 August 2020, 20 September 2020.

2. Methodology

Multitemporal remotely sensed data along with ancillary data such as cartographic maps and ground truth data were processed and analysed using image processing and GIS techniques.

Two methodologies were used for the analysis of data as shown in Figures S3 and S4. The hydrological analysis was based on Reuter, H.I. et al. [5], who suggested a comprehensive approach for DEM preprocessing and hydrological analysis. Initially, administrative boundaries of Greece were downloaded from GEODATA [6] to select the pilot project areas. The DEM of 5 m resolution of the Greek Cadastre was also used. Various image processing and vector GIS techniques were used in order to define watersheds and streams of the pilot areas.

Concerning the methodology for identifying changes to land cover due to flood events in the three pilot areas, two machine learning techniques were applied, namely Self Organising Maps (SOM) [7,8] and isodata clustering [9].

For the Evia Politika area and the Thessalia plain, both the spectral index and the Image-Unsupervised Classification Self-Organizing Map were used on the S2 images in order to discriminate all inherent land/flood cover classes of the satellite images.

Classification results were further analysed in GIS along with various maps and information acquired from ortho-photos. Land cover changes were identified by comparing land cover classification results based on satellite data before and after the flood event.

For the Kefalonia pilot area, the isodata unsupervised classification technique [9] was applied on S2 images. The areas classified as flooded areas were converted from raster to vector and combined with agricultural data from the Greek Payment Authority of Common Agricultural Policy to identify the land cover types affected by floods.

An area of 10 km^2 of Evia Politika basin had been affected by the flood event. The analysis provides satellite evidence that the disaster was caused by a large mass of rock/sediment/mud/debris dislodged from the slopes of/higher altitude areas of the basin. It started as erosion, slope instabilities and even landslides and then transformed into a mud and debris flow, causing destruction along its path. Fatalities were due to severe stream flow discharge from Polititika village towards the plain (Figure S5). The analysis of satellite images acquired before and after the event can be used to quickly determine and quantify key measures of the event, e.g., elevation differences and travel distances of erosion products/water. This study clearly shows that satellite data could play a significant role in future high mountain hazard assessments, in particular for evaluating large and relatively inaccessible areas. It is suggested that due to climate change, such events might be happening more frequently, and that the full potential of satellite data and knowledge should be utilized to identify possibly dangerous regions (Figure S6).

For the Thessalia plain, the flooded area was about 35% of the total area, while 15% of summer cultivations were affected. Changes caused by the flood event were mapped even at the field level, and this can be useful during field inspections. Various summer cultivations and mainly the ready-for-harvesting cotton fields were affected. Figure S2 images provided precise data for tracking the spatial footprint of surface water changes/flood waters at regional and local scales, i.e., a field of 23.7 ha (Figures S7 and S8).

2.1. Processing Techniques

Various image processing and vector GIS techniques were used for the analysis of both the satellite imagery and the collected map data and field information, such as Georeferencing, Resampling, Water/Vegetation spectral features, Colour Composites, Intensity Hue Saturation (HIS) Images, Identification of Areas of Interest (AOI), Automatic combination of the classification result of multi-temporal imagery, Automatic conversion of raster to vector data, Collection/Input/Coding, Storage/Management, Retrieval of various data and Processing/Analysis.

2.2. Visualization and Mapping

In terms of Image Classification-Unsupervised Classification techniques using neural networks, Artificial Neural Networks (ANNs) are generally quite effective for the classification of remotely sensed data. For classification purposes, the Self Organized ANN method was used on the Sentinel 2 images in order to discriminate all inherent land/flood cover classes of the satellite images using automated conversion of raster to vector data: the raster output of the classification and/or interpretation process was converted to vector data and these data were analyzed with the corresponding map data and observations acquired in the ortho-photos. Further processing and analysis was performed to derive information concerning land cover changes due to flooding in the pilot study areas (Table 2).

Table 2. Total areas calculated.

Land Use (LU)	Sum Area of LU (m^2)	LU Flooded (m^2)	LU Remain (m^2)	LU Remain (%)	LU Affected (LC) (%)	Total Percentage (%)
Forest	3,043,611	210,799	2,832,812	93%	7%	100%
Vineyard	229,634	18,240	211,394	92%	8%	100%
Vineyard Mix	181,007	27,853	153,154	85%	15%	100%
Arable	905,322	144,450	760,872	84%	16%	100%
Arable Mix	3,397,804	475,221	2,922,583	86%	14%	100%
Olive Growing	9,266,342	412,194	8,854,148	96%	4%	100%
Olive Growing Mix	3,354,840	384,223	2,970,617	89%	11%	100%

3. Conclusions

In summary the following areas were identified during the aforementioned process: flooding, erosion, abrasion surfaces, areas of sediment transport due to flood events, areas of slope instability, landslides and land cover types affected by the flood events.

In this work, we evaluated a methodology to automatically map flooded areas from multispectral Figure S2 images. The methodology enables the identification, delineation, and monitoring of floods and estimates of changes in surface land cover/use. The techniques involved in the developed methodology could be applied for the monitoring of aspects of floods and, eventually, could be used for the mitigation of their environmental, social and economic footprints. The methodology targets regional and local agencies that are in charge of managing rescue operations and assessing damage effectively.

Supplementary Materials: The following are available online at: https://tinyurl.com/efita178, Figure S1: Pilot Project Areas, Figure S2a: Evia Politika area a, Figure S2b: Evia Politika area b, Figure S3: Hydrological analysis, Figure S4: Change detection due to flood disaster, Figure S5: Mapping changes due to the 9th August Evia flood event, Figure S6: Example of classification of flood surface waters of Thessaly using multitemporal Sentinel 2 images, Figure S7: River basins and Hydrological analysis of the Pilareon municipality, Figure S8: Land use classes due to flood disaster.

Author Contributions: Conceptualization, L.T. and M.S.; methodology, Y.V.; software, P.M.; validation, E.C.; data curation L.T., M.S. and Y.V.; supervision P.M. and E.C. All authors have read and agreed to the published version of the manuscript.

Funding: This research was funded by the European Union and Greece (Partnership Agreement for the Development Framework 2014–2020) under the Regional Operational Programme Ionian Islands 2014–2020 for the project "Assessment of the impact on biodiversity of High Nature Value Areas in the Region of Ionian Islands due to the invasion of the allocthonus plant species *Ailanthus altissima* (HNV-Threat)" (MIS code 5034911).

Data Availability Statement: The data presented in this study are available on request from the corresponding author. The data are not publicly available due to the reason that this is an ongoing study.

Conflicts of Interest: The authors declare no conflict of interest.

References

1. Jonkman, S.N. Global perspectives on loss of human life caused by floods. *Nat. Hazards Dordr.* **2005**, *34*, 151–175. [CrossRef]
2. Kharin, V.V.; Zwiers, F.W.; Zhang, X.; Hegerl, G.C. Changes in temperature and precipitation extremes in the IPCC ensemble of global coupled model simulations. *J. Clim.* **2007**, *20*, 1419–1444. [CrossRef]
3. Bresciani, M.; Stroppiana, D.; Odermatt, D.; Morabito, G.; Giardino, C. Assessing remotely sensed chlorophyll-a for the implementation of the water framework directive in European perialpine lakes. *Sci. Total Environ.* **2011**, *409*, 3083–3091. [CrossRef] [PubMed]
4. Available online: https://earth.esa.int/web/sentinel/home (accessed on 1 July 2021).
5. Reuter, H.I.; Hengl, T.; Gessler, P.; Soille, P. Preparation of DEMs for geomorphometric analysis. *Dev. Soil Sci.* **2009**, *33*, 87–120.
6. Available online: https://geodata.gov.gr/ (accessed on 1 July 2021).
7. Kohonen, T. *Self-Organization and Associative Memory*, 3rd ed.; Springer: Berlin/Heidelberg, Germany; New York, NY, USA, 1989.
8. Vassilas, N.; Charou, E. A New Methodology for Efficient Classification of Multispectral Satellite Images Using Neural Network Techniques. *Neural Process. Lett.* **1999**, *9*, 35–43. [CrossRef]
9. Elkhrachy, I. Assessment and management flash flood in Najran Wady using GIS and remote sensing. *J. Indian Soc. Remote Sens.* **2018**, *46*, 297–308. [CrossRef]

Proceeding Paper

A Review of a Teaching–Learning Strategy Change to Strengthen Geomatic Concepts and Tools in the Biosystems Engineering Academic Studies at the Universidad de Costa Rica [†]

José Aguilar * and Matías Cháves

Escuela de Ingeniería de Biosistemas, Universidad de Costa Rica, San Jose 11501-2060, Costa Rica; matias.chaves@ucr.ac.cr

* Correspondence: jose.aguilar@ucr.ac.cr

† Presented at the 13th EFITA International Conference, online, 25–26 May 2021.

Abstract: A different teaching–learning strategy was implemented in the geomatic applications in the Biosystems Engineering course. The strategy, which was based on the educational theory of constructivism, promoted the strengthening of geomatic concepts and tools through case studies that aim to balance lectures and practices to achieve an efficient teaching–learning level. The evaluation and follow-up of this case study review the project's progress and include students' opinions, as collected in a survey over multiple semesters. The strategy allowed the classroom to become a space that promoted the benefit of the student's experience in their professional training.

Keywords: case studies; geoscience; geomatic; curricular design; constructivism; software engineering

Citation: Aguilar, J.; Cháves, M. A Review of a Teaching–Learning Strategy Change to Strengthen Geomatic Concepts and Tools in the Biosystems Engineering Academic Studies at the Universidad de Costa Rica. *Eng. Proc.* **2021**, *9*, 44. https://doi.org/10.3390/engproc2021009044

Academic Editors: Dimitrios Aidonis and Aristotelis Christos Tagarakis

Published: 25 January 2022

Publisher's Note: MDPI stays neutral with regard to jurisdictional claims in published maps and institutional affiliations.

1. Introduction

As a public higher education institution, the Universidad de Costa Rica (UCR) is committed to improving teaching, research, and social action. Academics and students use humanistic values to seek sustainable solutions to national problems in a global context. Career preparations related to agriculture and natural resources need transformation for higher learning goals in modern times. In 2013, the UCR Agricultural Engineering career changed to Biosystem Engineering, aligning with the digital revolution incorporating new subjects such as information and communication technologies (ICT). Viewed as a key and transversal part of the academic program, one of the areas of modernization was the informatics class in agriculture, which was renamed Geomatic. Geomatic is an essential subject in the biosystems program that aims to solve society's complex problems in a challenging way. Teachers also needed to change their way of teaching. One change made was balancing lectures and practices to achieve an efficient teaching–learning level. Students wanted to address social problems in their learning experience. Jean Piaget [1] developed the theory of constructivism and hypothesized that "humans create knowledge through the interaction between their experiences and ideas". With this approach, teacher and student develop a much closer learning strategy connected to life experiences, assuring that the knowledge is learned well. Holley [2] presented an example of the methodology success where she found there is a significant increase in positive attitudes toward the subject, and the results from tests demonstrated a significant improvement. Holley indicates that "the comparison of pretest and posttest shows a significant change" and that it helped the students in their learning and especially in their confidence. A case study methodology can be an effective way for the teacher who wants to improve the learning experience by developing higher cognitive skills in the students. However, this requires a much longer dedication to class preparations to have an effective methodology [3]. In an engineering career, where the context with the real world is fundamental, every engineering course

Eng. Proc. **2021**, *9*, 44. https://doi.org/10.3390/engproc2021009044

should have as many exposures to the world as possible. More specifically, adding agriculture to the engineering-based career and combining it with ICT produces a much more positive impact on the student [4]. Other authors believe that the modernization of the learning processes and development of information processing competencies and cognitive skills in students through ITC enables greater efficiency in the institutional and academic management processes of schools [5].

2. Materials and Methods

One of the teacher's roles is to design case studies based on actual problems that need potential solutions to promote two learning abilities. The first ability is using the appropriate knowledge and skills to identify, formulate, research (literature), and analyze engineering problems with geomatic solutions. This results in significant conclusions, using engineering principles. The second ability is applying the knowledge and fundamentals of geomatics tools to add to the space-time phenomena to engineering solutions. Case studies create learning in spaces that generate knowledge with a methodological logic of background, where the following possibilities are considered:

1. Problem planning, design, and execution of solutions;
2. Transformation of content into procedural, attitudinal, or conceptual tools;
3. Strengthened relations between teachers, students, and agricultural producers; and
4. Ability to work cooperatively to build knowledge among work teams.

3. Results

3.1. Case Study Portfolio Design

Both the biosystems engineering career and geomatic have broad and diverse concentrations. ICT applications are the critical link between the two. The ability to articulate them requires an adaptive and gradual teaching design. The case study methodology facilitates a real problem using an appropriate subset of both areas (Figure 1a). The case study portfolio design considers problems where temporal and spatial variability is necessary for decision making. Government, society, and academia are essential to strengthen the timeliness conception. The resulting cooperation promotes public and private articulations for their involvement and definition. The coordination required between the distinct groups is broad and requires effective communication, collaboration, and committed cooperation. It is essential to show the benefits of the parties by achieving win–win scenarios. The design and definition of the case studies were established from a national project for the digital transformation of agriculture, called Agroinnovación 4.0, established by the Ministry of Agriculture and Livestock. This project included both the private and public sectors, where the potential of geomatics tools for decision-making processes through models were discussed. A preliminary list of cases was established from these meetings. The case studies were prioritized according to importance and data set availability. The terms of reference were established with the facilitators of the government, producers, and academia sectors.

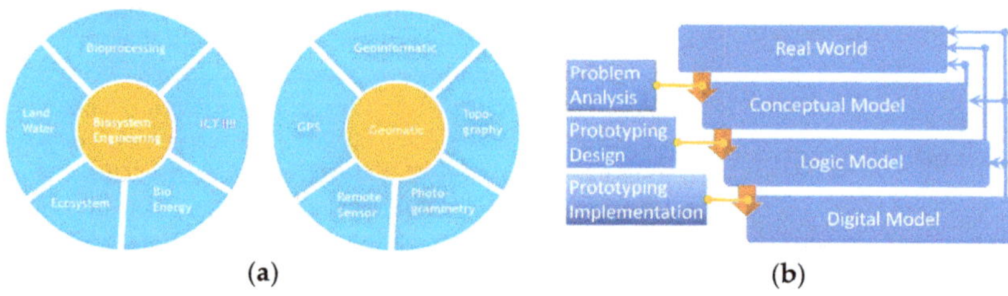

(a)   (b)

Figure 1. (**a**) Both the biosystems engineering career and geomatics have broad and diverse concentrations linked by ICT; (**b**) process flow diagram used to build the prototype and follow-up of the case studies.

3.1.1. The Evaluation and Follow-Up of the Case Studies

Each project's progress has a scope delimitation according to the terms of reference with the help of the facilitators. The final project integrates the three processes with feedback and maturity of the prototype reached. Figure 1b shows the process flow diagram used to build the prototype. It starts with analyzing a real-world problem with its challenges and the decision-making models required to specify a conceptual model. The prototyping design implies an investigation accompanied by expert criteria to define the necessary datasets, the design of the geographic database, and the development of a cartographic model that enables the definition of the logic model. Finally, the prototyping implementation is based on its digital model.

The problem analysis allows students to have a practical approach to problems by counting the perspectives of the farmers, the Ministry of Agriculture and Livestock official facilitator, and academia. The background and considerations of the case study allow the conceptual model to be set up through systematization tools such as concept maps, process flow diagrams, and the quality evaluation of temporal and spatial databases. The definition of the decision scale and delimitation are fundamental in this process. Prototyping design implies a review of the state of application in a Costa Rican context. Expert criteria are relevant and investigated to define the data set, organization, the model to be used, and the validation criteria according to the work scale. The prototype implementation seeks to specify the outputs of the proposed model based on the progress project. Finally, the students share the preliminary results with the target groups to gain their feedback and use it for the calibration and criteria validation processes. Once settings and validation considerations are in place, the students share their final experiences with colleagues and facilitators.

3.1.2. General Perception of Students

Figure 2 shows the students' responses on how their interest was before receiving the course and after it. The highest values are in the category that initially had interest and increased it at the end of the course. It is essential to highlight the first semester of 2020; the values of interest remain high despite the COVID-19 pandemic effect. Table 1 shows the student's perception responses related to learning method aspects of the balance of theory and practice, critical thinking, examples of the country's reality, research, and ethics. Whereby graphing these elements, as seen in Figure 2, the values are high. At the end of the course, the students' general opinions indicate that the teaching approach has an approval above 80% over four different semesters.

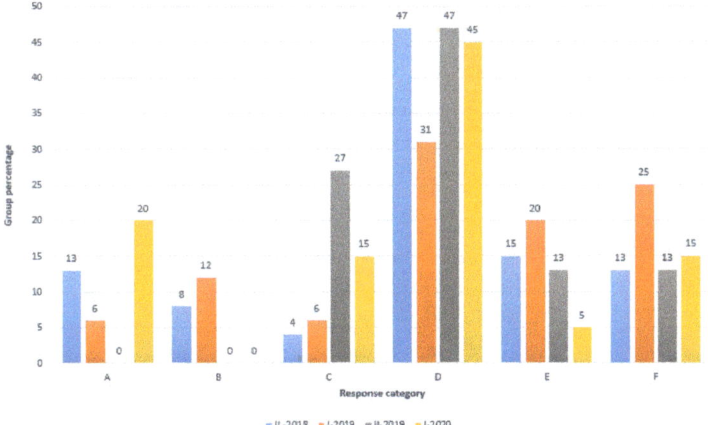

Figure 2. A: At first, I had interest and lost it; B: At first, I had no interest and still do not have it: C: At first, I had no interest, and now I do; D: At first, I had interest and kept it; E: At first, I had interest and increased it; F: Teacher has no influence.

Table 1. Students' perception responses related to learning method.

The Learning Method Provides	II-2018 (%)	I-2019 (%)	II-2019 (%)	I-2020 (%)
A balance between theory and practice	72	74	93	93
Activities that allow students to think in a diverse and innovative critical way	70	83	90	90
Examples related to future profession	94	97	96	97
Discussions with the reality of the country	93	94	97	97
Activities involving research	-	95	97	97
Topics of ethics, values, and responsibility	91	95	97	98

4. Discussion

The top value from the proposed teaching–learning strategy is encouraging collaborative work between students detecting deficiencies in the group and promoting teamwork. To support leadership attitudes, students identify how they can work to benefit all, helping the group analysis. The case study strategy allows the student to develop knowledge and skills, while the teacher provides strategical guidance by the facilitators. Values from the students' experience also accompany the proposed solutions from the theory and practical support. Accordingly, geomatic concepts and their tools applied in the Biosystems Engineering career are more effective if the connection with the real-world case studies is relevant. It is important to highlight that problem analysis and research abilities are fundamental for solutions, creativity, negotiation, teamwork, leadership, attitudes, and skills. Case studies allow the classroom to become a space that promotes the benefit of the students' life experiences in their professional training. The success factor in achieving the learning goals rely on the progress project that encourages and facilitates the gradual and adaptive follow-up from the theory and practice of geomatics.

Author Contributions: Conceptualization, J.A.; methodology, J.A.; validation, J.A. and M.C.; formal analysis, J.A. and M.C.; investigation, J.A. and M.C.; resources, J.A. and M.C.; writing—original draft preparation, J.A.; writing—review and editing, J.A.; visualization, J.A.; supervision, J.A.; project administration, J.A. All authors have read and agreed to the published version of the manuscript.

Funding: This research received no external funding.

Institutional Review Board Statement: Not applicable.

Informed Consent Statement: Not applicable.

Acknowledgments: We are grateful to Ing. Jairo González from MAG and lead of Agroinnovación 4.0 for his support given in the portfolio cases studies and Ing. Alvaro Brenes and Ing. Fernando Vázquez, professors of Facultad de Ciencias Agroalimentarias, as the facilitators in the case studies, and especially the farmer involved. We also thank all the students from the course Geomática aplicada en Ingeniería de Biosistemas IB-0011 and IB-0050 at UCR. Finally, we thank Luke Reese from MSU for his writing review.

Conflicts of Interest: The authors declare no conflict of interest.

References

1. Bru, B. Constructivism. In *Education Research: Across Multiple Paradigms*; Kimmons, R., Ed.; EdTech Books, 2021. Available online: https://edtechbooks.org/education_research/constructivismy (accessed on 1 May 2021).
2. Holley, E.A. Engaging engineering students in geoscience through case studies and active learning. *J. Geosci. Educ.* **2017**, *65*, 240–249. [CrossRef]
3. Yadav, A.; Beckerman, J.L. Implementing case studies in a plant pathology course: Impact on student learning and engagement. *J. Nat. Resour. Life Sci. Educ.* **2009**, *38*, 50–55. [CrossRef]
4. Sankar, C.S.; Varma, V.; Raju, P.K. Use of case studies in engineering education: Assessment of changes in cognitive skills. *J. Prof. Issues Eng. Educ. Pract.* **2008**, *134*, 287–296. [CrossRef]
5. Lindsey, A.J. Professional development participation by teachers facilitated student exposure to agriculture. *Nat. Sci. Educ.* **2020**, *49*, e20034. [CrossRef]

Proceeding Paper

Wearable Sensors for Identifying Activity Signatures in Human-Robot Collaborative Agricultural Environments [†]

Aristotelis C. Tagarakis [1], Lefteris Benos [1], Eirini Aivazidou [1], Athanasios Anagnostis [1], Dimitrios Kateris [1] and Dionysis Bochtis [1,2,*]

[1]   Centre for Research and Technology-Hellas (CERTH), Institute for Bio-Economy and Agri-Technology (IBO), 6th km Charilaou-Thermi Rd., GR 57001 Thessaloniki, Greece; a.tagarakis@certh.gr (A.C.T.); e.benos@certh.gr (L.B.); e.aivazidou@certh.gr (E.A.); a.anagnostis@certh.gr (A.A.); d.kateris@certh.gr (D.K.)
[2]   FarmB Digital Agriculture P.C., Doiranis 17, GR 54639 Thessaloniki, Greece
*   Correspondence: d.bochtis@certh.gr
†   Presented at the 13th EFITA International Conference, online, 25–26 May 2021.

Abstract: To establish a safe human–robot interaction in collaborative agricultural environments, a field experiment was performed, acquiring data from wearable sensors placed at five different body locations on 20 participants. The human–robot collaborative task presented in this study involved six well-defined continuous sub-activities, which were executed under several variants to capture, as much as possible, the different ways in which someone can carry out certain synergistic actions in the field. The obtained dataset was made publicly accessible, thus enabling future meta-studies for machine learning models focusing on human activity recognition, and ergonomics aiming to identify the main risk factors for possible injuries.

Keywords: accelerometer; magnetometer; gyroscope; situation awareness; human–robot interaction; safety; ergonomics

Citation: Tagarakis, A.C.; Benos, L.; Aivazidou, E.; Anagnostis, A.; Kateris, D.; Bochtis, D. Wearable Sensors for Identifying Activity Signatures in Human-Robot Collaborative Agricultural Environments. *Eng. Proc.* **2021**, *9*, 5. https://doi.org/10.3390/engproc2021009005

Academic Editors: Charisios Achillas and Dimitios Aidonis

Published: 19 November 2021

Publisher's Note: MDPI stays neutral with regard to jurisdictional claims in published maps and institutional affiliations.

1. Introduction

An emerging field in agriculture is collaborative robotics that take advantage of the distinctive human cognitive characteristics and the repeatable accuracy and strength of robots. One important factor, which is at the top of the priority list of Industry 5.0, is the safety of workers, due to the simultaneous presence of humans and autonomous vehicles within the same workplaces [1]. This human-centric approach is very challenging, especially in the agricultural sector, since it deals with unpredictable and complex environments, which contrast with other industries' structured domains. A crucial element in accomplishing a safe human–robot interaction is human awareness. Human activity recognition relying on wearable sensor data has received remarkable attention as compared with vision-based techniques, as the latter are prone to visual disturbances. To that end, sensors, including accelerometers, magnetometers and gyroscopes, are often utilized, either alone or in a synergistic manner. In general, multi-sensor data fusion is considered to be more trustworthy than a single sensor, because the potential information losses from one sensor can be offset by the presence of the others [2].

In the present study, a collaborative human–robot task was designed, while data were collected from field experimental sessions involving two different types of Unmanned Ground Vehicles (UGVs) and twenty healthy participants wearing five Inertial Measurement Units (IMUs). Consequently, the workers' activity "signatures" were obtained and analyzed, providing the potential to increase human awareness in human–robot interaction activities and provide useful feedback for future ergonomic analyses.

2. Materials and Methods

Experimental Setup and Signal Processing

A total of 13 male and 7 female participants, with neither recent musculoskeletal injury nor history of surgeries, took part in these outdoor experiments. Their average age, weight and height were 30.95 years (standard deviation $\approx$ 4.85), 75.4 kg (standard deviation $\approx$ 17.2) and 1.75 m (standard deviation $\approx$ 0.08), respectively. An informed consent form was filled out prior to any participation, which had been approved by the Institutional Ethical Committee. Moreover, a five-minute instructed warm-up was executed to avoid any injury. The aforementioned task, which has to be carried out three times by each person, included: (a) walking a 3.5 m unimpeded distance; (b) lifting a crate (empty or with a total mass of 20% of the mass of each participant); (c) carrying the crate back to the departure point; (d) placing the crate on an immovable UGV (Husky; Clearpath Robotics Inc. or Thorvald; SAGA Robotics, Oslo, Norway). A common plastic crate (height = 31 cm, width = 53 cm, depth = 35 cm) was used, with a tare weight of 1.5 kg and two handles at 28 cm above the base. Finally, the loading heights for the cases of Husky and Thorvald were approximately 40 and 80 cm, respectively.

Five IMU sensors (Blue Trident, Vicon, Nexus, Oxford, UK) were utilized in the present experiments, which are widely used in such studies. Each wearable sensor contains a tri-axial accelerometer, a tri-axial magnetometer and a tri-axial gyroscope. These sensors were attached via double-sided tape at the regions of chest (breast bone), the first thoracic vertebra, T1, (cervix), and the fourth lumbar vertebrae, L4, (lumbar region). In contrast, special velcro straps were used to attach the sensors at the left and right wrists (Figure 1). The sampling frequency was set equal to 50 Hz, while, for the purpose of synchronizing the IMUs and gathering the data, the Capture.U software, provided by VICON, was deployed.

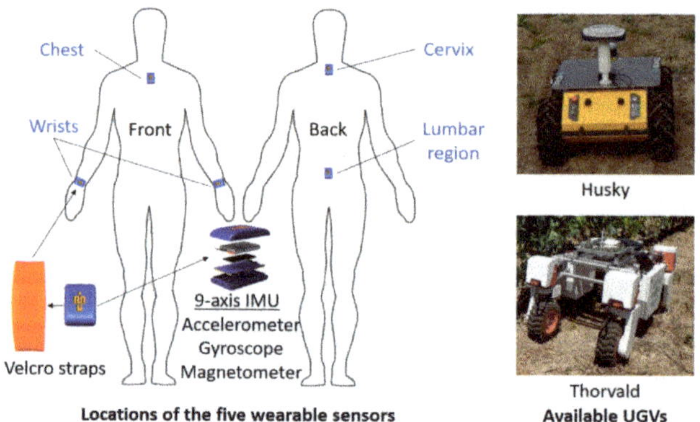

Figure 1. Body locations of the five IMU sensors and the available UGVs used in the experiments.

Distinguishing the sub-activities through carefully analyzing the video records was a particularly challenging task, since each participant performed the predefined sub-activities in their own pace and manner to increase the variability of the dataset. The most observed difference among the participants was definitely the technique they used to lift the crate from the ground. These techniques can be divided into the following main lifting postures [3]: (a) Stooping: bending the trunk forward from an erect position without kneeling; (b) Squat: bending knees by keeping the back straight and then standing back up; (c) Semi-squat: an intermediate posture between stooping and squat.

The sub-activities were continuous to obtain realistic measurements, hence increasing the degree of difficulty in labeling. To determine the critical instant where the transition occurred, the following well-defined criteria were imposed: (i) "Standing still" until the signal is given to begin; (ii) "Walking without the crate": One of the feet leaves the ground, corresponding to the start of the stance phase of gait cycle; (iii) "Bending": Starting bending the trunk forward (stooping), or kneeling (squat), or performing both simultaneously (semi-squat); (iv) "Lifting crate": Starting lifting the crate; (v) "Walking with the crate": The stance phase starts as analyzed above, but carrying the crate this time; (vi) "Placing crate": Starting stooping, squat or semi-squat and ending when the entire crate is placed to either Husky or Thorvald.

3. Results and Discussion

For the sake of brevity, only indicative raw signals in z direction are presented in this study (Figure 2), while the full dataset was made publicly available in [4]. Moreover, labels were assigned to the sub-activities, from 0 (standing still) to 5 (placing crate), with the intention of rendering them adequate for future machine learning studies.

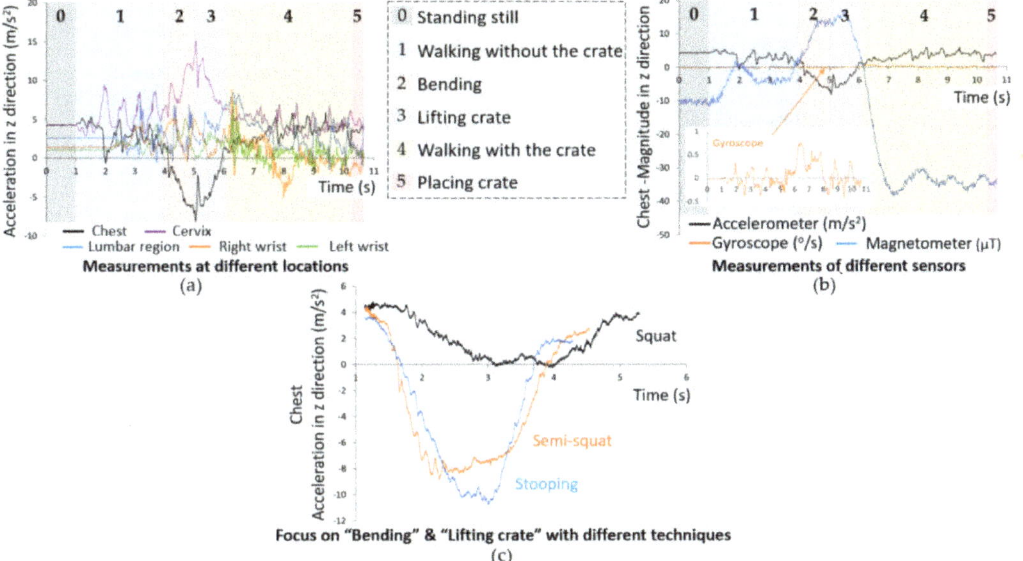

Figure 2. Indicative raw signals in z direction, considering the case of loading Thorvald with a crate of a total mass equal to 20% of the participant's mass, representing: (**a**) acceleration at different body locations, (**b**) measurements of different sensors at the chest, and (**c**) acceleration at the chest using different techniques.

As expected, the sub-activities demanding more time were those involving walking with and without the crate. In contrast, transitional sub-activities, including bending to approach the crate and lifting it, as well as placing the crate onto the robot, were considerably less time-consuming. Consequently, more effort was needed to capture the critical transitional instant through carefully analyzing the video records in accordance with the aforementioned criteria. In Figure 2a,b, the distinction of the sub-activities is clearly shown. More specifically, Figure 1a depicts the raw signals acquired by accelerometers at the five body locations. The signals originating from the wrists and lower back were quite complicated, while those from the chest and cervix presented local maxima or minima when a transition took place. Focusing on the acceleration measurements of the chest (Figure 2b), for instance, the nearly flat signal (corresponding to the standstill state) starts to fluctuate after t = 1 s, almost periodically indicating the repetitive parts of the gait cycle. This state is abruptly interrupted by an "indentation" in acceleration (or, equivalently, a "bulge" regarding the magnetometer signal). Within this indentation, the bending and lifting of the crate occur, requiring an approximately equal time. Subsequently, the signal indicating gait follows, while the sub-activity of placing the crate cannot easily be distinguished; it looks like a part of the previous sub-activity.

One very interesting feature extracted from the analysis of signals was their different forms, especially when bending and lifting the crate. This differentiation can be attributed to the several variations in the experiments (e.g., full/empty crate, different depositing heights, performing tasks at the individual's own pace). However, a more careful examination of video records revealed that the difference in these time series was mainly due to the lifting technique used. As can be seen in Figure 2c, when the participant used the squat technique, a relatively small indentation was observed, lasting longer than the other two techniques. The squat style was used by a minority of subjects, whereas stooping and semi-squat were used in the majority of cases, demonstrating a deeper indentation. In general, a squat lift leads to less stress on the spine while stooping; although more natural, this is considered to be the primary risk factor for lower back disorders. This type of lifting, especially when performed in a repetitive manner, appears to be the most common technique in agricultural activities, justifying the epidemic proportion of low back injuries in this sector [3,5]. The semi-squat is an alternative posture between squat and stooping, which avoids the deep kneeling of squat and the full lumbar flexion of stooping. There is considerable controversy regarding the best lifting posture, since all of them have drawbacks regarding oxygen consumption and fatigue in spine and knees [3].

In summary, a field experiment, involving well-defined continuous sub-activities, was designed, which collected data via wearable sensors. This dataset is characterized by a large variability, due to the inclusion of a plethora of different parameters. Finally, it was made publicly accessible [4] and is expected to be particularly useful for future research regarding machine learning and ergonomic studies.

Author Contributions: Conceptualization, A.C.T., L.B., D.B.; methodology, L.B., E.A., A.A.; writing—original draft preparation, A.C.T., L.B., D.K.; writing—review and editing, D.K., D.B.; visualization, L.B., A.A., E.A.; supervision, D.B. All authors have read and agreed to the published version of the manuscript.

Institutional Review Board Statement: The study was conducted according to the guidelines of the Declaration of Helsinki, and approved by the Ethical Committee under the identification code 1660 on 3 June 2020.

Informed Consent Statement: Informed consent was obtained from all subjects involved in the study.

Data Availability Statement: The dataset used in this work is publicly available in [4] which also provides additional information about the experimental procedure, file naming and sub-activity mapping.

Acknowledgments: This research has been partly supported by the Project: "Human-Robot Synergetic Logistics for High Value Crops" (project acronym: SYNERGIE), funded by the General Secretariat for Research and Technology (GSRT) under reference no. 2386.

Conflicts of Interest: The authors declare no conflict of interest.

References

1. Benos, L.; Bechar, A.; Bochtis, D. Safety and ergonomics in human-robot interactive agricultural operations. *Biosyst. Eng.* **2020**, *200*, 55–72. [CrossRef]
2. Anagnostis, A.; Benos, L.; Tsaopoulos, D.; Tagarakis, A.; Tsolakis, N.; Bochtis, D. Human activity recognition through recurrent neural networks for human-robot interaction in agriculture. *Appl. Sci.* **2021**, *11*, 2188. [CrossRef]
3. Benos, L.; Tsaopoulos, D.; Bochtis, D. A review on ergonomics in agriculture. Part I: Manual operations. *Appl. Sci.* **2020**, *10*, 1905. [CrossRef]
4. Open Datasets—iBO. Available online: https://ibo.certh.gr/open-datasets/ (accessed on 1 March 2021).
5. Benos, L.; Tsaopoulos, D.; Bochtis, D. A Review on Ergonomics in Agriculture. Part II: Mechanized Operations. *Appl. Sci.* **2020**, *10*, 3484. [CrossRef]

engineering proceedings

MDPI

Proceeding Paper

How Agricultural Digital Innovation Can Benefit from Semantics: The Case of the AGROVOC Multilingual Thesaurus [†]

Esther Mietzsch [1,*], Daniel Martini [1], Kristin Kolshus [2,‡], Andrea Turbati [2,‡] and Imma Subirats [2,‡]

[1] Association for Technology and Structures in Agriculture (KTBL), Bartningstr. 49, 64289 Darmstadt, Germany; d.martini@ktbl.de

[2] Food and Agriculture Organization of the United Nations, Viale delle Terme di Caracalla, 00153 Rome, Italy; Kristin.Kolshus@fao.org (K.K.); turbati@info.uniroma2.it (A.T.); Imma.Subirats@fao.org (I.S.)

[*] Correspondence: e.mietzsch@ktbl.de; Tel.: +49-6151-7001-230

[†] Presented at the 13th EFITA International Conference, online, 25–26 May 2021.

[‡] Group Email of Food and Agriculture Organization of the United Nations: agrovoc@fao.org.

Abstract: AGROVOC is the multilingual thesaurus managed and published by the Food and Agriculture Organization of the United Nations (FAO). Its content is available in more than 40 languages and covers all the FAO's areas of interest. The structural basis is a resource description framework (RDF) and simple knowledge organization system (SKOS). More than 39,000 concepts identified by a uniform resource identifier (URI) and 800,000 terms are related through a hierarchical system and aligned to knowledge organization systems. This paper aims to illustrate the recent developments in the context of AGROVOC and to present use cases where it has contributed to enhancing the interoperability of data shared by different information systems.

Keywords: AGROVOC; linked open data; FAIR principles; thesaurus; controlled vocabulary; food and agriculture; collaborative editing; multilinguality

Citation: Mietzsch, E.; Martini, D.; Kolshus, K.; Turbati, A.; Subirats, I. How Agricultural Digital Innovation Can Benefit from Semantics: The Case of the AGROVOC Multilingual Thesaurus. *Eng. Proc.* **2021**, *9*, 17. https://doi.org/10.3390/engproc2021009017

Academic Editors: Dimitrios Aidonis and Aristotelis Christos Tagarakis

Published: 24 November 2021

Publisher's Note: MDPI stays neutral with regard to jurisdictional claims in published maps and institutional affiliations.

1. Introduction

Since the 1980s, the Food and Agriculture Organization of the United Nations (FAO) has managed and published the AGROVOC Multilingual Thesaurus. It covers all of the FAO's areas of interest, mainly food, nutrition, agriculture, forestry, fisheries, and environment, as well as scientific and common names of organisms, biological notions, techniques of plant cultivation and environmental research, and other related subjects in more than 40 languages. Currently, around 25 national and international organizations worldwide support AGROVOC by contributing to the editorial community. These organizations provide language coverage by supplying terms in a specific language, and support thematic coverage by sharing their expertise in a specific discipline. AGROVOC also incorporates three specialized subsets: LandVoc, maintained by the LandPortal Foundation for concepts related to land governance; ASFA for aquatic sciences and fisheries; and FAOLEX for legislative and policy concepts in FAO areas.

2. AGROVOC as Linked Open Data

AGROVOC is available online as a linked open data set. The structural basis is the RDF [1] and the SKOS [2]. AGROVOC has more than 39,000 concepts formalized as SKOS concepts, identified by dereferenceable URIs. Each concept has at least one preferred term in a language. Optionally, a concept can have alternative or non-preferred terms. AGROVOC uses the SKOS extension for labels: SKOS-XL, which treats labels as full resources, thus allowing to assign to them further properties alongside the pure label text string. The predicates used are as follows: skosxl:prefLabel, used for preferred terms, and skosxl:altLabel, used for alternative or non-preferred terms.

The concepts are related through a hierarchical system based on skos:broader and skos:narrower and aligned to other vocabularies and thesauri mainly by skos:exactMatch and skos:closeMatch. Using this basis, AGROVOC is the key element to produce FAIR data in food and agricultural sciences. The FAIR principles state that data must be findable, accessible, interoperable, and reusable [3] and formulate measures to be taken to achieve these goals. AGROVOC fulfills these recommendations in itself, and thus its use in data sets and services contributes to the fulfillment of principle I2: (meta)data use vocabularies that follow FAIR principles. This is to support knowledge discovery and innovation, data and knowledge integration, and promote sharing and reuse of data, which focuses more on data intensive research and sharing data across the data value chain [4].

3. AGROVOC Ecosystem

The AGROVOC technical infrastructure is based on a comprehensive ecosystem of tools for users and editors to provide access to the data for both humans and machines. AGROVOC is a stable and reliable resource, which is continuously expanded by the activity of the curators and the editorial community. Improvements in the underlying technology have led to improvements in content representation. The Skosmos search interface allows users to search for terms and to browse through the hierarchy tree [5]. The data are also available in machine-readable formats like RDF/XML, Turtle, and JSON-LD. For targeted querying, selection and extraction of subsets, a public query endpoint is available [6]. It allows use of the SPARQL query language for semantic data [7] on the AGROVOC data set and includes sample queries. The results of the queries can be displayed in tables and can be downloaded as files in a number of formats, including comma-separated value (CSV).

Access for editors (who edit terms, add definitions, add concepts, and so on) is provided via a dedicated web access tool called VocBench [8]. Apart from providing a form-based frontend for data entry, VocBench supports the editorial workflow by offering features such as history, automatic capturing of modification metadata based on Dublin Core [9], and role-based validation of new entries and changes. Editorial rules and guidelines have been defined and are continuously evaluated and revised to facilitate the work delivered by the multilingual and distributed community of editors [10]. Currently, AGROVOC publishes new releases on a monthly basis.

4. AGROVOC Use Cases

AGROVOC can be used to enable agricultural digital innovation in different ways; i.e., linking AGROVOC concept URIs from data sets or other resources like bibliographic records and annotations of research data and text corpora allows users to unambiguously define concepts. Reliance on this common URI set implicitly links all these resources to each other, effectively integrating them into a global, interoperable data space. Apart from data integration, concept labels available in multiple languages support the internationalization of applications and information systems. Possible areas of application include the following:

- Organization of knowledge for subsequent data retrieval;
- Metadata annotation of agricultural (research) data;
- Standardization of agricultural information data and services;
- Indexing of literature;
- Auto-tagging/annotation of text corpora and web sites;
- Thesaurus for international cooperation (translation purposes);
- Multilingual search engine discovery.

A current use case is the project HortiSem [11], where an agricultural advisory system focusing on horticulture is enhanced by semantic web technology. Data relevant for planning of pest control measures and developing plant disease management strategies are currently spread across different heterogenous data sources. Currently, three of them have been prioritized to be dealt with in the project and to be semantically annotated, enriched, and integrated into a knowledge graph. One is the German Federal Office of

Consumer Protection and Food Safety (BVL) registered pesticides relational database, which contains data on active ingredients, crops, pests, and pesticides, alongside their most important attributes like temporal and spatial application restrictions and relations, such as which pesticide can be used on which combination of crop and pest (so-called indications). Another source is the EU Pesticides Database on maximum residue levels (MRLs). The MRL data are available in XML format. Finally, a large text corpus comprising agricultural advisory alerting newsletters is used as a large sample of free-text unstructured data.

Each of these sources undergoes a preparatory processing step to convert (part of) the data into RDF and load them into a Fuseki Triple Store to create an integrated knowledge graph. Structured data sources are converted to RDF using specific tools, namely db2triples for the database and an XML2RDF converter for the XML files. Unstructured data (texts) are processed using the SpaCy toolkit for named entity recognition/linking. Using that, automatically generated entity annotations are produced and integrated into the graph using the Web Annotation Vocabulary [12]. Generating these annotations is currently a work in progress, but, depending on concept availability, they point on the one hand to entity occurrences in the documents, and on the other hand to AGROVOC concepts or additional concepts for pest control products derived from the BVL database data. Approaches to prepare the required knowledge base for the named entity linker with subsets of AGROVOC are currently being developed. All of the data sets processed are then linked and mapped among the data sources, as well as to AGROVOC as a central concept hub (see Figure 1). This merged data set is then accessible by several applications, among them three horticultural information portals. AGROVOC here is both a set of concepts and a hub to the global data space built by a multitude of vocabularies and knowledge systems, which are linked by alignments in AGROVOC.

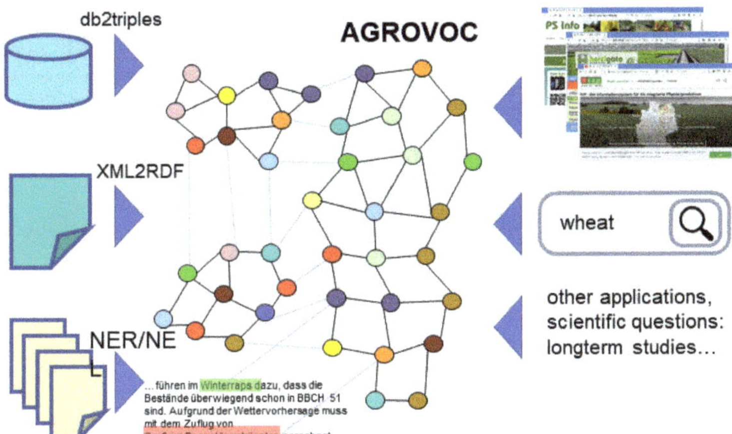

Figure 1. HortiSem overall architectural model.

5. Outlook on Further Development

In 2010, a set of properties describing agricultural and biological concept relations to be used in AGROVOC were devised and specified in the Agrontology. While that approach could have led to a semantic enrichment of AGROVOC, in practice, only some of these properties have been considered useful and others have been rarely or inconsistently used. An ongoing activity is thus revising the Agrontology, deprecating superfluous or unclear properties, better documenting the remaining ones, and giving guidance for editors on how to use it. Apart from using ontology relations in AGROVOC, approaches to using AGROVOC concepts in external ontologies are also addressed in initiatives with partners like CGIAR. Further work includes improvements to data quality and coverage like adding more language, enhancing language coverage, and closing thematic gaps in close collaboration with specific expert communities like fisheries or land governance.

Author Contributions: Writing—original draft preparation, E.M.; writing—review and editing, D.M., K.K., A.T. and I.S.; supervision, I.S.; project administration, I.S.; funding acquisition, D.M. All authors have read and agreed to the published version of the manuscript.

Funding: HortiSem is supported by funds of the Federal Ministry of Food and Agriculture (BMEL) based on a decision of the Parliament of the Federal Republic of Germany via the Federal Office for Agriculture and Food (BLE) under the innovation support programme (funding number: 2818508B18).

Data Availability Statement: Publicly available datasets were analyzed in this study. These data can be found here: https://agrovoc.fao.org/browse/agrovoc/en/ and https://hortisem.de/ (accessed on 1 June 2021).

Acknowledgments: The authors acknowledge the work of the 22 institutions in the AGROVOC Editorial Community.

Conflicts of Interest: The authors declare no conflict of interest.

References

1. Cyganiak, R.; Wood, D.; Lanthaler, M. RDF 1.1 Concepts and Abstract Syntax. World Wide Web Consortium (W3C). 2014. Available online: https://www.w3.org/TR/rdf11-concepts/ (accessed on 7 June 2021).
2. Miles, S. Bechhofer. SKOS Simple Knowledge Organization System Reference. World Wide Web Consortium (W3C). 2009. Available online: https://www.w3.org/TR/skos-reference/ (accessed on 8 June 2021).
3. The Fair Data Principles. Available online: https://www.force11.org/group/fairgroup/fairprinciples (accessed on 1 June 2021).
4. FAO. AGROVOC—Semantic Data Interoperability on Food and Agriculture. Rome. 2021. Available online: https://doi.org/10.4060/cb2838en (accessed on 1 June 2021).
5. AGROVOC: AGROVOC Multilingual Thesaurus. Available online: https://agrovoc.fao.org/browse/agrovoc/en/ (accessed on 1 June 2021).
6. AGROVOC SPARQL Endpoint. Available online: https://agrovoc.fao.org/sparql (accessed on 1 June 2021).
7. Harris, S.; Seaborne, A. SPARQL 1.1 Query Language. World Wide Web Consortium (W3C). 2013. Available online: https://www.w3.org/TR/sparql11-query/ (accessed on 8 June 2021).
8. Stellato, A.; Fiorelli, M.; Turbati, A.; Lorenzetti, T.; Van Gemert, W.; Dechandon, D.; Laaboudi-Spoiden, C.; Gerencsér, A.; Waniart, A.; Costetchi, E. VocBench 3: A collaborative Semantic Web editor for ontologies, thesauri and lexicons. *Semant. Web* **2020**, *5*, 855–881. [CrossRef]
9. DCMI Usage Board. DCMI Metadata Terms. 2020. Available online: https://www.dublincore.org/specifications/dublin-core/dcmi-terms/ (accessed on 8 June 2021).
10. FAO. The AGROVOC Editorial Guidelines 2020. Rome. 2020. Available online: http://www.fao.org/documents/card/en/c/cb2328en (accessed on 1 June 2021).
11. HortiSem. Available online: https://hortisem.de/ (accessed on 1 June 2021).
12. Sanderson, R.; Ciccarese, P.; Young, B. Web Annotation Vocabulary. World Wide Web Consortium (W3C). 2017. Available online: https://www.w3.org/TR/annotation-vocab/ (accessed on 8 June 2021).

engineering proceedings

Digital Agriculture Infrastructure in the USA and Germany [†]

Heinz Bernhardt [1,*], Leon Schumacher [2], Jianfeng Zhou [2], Maximilian Treiber [1] and Kent Shannon [2]

[1] TUM School of Life Sciences, Technical University of Munich, 85354 Freising, Germany; maximilian.treiber@wzw.tum.de
[2] College of Agriculture, Food & Natural Resources, University of Missouri, Columbia, MO 65211, USA; schumacherl@missouri.edu (L.S.); zhoujianf@missouri.edu (J.Z.); ShannonD@missouri.edu (K.S.)
* Correspondence: heinz.bernhardt@wzw.tum.de; Tel.: +49-8161-713086
† Presented at the 13th EFITA International Conference, online, 25–26 May 2021.

Abstract: The USA and Germany have compared the issues that surround the adoption of digital technology on the farm that will foster more environmentally sustainable food production/processing systems. Both countries lack robust broadband internet pathways to foster the adoption of these technologies. The problem is currently relevant to making this data technology available on every farm and field. The implementation of this infrastructure is even more important as society demands more and more information on the product and production process of agriculture and industry.

Keywords: data availability; rural regions; smart farming

Citation: Bernhardt, H.; Schumacher, L.; Zhou, J.; Treiber, M.; Shannon, K. Digital Agriculture Infrastructure in the USA and Germany. *Eng. Proc.* **2021**, *9*, 1. https://doi.org/10.3390/engproc2021009001

Academic Editors: Charisios Achillas and Lefteris Benos

Published: 19 November 2021

Publisher's Note: MDPI stays neutral with regard to jurisdictional claims in published maps and institutional affiliations.

1. Introduction

Agriculture worldwide is undergoing a transformation process toward the integration of digital process and production chains. The development of the necessary infrastructure for data transmission will be compared and analyzed using the examples of Germany and the USA (Corn Belt). The current development of smart farming is also strongly characterized by the fact that data and information can be easily shared between different partners. This can be control data for fully autonomous tractors, results from in situ sensors in the field, application rates for implements derived from satellite data, or the transmission of production data to trading partners.

The fact that data is collected directly on the working machine and processed there, and the information is converted directly into decisions is less and less the case, due to increasing complexities and amounts of data, on the one hand, and the need for third-party information and complex algorithms for decision making, on the other hand. Data is increasingly being collected at different points and times, blended with existing data, processed by different parties, then forwarded as information to different points for later analysis. To enable this development, the infrastructure for data transmission plays a crucial role.

2. Materials and Methods

In this study, we focus on three subareas of this infrastructure network. The first part is between the cloud and the farm. The second subarea is data transmission between farm and field/tractor. Here the discussion is between data rate, range, penetration, cost, and energy demand of different transmission technologies. Low-power, wide-area network (LPWAN) systems (e.g., LoRa, mioty, Sigfox, etc.) are on the rise. They combine low-power and low data rates with high range, therefore making them effective for many use cases of stationary sensor deployment in agriculture. A third category blends in between the mentioned technologies. In the US, these transmission technologies (e.g., TV whitespace) are suitable to build a bridge/backbone between regional locations of a farming operation. With medium data rates and very high ranges, they are suitable to connect LPWAN-based wireless sensor networks from remote locations to an internet backbone on a farmer's home base. Streaming technologies, such as what is used by Netflix, are being adopted in the USA by some precision agriculture vendors, but many farmers in the USA gather data via an iPad while in the field and upload this data after they return to their home office.

3. Digital Agriculture Infrastructure

3.1. USA

The deployment and adoption of digital technology by agriproducers is expected to enhance food production in the USA as in other parts of the world to meet the increasing population of the world with more than 9.5 billion people by 2050. Digital devices and sensors, such as satellite remote sensing, UAV imaging systems, Internet of Things (IoT) sensing systems, and ground-based robotic systems, are rapidly becoming more common in agriculture for collecting high-resolution temporal and spatial big data of crops, animals, environment, and farm equipment. With the advance of big data analytic technologies and artificial intelligence, digital data are able to be used to monitor production, improve efficiency, and increase agricultural sustainability. In addition, these advanced systems allow farms to be more profitable, efficient, safe, and environmentally friendly, conserving our natural resources.

To successfully adopt and implement these advancements, wireless connectivity is an enabling technology. However, the wireless network infrastructure for data transfer in agriculture is still lacking. We feel that there are three layers of networks that need consideration in the US: (i) backbone connectivity ("to the internet"), (ii) regional network infrastructure (wireless transfer of data from the farm to the nearest backbone, e.g., TV whitespace, private 4G LTE, or cellular hotspot), and (iii) local sensor networks (e.g., low-power, wide-area networks or LPWANs) for collection of sensor/machine data.

Companies, such as Farmobile, The Climate Corporation, Trimble, Farmers Edge, Ag Leader, and John Deere, are establishing proprietary networks, some with the assistance of companies, such as Trilogy, a company that advocates the use of a private 4G LTE network on the farm. Why 4G? Data would be more secure and faster, since a private 4G LTE network does not use the more traditional cell phone network. Data is collected directly via the CAN bus on the tractor, combine, or sprayer. Data transfer to the cloud varies as The Climate Corporation (Bayer) first transfers that data from the CAN bus via Bluetooth to the iPad. The iPad later connects to the Internet via Wi-Fi, and the data is transferred to The Climate Corporation cloud. Farmobile uses traditional cell phone networks (when available) to immediately transfer the data to the Farmobile cloud.

Data analysis is more efficient via artificial intelligence systems after being placed on cloud-based servers. Companies have also adopted data transfer systems similar to those used by Netflix and Hulu (for TV viewing). Why are multiple and different data transfer systems used and not a single common system? Because 80% of the 24 million American households do not have reliable, affordable, high-speed broadband in the rural areas (Federal Communications Commission's report). Digital technology and understanding this data will change the way the American farmer works; however, the USDA [1] recently noted that digital row crop technology is being adopted, yet the livestock and special crop production remains at the early innovator stages. According to the same article, "90 percent of the people do nothing with the data that they collect. They don't know what to do with it." It seems we have two challenges: (i) the USA needs an agricultural workforce that understands the best use of digital technologies, and (ii) as stated earlier, there is a lack of Internet access in rural communities. According to the Leichtman Research Group [2], broadband is penetrating the USA home in many cities. However, this is not the case for rural areas in US.

3.2. Germany

Citizens in Germany and Europe are increasingly demanding information on the product and process quality of their food. Smart farming is one way to satisfy these demands [3]. Data exchange is necessary for this. In Germany, the data rate between the farm and the Internet cloud is strongly influenced by the transmission technology. Germany is relatively densely populated and therefore has a historically good network infrastructure based on copper cables. In the meantime, however, significant problems can be observed with the upgrade to fiber-optic networks. The expansion rate is only 4.7% and is mostly limited to the conurbations [4]. In rural areas, the network infrastructure is therefore only weakly developed. The data rate for downloads in landline networks at the district level is relatively low, especially in rural areas. The median for the worst-served areas is 15.5 Mbps. Most rural areas are in the 35–55 Mbps range, but with sometimes large differences between providers. The available data download in the landline network is less than 15 Mbps for 20% of all users; in 50% of all users less, than 50 Mbps; and in 20% of all users, more than 100 Mbps [5].

For the second area, field-to-farm data exchange, the quality of the mobile data networks is crucial. Although the 4G network is fairly widespread, there are still gaps in rural and sparsely populated northeastern Germany. The 5G network is mostly only developed in densely populated regions and along long-distance routes. There are still large gaps in rural regions [6]. Due to this structure, download rates between only 6.8–8 Mbps are possible in some districts. The majority of agricultural districts have a download rate of 8–18 Mbps. In large cities, on the other hand, 82 Mbps is possible. If one analyzes one level deeper and looks at agricultural areas and villages separately, data availability drops even further. Here, if a network is available at all, download rates of only 3 Mbps are usually possible. The available data download in the mobile network is less than 3.8 Mbps for 20% of all users, less than 14.3 Mbps for 50% of all users, and greater than 44.1 Mbps for 20% of all users. Since many smart farming applications in the field also require data to be uploaded to the cloud, the upload data rate is also important. The available data upload in the mobile network is less than 0.9 Mbps for 20% of all users, less than 4.4 Mbps for 50% of all users, and greater than 14.8 Mbps for 20% of all users.

The analysis of the transmission technology favored in Germany shows that it is mostly oriented towards consumer applications in large cities and along long-distance traffic routes. In contrast, rural areas and fields still face selective availability of even 3G connectivity. However, the requirements of agriculture are more diverse. On the one hand, some mobile machinery requires a remote broadband connection from time to time. On the other hand, local wireless sensor networks prefer long functional life, low energy consumption, and transmission ranges typically closely related to the geographic extension of a farm, that can be provided by a local LPWAN gateway.

4. Discussion

In the studies in the USA and Germany, the various problems of the data transmission infrastructure for low, medium, and high data rates were analyzed. It is evident in both countries that the quality of the data network infrastructure is still insufficient. Ultimately, political support (social support) are needed before the USA invests in robust rural broadband, thus enabling the adoption of digital technology and the use of more sustainable/efficient food production and processing systems needed by society. In some ways, rural broadband needs to mirror the adoption of electricity in the USA. The electric light bulb was invented in 1880. On 3 June 1889, the first North American electric power transmission line went online. On 11 May 1935, President Roosevelt created the Rural Electrification Administration, and by 1950, most of rural America had electricity. However, things are much different now, especially after the COVID-19 pandemic; society is better informed of the value of rural broadband for the USA. We believe that rural broadband will be deployed much quicker than electricity in rural USA.

The development of network infrastructure in Germany is such that, in the case of the railroad, electricity, and telephone networks, this was also started by private companies at the beginning, but was then soon transferred by the state to state or semipublic companies. The reason for this development was to ensure network expansion in rural regions as well. From the 1980s onward, the state withdrew more and more from the infrastructure networks and privatized the networks. In the implementation of the individual data networks, it can now be seen very clearly that companies are concentrating on urban areas with a high return on sales. In the case of 4G and 5G licenses, the contracts stipulate that rural regions in particular are to be expanded, but this is not apparent in the analysis. It is therefore evident in both study areas that there is still a need for action.

Another aspect that the study has shown is that, in some cases, there are no systems that are adapted to the needs of agriculture. Many applications for smart farming in agriculture require continuous and long-term data transmission with only low data rates. Examples are moisture sensors in the soil or position data for cattle. Low-power, wide area networks (LPWAN), such as LoRaWAN, offer solutions for these applications. Although only limited amounts of data can be exchanged via this system, they can be exchanged within a radius of 10 to 20 km (depending mostly on topography) and with low energy requirements. The distribution of these networks in Europe is not uniform. In Switzerland, nationwide use is possible via the major telephone companies. In Germany, on the other hand, there is little interest on the part of commercial providers, and there is only 70% coverage, which is mostly organized via citizens' networks. The question remaining is whether technologies, such as LPWANs, will really become so user friendly that farmers deploy their own networks on a farm level in contrast to relying on network providers like in the past. Another interesting trend, to cover the network availability gaps worldwide, are satellite communication systems. With companies, such as Starlink and Inmarsat, expanding their satellite constellations and connectivity services, this could become an important pillar for farms in remote areas, not only as a broadband connection backbone, but also for the collection of low-power and low-data rate wireless sensor networks.

Author Contributions: Conceptualization, H.B. and L.S.; methodology, H.B., L.S. and M.T.; investigation, H.B., L.S., J.Z., M.T. and K.S.; writing—original draft preparation, H.B. and L.S.; writing—review and editing, J.Z., K.S. and M.T.; visualization, M.T. All authors have read and agreed to the published version of the manuscript.

Funding: This research received no external funding.

Conflicts of Interest: The authors declare no conflict of interest.

References

1. A Case for Rural Broadband. USDA Publication. Available online: https://www.usda.gov/sites/default/files/documents/case-for-rural-broadband.pdf (accessed on 7 June 2021).
2. Leichtman Research Group. Actionable Research on the Broadband, Media and Entertainment Industries Since 2002. Available online: https://www.leichtmanresearch.com/ (accessed on 7 June 2021).
3. Aulbur; Wilfried; Henske, R.; Uffelmann, W.; Schelfi, G. *Farming 4.0: How Precision Agriculture Might Save the World Precision Farming Improves Farmer Livelihoods and Ensures Sustainable Food Production*, 1st ed.; Roland Berger Focus: Munich, Germany, 2019.
4. Glasfasernetz. Available online: https://de.statista.com/statistik/daten/studie/415799/umfrage/anteil-von-glasfaseranschluessen-an-allen-breitbandanschluessen-in-oecd-staaten/ (accessed on 1 June 2021).
5. Bundesnetzargentur. Available online: https://www.bundesnetzagentur.de (accessed on 1 June 2021).
6. Mobilfunk Netzausbau. Available online: https://www.telekom.de/netz/mobilfunk-netzausbau (accessed on 1 June 2021).

MDPI

Proceeding Paper

Drought Periods Identification in Ecuador between 2001 and 2018 Using SPEI and MODIS Data [†]

César Sáenz [1,2,*], Javier Litago [1], Klaus Wiese [3,4], Laura Recuero [2], Victor Cicuéndez [1] and Alicia Palacios-Orueta [2,5]

[1] Departamento de Economía Agraria, Estadística y Gestión de Empresas, ETSIAAB, Universidad Politécnica de Madrid (UPM), Avda. Complutense 3, 28040 Madrid, Spain; javier.litago@upm.es (J.L.); victor.cicuendez.lopezocana@alumnos.upm.es (V.C.)

[2] Centro de Estudios e Investigación para la Gestión de Riesgos Agrarios y Medioambientales (CEIGRAM), Universidad Politécnica de Madrid (UPM), C/Senda del Rey 13 Campus Sur de prácticas de la ETSIAAB, 28040 Madrid, Spain; laura.recuero@upm.es (L.R.); alicia.palacios@upm.es (A.P.-O.)

[3] Escuela de Biología, Universidad Nacional Autónoma de Honduras, Tegucigalpa 11101, Honduras; klauswiese@hotmail.com

[4] Programa de Ingeniería y Gestión del Medio Natural, Escuela Técnica Superior de Ingeniería de Montes, Forestal Y del Medio Natural, Universidad Politécnica de Madrid, 28040 Madrid, Spain

[5] Departamento de Ingeniería Agroforestal, ETSIAAB, Universidad Politécnica de Madrid (UPM), Avda. Complutense 3, 28040 Madrid, Spain

*   Correspondence: cesar.saenzf@alumnos.upm.es

†   Presented at the 13th EFITA International Conference, Online, 25–26 May 2021.

Abstract: Drought is a natural phenomenon in which the precipitation amount is below normal in a specific region over a long period. The main objective of this study is to identify periods of drought in Ecuador between 2001 and 2018 using the Standardized Precipitation Evapotranspiration Index (SPEI) and the Normalized Difference Water Index (NDWI) derived from MODIS data. Firstly, the SPEI at a six-month scale and the Runs theory were used to identify periods of drought. Secondly, the NDWI from MOD09A1 MODIS product was used to identify the areas affected by drought.

Keywords: drought; SPEI; NDWI; remote sensing; Ecuador

Citation: Sáenz, C.; Litago, J.; Wiese, K.; Recuero, L.; Cicuéndez, V.; Palacios-Orueta, A. Drought Periods Identification in Ecuador between 2001 and 2018 Using SPEI and MODIS Data. *Eng. Proc.* **2021**, *9*, 24. https://doi.org/10.3390/engproc2021009024

Academic Editors: Dimitrios Kateris and Maria Lampridi

Published: 25 November 2021

Publisher's Note: MDPI stays neutral with regard to jurisdictional claims in published maps and institutional affiliations.

1. Introduction

More frequent and intense drought events during the last few years are a clear consequence of climate change worldwide. Drought is defined as a natural phenomenon in which rainfall values are below normal in a specific region over a prolonged period [1]. However, drought is a complex phenomenon and presents different ways with diverse features [2]. The American Meteorological Society classifies it into four categories: meteorological or climatological drought, hydrological drought, agricultural drought and socioeconomic drought [3].

The scientific community has developed several indices in order to assess this complex phenomenon using, two main different methods [4]: (a) indices developed with a traditional meteorological method, taking into account meteorological variables, such as the Standardized Precipitation Evapotranspiration Index (SPEI) [5] and (b) indices computed using multispectral information from remote sensors, such as the Normalized Difference Water Index (NDWI) that evaluates the water content in vegetation and soils [6]. This index has been widely used for drought monitoring, especially when using satellite imagery offering a frequent and continuous spatial observation, even in areas where access is difficult. Resolution Imaging Spectroradiometer (MODIS), available since the beginning of the 2000s, has been widely used for drought monitoring at national scales [7] being especially interesting when analyzing long time series.

The main objective of this study is to develop a methodology to assess drought periods in Ecuador between 2001 and 2018 using the SPEI and NDWI indices.

63

2. Materials and Methods

2.1. Study Region

Ecuador is located between latitudes $1°28'$ N–$5°01'$ S and longitudes $75°11'$ W–$81°01'$ W and it is bordered to the west by the Pacific Ocean. The average temperature in Ecuador varies between regions, fluctuating between 20 °C and 26 °C. The total annual precipitation varies with respect to temporal and spatial distribution from 250 mm on the Coast to 4000 mm in the Amazon. According to the Köppen classification, Ecuador has a humid tropical climate (Af), without defined climatic seasons, although two well-marked periods: dry season (June–November) and rainy season (December–May). This research is focused on continental Ecuador, which is divided into three regions: Coast, Andean and Amazon. This area expands across 256,370 km^2 with a population of approximately 17.6 million people.

2.2. Data Source and Processing

Meteorological and MODIS Data

The Climate Research Unit (CRU) Time-Series (TS), version 4.03 of high-resolution gridded climate dataset with a spatial resolution of 0.5 degrees [8], was used to obtain monthly temperature and precipitation observations for the period 2001–2018.

The MOD09A1 version 6 product from MODIS sensor [9] was used to calculate NDWI spectral index. This product consists of 8 days temporal composites with a 500 m spatial resolution containing reflectance values at seven spectral bands: 620–670 nm (Red), 841–876 nm (Nir), 459–479 nm (Blue), 545–565 nm (Green), 1230–1250 nm (SWIR$_1$), 1628–1652 nm (SWIR$_2$), 2105–2155 nm (SWIR$_3$). To cover all continental Ecuador, it was necessary to download and process information from 4 tiles (h09v08, h09v09, h10v08, h10v09). The data were downloaded from the NASA website (https://lpdaac.usgs.gov, accessed on 20 April 2021) and for the period 2001–2018.

2.3. Research Methods

2.3.1. Calculation of the SPEI

The SPEI proposed by V. Serrano in 2010, is an index focused on detecting drought from climatic data calculated from the difference in precipitation and potential evapo-transpiration for different spatial and temporal scales (1 to 48 months). In this study, a six-month scale was used. The SPEI values range from -2 to 2 and the categories are: extremely wet (≥ 2), severely wet (1.50 to 1.99), moderately wet (1 to 1.49), normal (-0.99 to 0.99), moderate drought (-1 to -1.49), severe drought (-1.50 to -1.99) and extreme drought (≤ -2).

2.3.2. Calculation of the NDWI

The NDWI proposed by Gao in 1996 uses the bands of NIR and SWIR$_1$ to identify moisture (Equation (1)) [6]. The range of the NDWI is from -1 to 1. When the value is closer to 1, vegetation is considered to contain high moisture, and if it approaches 0, it is associated with surfaces without the presence of vegetation or bodies of water [10].

$$\text{NDWI} = (\text{NIR} - \text{SWIR}_3)/(\text{NIR} + \text{SWIR}_3), \tag{1}$$

The ranges we use for the different NDWI categories were: very high moisture (>0.7), high moisture (0.61 to 0.7), moderate moisture (0.51 to 0.6), low moisture (0.41 to 0.5), weak moisture (0.31 to 0.4), moderate drought (0.21 to 0.3), strong drought (0.01 to 0.2) and very strong drought (≤ 0).

2.4. Methodology

2.4.1. Runs Theory

Once the six-month SPEI was calculated, it was analyzed using the Runs theory [11] that allows finding the duration, severity, intensity and identifying the dates of the drought periods.

2.4.2. NDWI Images

To map and calculate the area affected by drought, the dry period was identified with the Runs theory. The pixels are classified with the proposed NDWI values and those that are less than or equal to 0.3 are considered as affected surfaces.

3. Results

The years that were identified as dry periods were 2005 and 2018. Although 2005 was the year most affected at national level, the Coastal region had the highest severity value of −1.43 and a duration of 10 months, following the Andean and Amazon region in Table 1.

Table 1. The most important dry periods between 2001 to 2018 of continental Ecuador regions.

Region	Coast		Andean		Amazon	
Year	2005	2018	2005	2018	2005	2018
Duration	10	6	13	6	6	6
Severity	−14.25	−6.79	−17.80	−7.16	−7.57	−7.35
Intensity	−1.43	−1.13	−1.37	−1.19	−1.26	−1.23
Date	January–October	April–September	February–February	April–September	May–October	March–August

In Figure 1, the drought period of 2005 using the NDWI and pixels with values less than 0.3 were classified as affected by drought and the most affected month was October, which affected 32.36% of the total surface of continental Ecuador.

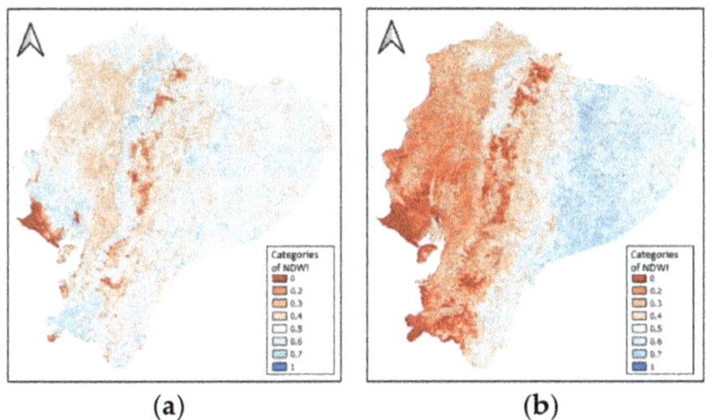

(a)　　　(b)

Figure 1. Surface of the month least and the most affected by drought in 2005: (**a**) May 4.74%; (**b**) October 32.36%.

4. Discussion and Conclusions

The combination of SPEI on a six-month scale and NDWI from MODIS provided relevant information tools for the identification, assessment and monitoring of dry periods and estimation of the affected surface. It was possible to identify the dates of duration of the dry periods and their levels of affectation based on the SPEI on a six-month basis, and the surface affected based on the NDWI information.

The year 2005 was the most affected by drought in all regions of Ecuador, followed by 2018, which also has high intensity values, but with a shorter duration than 2005. The year 2005 has been considered the driest year in the last 100 years in the southwestern part of the Amazon [12]. It was important to identify the month (October 2005) and the region most affected (Coast region) by this natural threat to be able to take preventive measures in the future, knowing that the sector of first necessity is the agricultural sector.

Author Contributions: Conceptualization, C.S. and J.L.; methodology, C.S. and J.L.; software, K.W. and L.R.; validation, K.W. and L.R.; formal analysis, C.S., J.L. and A.P.-O.; investigation, C.S., J.L., A.P.-O., V.C., L.R. and K.W.; resources, C.S., L.R. and V.C.; data curation, C.S. and L.R.; writing—original draft preparation, C.S., V.C. and L.R.; writing—review and editing, C.S., V.C., J.L. and A.P.-O. All authors have read and agreed to the published version of the manuscript.

Funding: César Sáenz was supported by a predoctoral scholarship awarded by the Community of Madrid (No. IND2020/AMB-17747).

Data Availability Statement: Publicly available datasets were analyzed in this study. This data can be found here: https://lpdaac.usgs.gov (accessed on 20 April 2021).

Acknowledgments: The authors would like to thank NASA for providing MODIS data. We are especially grateful to the anonymous reviewers and editors for reviewing our manuscript and for offering feedback.

Conflicts of Interest: The authors declare no conflict of interest.

References

1. Van Loon, A.F.; Laaha, G. Hydrological drought severity explained by climate and catchment characteristics. *J. Hydrol.* **2015**, *526*, 3–14. [CrossRef]
2. West, H.; Quinn, N.; Horswell, M. Remote sensing for drought monitoring & impact assessment: Progress, past challenges and future opportunities. *Remote Sens. Environ.* **2019**, *232*, 111291. [CrossRef]
3. Wilhite, D.A.; Glantz, M.H. Understanding: The Drought Phenomenon: The Role of Definitions. *Water Int.* **1985**, *10*, 111–120. [CrossRef]
4. Du, L.; Tian, Q.; Yu, T.; Meng, Q.; Jancso, T.; Udvardy, P.; Huang, Y. A comprehensive drought monitoring method integrating MODIS and TRMM data. *Int. J. Appl. Earth Obs. Geoinf.* **2013**, *23*, 245–253. [CrossRef]
5. Vicente-Serrano, S.M.; Beguería, S.; López-Moreno, J.I. A multiscalar drought index sensitive to global warming: The standardized precipitation evapotranspiration index. *J. Clim.* **2010**, *23*, 1696–1718. [CrossRef]
6. Gao, B. NDWI—A Normalized Difference Water Index for Remote Sensing of Vegetation Liquid Water From Space. *Remote Sens. Environ.* **1996**, *58*, 257–266. [CrossRef]
7. Xie, F.; Fan, H. Deriving drought indices from MODIS vegetation indices (NDVI/EVI) and Land Surface Temperature (LST): Is data reconstruction necessary? *Int. J. Appl. Earth Obs. Geoinf.* **2021**, *101*, 102352. [CrossRef]
8. Harris, I.; Osborn, T.J.; Jones, P.; Lister, D. Version 4 of the CRU TS monthly high-resolution gridded multivariate climate dataset. *Sci Data* **2020**, *7*, 109. [CrossRef] [PubMed]
9. Vermote, E. MOD09A1 MODIS/Terra Surface Reflectance 8-Day L3 Global 500m SIN Grid V006. 2015. Distributed by NASA EOSDIS Land Processes DAAC. Available online: https://doi.org/10.5067/MODIS/MOD09A1.006 (accessed on 20 April 2021).
10. Gulácsi, A.; Kovács, F. Drought monitoring with spectral indices calculated from modis satellite images in hungary. *J. Environ. Geog.* **2015**, *8*, 11–20. [CrossRef]
11. Yevjevich, V. *An Objective Approach to Definitions and Investigations of Continental Hydrologic Drought; Hydrology*; Paper No. 23; Colorado State University: Fort Collins, CO, USA, 1967.
12. Marengo, J.A.; Nobre, C.A.; Tomasella, J.; Oyama, M.D.; Sampaio de Oliveira, G.; de Oliveira, R.; Camargo, H.; Alves, L.M.; Brown, I.F. The Drought of Amazonia in 2005. *J. Clim.* **2008**, *21*, 495–516. [CrossRef]

Proceeding Paper

Integrating Ambient Intelligence Technologies for Empowering Agriculture [†]

Christos Stratakis [1], Nikolaos Menelaos Stivaktakis [1], Manousos Bouloukakis [1], Asterios Leonidis [1,*], Maria Doxastaki [1], George Kapnas [1], Theodoros Evdaimon [1], Maria Korozi [1], Evangelos Kalligiannakis [1] and Constantine Stephanidis [1,2]

[1] Institute of Computer Science, Foundation for Research and Technology—Hellas (FORTH), 70013 Heraklion, Crete, Greece; stratak@ics.forth.gr (C.S.); nstivaktak@ics.forth.gr (N.M.S.); mboulou@ics.forth.gr (M.B.); mdoxasta@ics.forth.gr (M.D.); gkapnas@ics.forth.gr (G.K.); evdemon@ics.forth.gr (T.E.); korozi@ics.forth.gr (M.K.); vaggeliskls@ics.forth.gr (E.K.); cs@ics.forth.gr (C.S.)

[2] Department of Computer Science, University of Crete, 70013 Heraklion, Crete, Greece

* Correspondence: leonidis@ics.forth.gr

[†] Presented at the 13th EFITA International Conference, online, 25–26 May 2021.

Abstract: This work blends the domain of Precision Agriculture with the prevalent paradigm of Ambient Intelligence, so as to enhance the interaction between farmers and Intelligent Environments, and support their various daily agricultural activities, aspiring to improve the quality and quantity of cultivated plants. In this paper, two systems are presented, namely the Intelligent Greenhouse and the AmI seedbed, targeting a wide range of agricultural activities, starting from planting the seeds, caring for each individual sprouted plant up to their transplantation in the greenhouse, where the provision for the entire plantation lasts until the harvesting period.

Keywords: precision agriculture; ambient intelligence; smart greenhouse; smart seedbed

Citation: Stratakis, C.; Stivaktakis, N.M.; Bouloukakis, M.; Leonidis, A.; Doxastaki, M.; Kapnas, G.; Evdaimon, T.; Korozi, M.; Kalligiannakis, E.; Stephanidis, C. Integrating Ambient Intelligence Technologies for Empowering Agriculture. *Eng. Proc.* **2021**, *9*, 41. https://doi.org/10.3390/engproc2021009041

Academic Editors: Dimitrios Aidonis and Aristotelis Christos Tagarakis

Published: 11 January 2022

Publisher's Note: MDPI stays neutral with regard to jurisdictional claims in published maps and institutional affiliations.

1. Introduction

Throughout their cultivation activities, farmers often face complex decision-making problems [1], which do not necessarily have straightforward solutions. This complexity becomes apparent when considering the amount and type of information that they have to process on a daily basis. In more detail, farmers must be aware of various parameters of their cultivated area (e.g., soil moisture levels, weather conditions), while they should also be able to assess the status of each plant separately [2]. The advancements in technological components, such as sensors, actuators, robotics, and mechanized equipment, have contributed towards the alleviation of several complex tasks and the emergence of the term Precision Agriculture (PA). PA depends inherently on Information and Communication Technologies (ICTs), so as to analyze data collected from multiple resources and make informed decisions associated with plant cultivation [3]. However, despite the fact that PA undoubtedly contributes to the improvement of the quantity and quality of agricultural products and the optimization of resource utilization [4], there is still limited research towards creating human-oriented agriculture-related intelligent environments (i.e., farms, greenhouses) whose primary goal is to satisfy the needs of the people working in them. The term "Intelligent Greenhouse" [5], became prevalent in the past quinquennium, and refers to greenhouses that focus on minimizing their carbon footprint [6], permitting the creation of self-regulating cultivation environments, automating several agricultural activities [7]. They are usually equipped with several I/O devices and control devices [8], and permit users to monitor and control them remotely. Apart from technologically enhanced greenhouses, there are several approaches concentrating on seedbeds [9], so as to streamline the process of planting the seeds and caring for the seedling before transplanting them elsewhere. However, these approaches do not support automated daily care and precise seedling measurement, while there is no provision for the post-grow period of seedlings, especially for climbing and brassica vegetables. In general, existing solutions targeting either greenhouses or seedbeds mainly focus on advancing the hardware infrastructure rather than focusing on the needs of users and enhancing User Experience (UX). Conversely, this work takes advantage of the Ambient Intelligence (AmI) paradigm [10] towards creating a user-centric [11] Intelligent Environment (IE) consisting of a greenhouse and a seedbed. Both systems target the domain of PA, deliver novel interaction techniques, and respond to user needs.

2. Design Process

A user-centered design approach was followed, based on the Design Thinking Methodology [12], which is an iterative process consisting of five phases: Empathize, Define, Ideate, Prototype, and Test. For the first two phases, the design team studied the related work on the domain of Intelligent Systems for Agriculture and examined relevant commercial solutions. Next, in order to gain an empathic understanding of the user needs, the team defined specific personas (i.e., Agronomist, Professional Farmer, and Hobby/Amateur Farmer) and created several motivating scenarios under the guidance of an agronomist familiar with AmI Technologies. Then, for the Ideate phase, the team organized several brainstorming sessions with the participation of several stakeholders, that resulted in an abundance of ideas which were subsequently filtered, so as to remove unfeasible solutions and to prioritize the most promising ones. This process led to the decision of firstly investigating the potentials of AmI in the domain of closed-type Agriculture, and to the formulation of a list with high-level requirements (Table 1) for an AmI Seedbed and an Intelligent Greenhouse. For the Prototype phase, the design team, along with specialists from other relevant fields (e.g., automation and robotics engineers), collaborated during focus group meetings in order to (i) decide the type of sensors, actuators, and other technological components that should be used, (ii) identify any custom-made artefacts that should be designed and developed, (iii) discuss the specifics regarding the installation of the technological equipment (e.g., where each sensor should be placed), and (iv) pinpoint the requirements that the greenhouse construction should satisfy so as to accommodate the envisioned scenarios. As a next step, User Interface (UI) and UX experts undertook the task of prototyping the identified end-user applications.

Table 1. High-level requirements.

	Intelligent Greenhouse		AmI Seedbed		Common
A.	Detailed information regarding the cultivated plants	I.	On-demand dibbling and planting of seeds	M.	Collaboration between farmers and agronomists
B.	Detailed information regarding the interior/exterior conditions	J.	Precise irrigation of the seedlings on a daily basis	N.	Intercommunication with other IEs
C.	Remote and on-site monitoring of one or more greenhouses	K.	Daily image capturing	O.	Natural and intuitive interaction
D.	Remote and on-site management of any controllable equipment	L.	Measurement of air temperature and humidity		
E.	Remote and on-site access to details for pending activities				
F.	Automations				
G.	Context-sensitive guidelines for planting/caring/harvesting plants				
H.	Notifications and warnings				

3. AmI Seedbed and Intelligent Greenhouse

The AmI Seedbed (Figure 1a) is a PA Computer Numerical Control (CNC) project that comprises a metal bench on which there are various sensors to measure air temperature and humidity and a Cartesian (xyz) coordinate robotic "arm" made of high-quality aluminum linear rail profiles. The robotic system's head is equipped with a commercial Raspberry Pi camera for image capturing and custom-made 3D printed planting and irrigation units. A small air pump through a tiny tube that ends to a trunk and is used for planting seeds into well-prepared pots with soil substrate, while a thin water tube pours water precisely into each pot.

The Intelligent Greenhouse (Figure 1a) is a small-scale (approximately 25 m^2) experimental gable-type greenhouse covered with polycarbonate cover sheets and equipped with various sensors, actuators, as well as custom-made artifacts (Table 2). The installed sensors permit the monitoring of internal and external conditions (e.g., air and soil temperature and humidity, solar radiation, external weather conditions), as well as the assessment of specific parameters of each plant (e.g., weight). Additionally, the employed actuators can be controlled via appropriate software, permitting users to manually control them, and enabling the greenhouse to automatically change its status so as to optimize its interior conditions based on Artificial Intelligence (AI). Finally, custom-made control panels are installed on the pathway in front of each cultivation zone, allowing users to view the zone's current status and, if needed, adjust its actuators.

Table 2. The technological infrastructure of the Intelligent Greenhouse.

Sensors	Actuators	Custom-Made Artefacts
• Soil moisture and soil temperature sensor • Solar radiation sensor • Internal air temperature sensor • External weather station • Fruit weight sensor (custom-made for climbing vegetables) • IP cameras	• Roof window • Door • Fan for air circulation • Irrigation valves • Grow lights • Work lights	• Control panel embedding a touchscreen, a LED display, metal buttons, flip switches, IR receiver, and an NFC reader

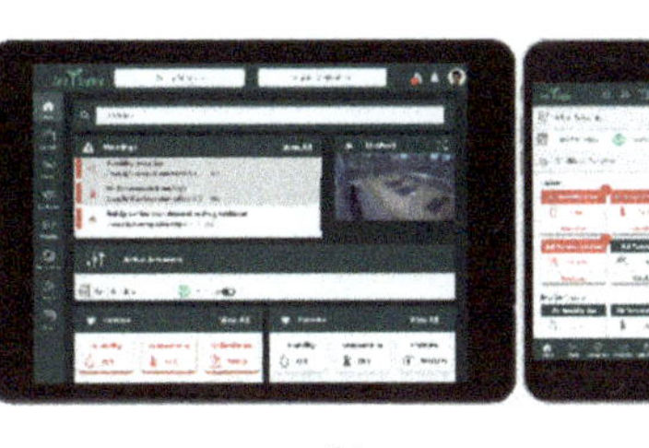

(a) (b) (c)

Figure 1. (**a**) AmI Seedbed and Intelligent Greenhouse, (**b**) snapshots of the web applications, (**c**) AR plant growth simulation.

4. End-User Applications

Intuitive and user-friendly applications [13] were designed and developed targeting PCs, tablets, smartphones (Figure 1b), and other technologically-enhanced artifacts (e.g., billboards, smart TVs, the smart refrigerator of the Intelligent Home [14]) so as to enable remote and on-site management of multiple dispersed greenhouses and seedbeds. The applications satisfy the requirements A-H mentioned in Table 1 and deliver personalized content, keeping in mind the characteristics and needs of each user. Additionally, following the advancements in the domain of Mixed Reality (MR), and considering their benefits towards PA, the mobile applications foster such technologies so as to offer advanced visualizations regarding the greenhouse's conditions. For example, the application features a simulation tool that offers an animated visualization of the plant growth (Figure 1c).

5. Conclusions and Future Work

Currently, the end-user applications are accessible via a PC in the greenhouse's storehouse. Through that PC, the in-house agronomist is able to monitor and control the greenhouse, as well as analyze the collected data, so as to suggest automations and improvements in the system's decision-making logic. In the past five months, the AmI Seedbed undertook the tasks of automatically (i) planting several pots with cucumber seeds (which are of great importance for the local (Cretan) agricultural economy), (ii) capturing an image of each pot to calculate the seedlings' growth stage according to the percentage of their leaf area, and (iii) irrigating them on a daily basis. Each column of the seedbed was irrigated with a different amount of water, starting from 0 mL and ending at 18 mL, so as to investigate the optimal amount of irrigation. The latter revealed that 4–6 mL of water per day lead to normal grown seedlings up to the stage of four leaves and a height of approximately 120–140 mm. Next, the young cucumbers were transplanted into the greenhouse, where the intelligent infrastructure was used to reach optimal internal conditions (i.e., temperature, humidity, and soil moisture) automatically. The results were quite positive, and we were able to harvest almost 100 kg of cucumbers out of a net cultivation area of 2 m^2 (nine plants distributed in two rows 1 m away, and a 40 cm gap between each plant). Regarding future work, we are planning a full-scale user-based evaluation with the participation of agronomists, professional farmers, and hobbyists so as to assess the UX, while apart from improving the current infrastructure, we are already investigating the potentials of AmI in open-area fields.

Author Contributions: Conceptualization, C.S. (Christos Stratakis), M.B. and A.L.; Methodology, A.L. and M.K.; Project administration, C.S. (Constantine Stephanidis); Software, N.M.S., M.D., G.K., T.E. and E.K.; Supervision, C.S. (Christos Stratakis), M.B. and A.L.; Writing—original draft, M.K. and C.S. (Christos Stratakis); Writing—review & editing, A.L. All authors have read and agreed to the published version of the manuscript.

Funding: This research received no external funding.

Institutional Review Board Statement: Not applicable.

Informed Consent Statement: Not applicable.

Data Availability Statement: Data is contained within the article.

Acknowledgments: This work has been supported by the FORTH-ICS internal RTD Programme 'Ambient Intelligence and Smart Environments'.

Conflicts of Interest: The authors declare no conflict of interest.

References

1. Aubert, B.A.; Schroeder, A.; Grimaudo, J. IT as Enabler of Sustainable Farming: An Empirical Analysis of Farmers' Adoption Decision of Precision Agriculture Technology. *Decis. Support Syst.* **2012**, *54*, 510–520. [CrossRef]
2. Atherton, B.C.; Morgan, M.; Shearer, S.A.; Stombaugh, T.S.; Ward, A.D. Site-Specific Farming: A Perspective on Information Needs, Benefits and Limitations. *J. Soil Water Conserv.* **1999**, *54*, 455–461.
3. Council, N.R. *Precision Agriculture in the 21st Century: Geospatial and Information Technologies in Crop Management*; The National Academies Press: Washington, DC, USA, 1997; ISBN 0-309-05893-7.
4. McBratney, A.; Whelan, B.; Ancev, T.; Bouma, J. Future Directions of Precision Agriculture. *Precis. Agric.* **2005**, *6*, 7–23. [CrossRef]
5. Muñoz, A.; Park, J. *Agriculture and Environment Perspectives in Intelligent Systems*; IOS Press: Amsterdam, The Netherlands, 2019; Volume 24, ISBN 1-61499-969-4.
6. Smart Greenhouse. Available online: https://myfood.eu/our-technology/smart-greenhouse/ (accessed on 8 March 2021).
7. Belgibaev, B.; Nikulin, V.; Umarov, A. Designing Smart Greenhouses, Satisfactory Price-Quality. *J. Math. Mech. Comput. Sci.* **2020**, *105*, 174–190. [CrossRef]
8. Mirabella, O.; Brischetto, M. A Hybrid Wired/Wireless Networking Infrastructure for Greenhouse Management. *IEEE Trans. Instrum. Meas.* **2010**, *60*, 398–407. [CrossRef]
9. FarmBot | Open-Source CNC Farming. Available online: https://farm.bot/ (accessed on 27 November 2020).
10. Aarts, E.; Wichert, R. Ambient intelligence. In *Technology Guide*; Springer: Berlin/Heidelberg, Germany, 2009; pp. 244–249.
11. Stephanidis, C. Human Factors in Ambient Intelligence Environments. *Handb. Hum. Factors Ergon.* **2012**, 1354–1373.
12. Plattner, H.; Meinel, C.; Weinberg, U. *Design-Thinking*; Springer: Berlin/Heidelberg, Germany, 2009; ISBN 3-86880-013-1.

13. Bekiaris, I.; Leonidis, A.; Korozi, M.; Stratakis, C.; Zidianakis, E.; Doxastaki, M.; Stephanidis, C. GRETA: Pervasive and AR Interfaces for Controlling Intelligent Greenhouses. In Proceedings of the 2021 17th International Conference on Intelligent Environments (IE), Dubai, United Arab Emirates, 21–24 June 2021.
14. Leonidis, A.; Korozi, M.; Kouroumalis, V.; Poutouris, E.; Stefanidi, E.; Arampatzis, D.; Sykianaki, E.; Anyfantis, N.; Kalligiannakis, E.; Nicodemou, V.C. Ambient Intelligence in the Living Room. *Sensors* **2019**, *19*, 5011. [CrossRef] [PubMed]

engineering
proceedings

MDPI

Proceeding Paper

An Ontology Based Approach for Regulatory Compliance of EU Reg. No 995/2010 in Greece [†]

Efthymios Lallas *, Anthony Karageorgos and Georgios Ntalos

Applied Informatics and Digital Technologies Laboratory, Department of Forestry, Wood Sciences and Design, School of Technology, University of Thessaly, 43100 Karditsa, Greece; karageorgos@uth.gr (A.K.); gntalos@uth.gr (G.N.)
* Correspondence: elallas@uth.gr
† Presented at the 13th EFITA International Conference, online, 25–26 May 2021.

Abstract: Illegal logging has always been considered as a major environmental and social global concern, as it is directly associated with deforestation and climate change. Nowadays, EU Regulation No 995/2010 has been successfully enforced to impede the placement of illegally produced timber within the EU market and therefore to efficiently enhance sustainable forest management and restore ecosystem balance. However, EU 995 regulatory compliance and enforcement itself is quite complex, since it requires long-term conformity, on a common basis for various heterogeneous groups and communities of stakeholders, in a global, even beyond EU, rule regulation framework. To make things worse, such a framework must be applied to the entire supply distribution chain and a wide variety of wood products, ranging from paper pulp to solid wood and flooring. Hence, in such complex and multivariate information environments, an ontological approach can more efficiently support regulatory compliance and knowledge management, due to its openness and richness of semantics for representing, analyzing, interpreting and managing such kind of information. In this paper, a rule-based regulatory compliance ontology is proposed, which fully captures EU Regulation No 995/2010 concepts and compliance rules and guidelines, as well as Greek legislations governing wood trade. The proposed ontology can be the basis for a computerized system providing automated support for illegal wood trade and monitoring EU regulation information provision and audit information storage and analysis.

Keywords: illegal wood trade; EU Regulation No 995/2010; regulatory compliance; ontology engineering

Citation: Lallas, E.; Karageorgos, A.; Ntalos, G. An Ontology Based Approach for Regulatory Compliance of EU Reg. No 995/2010 in Greece. *Eng. Proc.* **2021**, *9*, 38. https://doi.org/10.3390/engproc2021009038

Academic Editors: Dimitrios Kateris and Maria Lampridi

Published: 31 December 2021

Publisher's Note: MDPI stays neutral with regard to jurisdictional claims in published maps and institutional affiliations.

1. Introduction

Illegal logging has always been considered as a major environmental and social global concern, as it is directly associated with deforestation and climate change. Over the past few decades, there has been a worldwide attempt to monitor and eliminate illegal logging motivations and activities via appliance of appropriate regulation frameworks and by establishing monitoring organizations for assuring conformity behavior. The EU community has successfully enforced EU Regulation No 995/2010 (EUTR) [1] to its member states for efficiently dealing with this matter.

The main goal of EU Regulation No 995/2010 is to prohibit placing illegally harvested timber and products within the EU market. For this purpose, it requires that traders maintain valid formal records of who they buy from and who they sell to and that operators apply a due diligence system including risk assessment and risk mitigation procedures [2]. Unfortunately, supply chain traceability is a challenging task for monitoring organizations, as there is commonly not sufficient evidence of wood product originality and of the intermediate merchant transactions. Furthermore, supply chain complexity is an additional issue, since it impedes the long-term conformity of heterogeneous groups and communities involved, such as stakeholders and suppliers. Several research works have focused on solutions targeting supply chain traceability and complexity [3–9].

In such complex regulation environments comprising multivariate information, an ontology-based data model could better support compliance and knowledge management due to its openness and richness of semantics.

The work described in this paper focuses on a regulatory compliance ontology model, aiming to facilitate meeting EU Regulation No 995/2010 compliance guidelines and therefore to efficiently tackle the tremendous consequences of illegal logging. A case study in Greece was considered. The proposed ontology can be the basis for the implementation of a computerized regulation compliance system, providing automated support for illegal wood trade monitoring and audit information storage and analysis.

2. Description of the Regulatory Compliance Ontology of EU Regulation No 995/2010 in Greece

In this section the main ontology entities are identified based on the EU Regulation No 995/2010 guidelines, applied to a case study in Greece. The main building blocks of constructing such an ontology are ontological subdomains and class objects with common characteristics, which represent application-domain concepts. Class individuals are used for accurate ontology description, which are the instances of a class object.

More specifically, the proposed ontology comprises six regulatory compliance domains, which are deemed as necessary for fully representing the EUTR regulation domain in Greece. Each domain is associated with a number of class entities. The proposed domains are as follows: (i) the administration authority domain including classes corresponding to the respective authorities of Greece and the EU. The central competent authority in Greece is the Department of Planning and Forest Policy of the General Secretariat of Development and Protection of Forests and Agri-Environment of the Ministry of Environment and Energy, with the support of Distribution and Trade Department of Forest Products and Species (Convention on International Trade in Endangered Species (CITES) of Wild Fauna and Flora). There are also regional competent delegated authorities for each region of Greece, such as the Forest Secretariat of the Greece Regional Administrations or the Secretariat of Forest Coordination and Inspection. Regional competent authorities usually cooperate with regional audit services, for carrying out compliance audit checks, and inter-ministerial technical expertise working groups, for technical assistance, whenever required; (ii) the supply chain domain, including trader and operator classes, along with their attributes, and additional general classes, such as merchants, retailers, manufacturers, suppliers, commercial customers, end (final) consumers, agents, buyers, and sellers; (iii) the directory/registry domain, including all relevant national and regional directories of traders, operators, and monitoring organizations; (iv) the product domain including class individuals related to wood products, such as code, name, description, batch order, origin country, quantity, value, invoice, packing list, delivery note, life cycle, and product list within, and out of the scope of EUTR; (v) the audit control domain, including class individuals related to audit control status (premises, field audit control, and due-diligence service audit) and associated attributes, such as risk assessment, risk mitigation, risk status, risk assessment report, and penalty/fines; (vi) the documentation domain comprising documentation record repository and maintain record period classes, and two subdomains—(a) certificate and license validation subdomain, including classes related to validation documentation (forest law enforcement, governance, and trade-FLEGT license, voluntary partnership agreements (VPAs) CITES certificate, 3rd-party-verified scheme, forest stewardship council (FSC), programs for the endorsement of forest certification(PEFC), and chain of custody(CoC) certification) [10] and (b) supply chain documentation subdomain, including classes related to product documentation, with some of them also included in the product domain.

3. Regulatory Compliance Ontology Design of EU Regulation No 995/2010 in Greece

The next ontology design step involves the identification of rule compliance constraints based on EUTR articles and related law frameworks [11–14], as well as regulation rules that govern traders and operators, competent authorities, and monitoring organizations. Such constraint requirements completely define the roles and the relationships among ontology class individuals, and they are reflected on the overall regulatory compliance ontology layout. Specifically, each EU country, including Greece, has its own delegated central and regional competent authorities responsible for the correct application of EUTR regulations on all parties involved. This is achieved by the consulting support of monitoring organizations for properly executing these relevant control checks. Monitoring organizations are EU-based industry or trade associations, certification bodies, or other service providers, which offer consulting services to assist wood enterprises in achieving compliance with due diligence requirements. In general, regulation compliance audit checks require, for both traders and operators, the certified record maintenance of particular transaction information for a certain period, in order to enable supply chain traceability. Moreover, operators should be able to apply a due-diligence service audit check, with the guidance of monitoring organizations, in order to assess the risk of illegal timber in the supply chain, and to provide additional information and verification documentation in case there is a risk assessment that needs to be mitigated.

Due to the large number of classes and their associated relationships included in the EUTR compliance ontology, it is hard to show and clearly discriminate all of them in a single layout. It would be more practical to represent a subset of interdomain class relationships as a sample for the proposed regulatory compliance ontology. For this reason, we have chosen the interdomain class entity relationships of a typical due-diligence audit check. A due-diligence audit check is a three-step procedure: The first step requires access to relevant information maintained in the compliance documentation repository. This step is mandatory for both traders and operators, while the next two steps are mandatory only for operators. The second step, termed risk assessment, aims for determining and classifying risk status. There could be either "no risk" status, in case a FLEGT or CITES license is provided, "negligible" (low/medium) risk, or "non-negligible" (high) risk status. Risk mitigation measures need to be performed for the latter two risk status cases, which necessitates the initiation of the third step. Its basic goal is to mitigate all risk status cases, down to "negligible" risk status and finalize successfully the audit control procedure. In any case, a risk assessment report is required, in addition, so as to officially finalize the procedure, by writing down its results in detail. The regulatory compliance ontology layout for highlighting class relationship representation of a typical due-diligence audit check is depicted in Figure 1.The relations between class individuals based on their specified roles according to EUTR framework clauses are indicated by arrows.

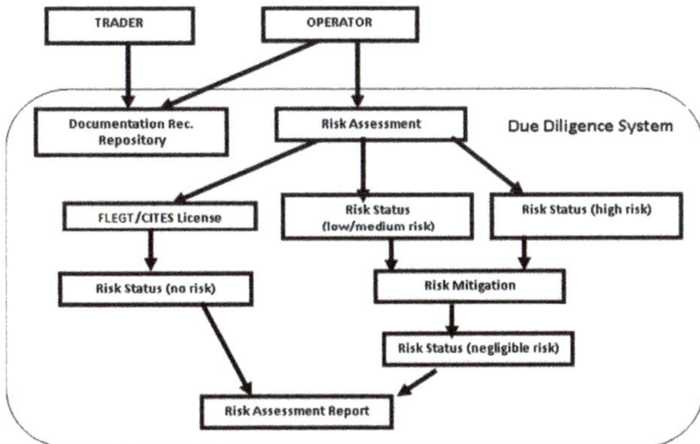

Figure 1. Regulatory compliance ontology layout of a due-diligence audit check.

4. Discussion and Conclusions

For over a decade, the EU community has successfully enforced EU Regulation No 995/2010 for tackling illegal harvesting and its tremendous consequences in forest management, ecosystem balance, and climate change. However, supply chain traceability and related data complexity are quite challenging issues that require efficient support for fast audit checks and effective risk assessment and risk mitigation procedures. In addition, each EU member must convert the EUTR into national law standards with the least exceptions and law deviations from the basic EU law framework.

This paper proposes a regulatory compliance ontology to support EU Regulation No 995/2010 compliance and knowledge management. The proposed ontology is a complete representation of EUTR focusing on its prohibition requirements, the competent authorities, and the required audit controls, while paying particular emphasis to compliance requirements in the context of Greek legislations. Furthermore, the proposed ontology describes in full detail the concepts in the domain of illegal wood trade, as well as the relationships that hold between these concepts. Its analytical representation is appropriate for the support of decision making in EUTR critical risk assessment issues. Finally, the ontology implementation is independent of any software platform and could thus be the basis for developing computerized audit monitoring information systems.

Funding: This research received no external funding.

Institutional Review Board Statement: Not applicable.

Informed Consent Statement: Not applicable.

Conflicts of Interest: The authors declare no conflict of interest.

References

1. Available online: https://eur-lex.europa.eu/LexUriServ/LexUriServ.do?uri=OJ:L:2010:295:0023:0034:EN:PDF (accessed on 11 November 2021).
2. Available online: https://ec.europa.eu/environment/forests/timber_regulation.htm (accessed on 11 November 2021).
3. Holopainen, J.; Toppinen, A.; Perttula, S. Impact of European Union Timber Regulation on Forest Certification Strategies in the Finnish Wood Industry Value Chain. *Forests* **2015**, *6*, 2879–2896. [CrossRef]
4. Leipold, S. How to move companies to source responsibly? German implementation of the European Timber Regulation between persuasion and coercion. *For. Policy Econ.* **2016**, *82*, 41–51. [CrossRef]
5. Koch, G.; Haag, V. Control of Internationally Traded Timber—The Role of Macroscopic and Microscopic Wood Identification against Illegal Logging. *J. Forensic Res.* **2015**, *6*, 1000317. [CrossRef]
6. McDermott, C.L.; Sotirov, M. Apolitical economy of the European Union's timber regulation: Which member states would, should or could support and implement EU rules on the import of illegal wood? *For. Policy Econ.* **2018**, *90*, 180–190. [CrossRef]
7. Köthke, M. Implementation of the European Timber Regulation by German importing operators: An empirical investigation. *For. Policy Econ.* **2020**, *111*, 102028. [CrossRef]
8. Dudik, R.; Sisak, L. Placing Timber and Timber Products on the Market in the Czech Republic and Related Economic Impacts. In Proceedings of the IUFRO Symposium 2014, Sopron, Hungary, 26–28 November 2014; pp. 35–42.
9. Neshataeva, E.; Karjalainen, T. Impact of the EU Timber Regulation on Russian companies exporting wood and wood-based products. *Resour. Technol.* **2015**, *12*, 37–44. [CrossRef]
10. Available online: https://www.flegtlicence.org/flegt-certification-cites (accessed on 11 November 2021).
11. Available online: https://eur-lex.europa.eu/legal-content/EN/TXT/?uri=CELEX%3A32012R0607 (accessed on 11 November 2021).
12. Available online: https://eur-lex.europa.eu/legal-content/EN/TXT/?uri=CELEX%3A32012R0363 (accessed on 11 November 2021).
13. Available online: https://www.e-nomothesia.gr/kat-dasos-thera/kya-134627-5835-2015.html (accessed on 11 November 2021).
14. Available online: https://dasarxeio.com/wp-content/uploads/2015/12/134382_5229_2015.pdf (accessed on 11 November 2021).

engineering
proceedings

Proceeding Paper

The Common Greenhouse Ontology: An Ontology Describing Components, Properties, and Measurements inside the Greenhouse †

Roos Bakker [1,*], **Romy van Drie** [1], **Cornelis Bouter** [1], **Sander van Leeuwen** [2], **Lorijn van Rooijen** [2] and **Jan Top** [2]

[1] Department Data Science, The Netherlands Organisation for Applied Scientific Research, 2595 DA Den Haag, The Netherlands; romy.vandrie@tno.nl (R.v.D.); cornelis.bouter@tno.nl (C.B.)
[2] Department FFC Supply Chain & Information Management, Wageningen University & Research, 6708 WG Wageningen, The Netherlands; sander.vanleeuwen@wur.nl (S.v.L.); lorijn.vanrooijen@wur.nl (L.v.R.); jan.top@wur.nl (J.T.)
* Correspondence: roos.bakker@tno.nl
† Presented at the 13th EFITA International Conference, online, 25–26 May 2021.

Abstract: Modern greenhouses have systems that continuously measure the properties of greenhouses and their crops. These measurements cannot be queried together without linking the relevant data. In this paper, we introduce the Common Greenhouse Ontology, a standard for sharing data on greenhouses and their measurable components. The ontology was created with domain experts and incorporates existing ontologies, SOSA and OM. It was evaluated using competency questions and SPARQL queries. The results of the evaluation show that the Common Greenhouse Ontology is an innovative solution for data interoperability and standardization, and an enabler for advanced data science techniques over larger databases.

Keywords: ontology; semantic alignment; greenhouse observations; data interoperability; SOSA ontology; semantic sensor network

Citation: Bakker, R.; van Drie, R.; Bouter, C.; van Leeuwen, S.; van Rooijen, L.; Top, J. The Common Greenhouse Ontology: An Ontology Describing Components, Properties, and Measurements inside the Greenhouse. *Eng. Proc.* **2021**, *9*, 27. https://doi.org/10.3390/engproc2021009027

Academic Editors: Dimitrios Aidonis and Aristotelis Christos Tagarakis

Published: 2 December 2021

Publisher's Note: MDPI stays neutral with regard to jurisdictional claims in published maps and institutional affiliations.

1. Introduction

Greenhouses are used to grow vegetables and plants year-round. The climate in the greenhouse has a considerable influence on the growth of the crops. Therefore, it is important to optimally control the climate. Increasingly, more tools become available to help with this, such as climate computers, sensors, and other systems. All these systems produce different data, such as temperatures, crop growth, and weather statistics, which are stored in different databases. In this paper, we introduce the Common Greenhouse Ontology (CGO). It provides a semantic alignment of different databases, as well as a standard for high-tech greenhouses and their components.

The CGO was created in the framework of a national project "Data-Driven Integrated Greenhouse Systems" (DDINGS). In this project, a platform was created to connect databases and perform data analysis [1]. In that work, we also introduced a first version of the CGO. Since then, we fully developed the CGO with different modules, a broad set of classes that describe the components of greenhouses, and an integration with other ontologies such as the Semantic Sensor Network ontology [2] and the Ontology of units of Measure [3].

In the agricultural domain, multiple ontologies and linked open databases have been developed. A well-known example is AGROVOC [4], a large agricultural thesaurus. There are also search engines for agricultural ontologies such as AgroPortal [5]. Many ontologies can be found on these platforms, such as the Plant Phenology Ontology [6]. These ontologies provide an extensive collection of concepts in the agriculture domain, however, there does not yet exist an ontology on greenhouses. As such, we developed the CGO to provide a standard on high-tech greenhouses and their elements, and to stimulate data integration. Some concepts from greenhouses do exist in other ontologies, which the CGO uses where possible.

2. Materials and Methods

An ontology is a formal model of the structure of a domain. It captures relevant concepts and describes the relations between these concepts [7]. The CGO was developed with domain experts and focuses on greenhouse-related concepts and measurements. Several ontology development methods were consulted and applied at various stages of development, including SABiO [8], and Ontology 101 [9]. The CGO also makes use of other ontologies: the Semantic Sensor Network ontology (SSN), which includes the Sensor, Observation, Sample, and Actuator ontology (SOSA), and the Ontology of units of Measure (OM).

Observations are an important aspect of the CGO. In the CGO, the SOSA ontology is used to model observations. In Figure 1, a simplified version of SOSA's architecture of observations is shown in green. The observation in the example is the thickness of a stem (of a flower) and the stem is the part that is observed, the Feature of Interest. The property we are observing is its thickness, the Observable Property. The Feature of Interest and the Observable Property, but also the result, are linked to the Observation, the center of the architecture to which all elements are connected.

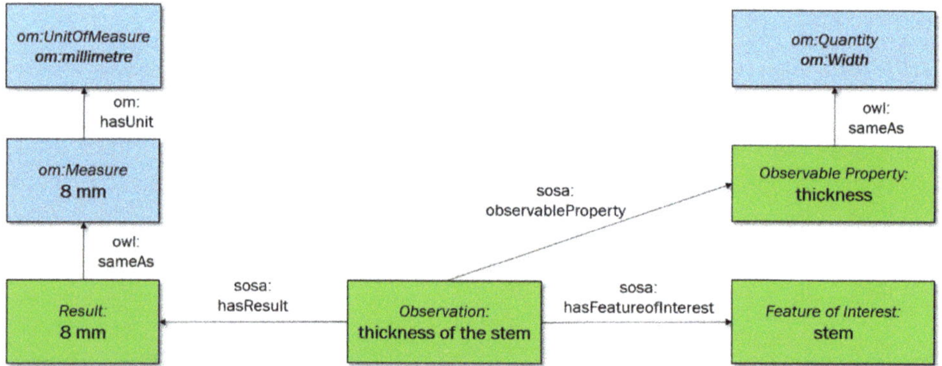

Figure 1. Design of measurements in CGO and its integration with SOSA and OM.

The Ontology of units of Measure (OM) is an OWL ontology for the domain of quantities, measurements and units of measure [3]. OM provides the possibility to link aspects of an observation to a quantity and a unit. A schematic overview of how this is implemented in CGO is given in Figure 1. From a cgo:Class (e.g., thickness), a link to an om:Quantity (e.g., om:Width) can be provided. The results are linked to a numerical value through an om:NumericalValue (this is not shown in Figure 1 for reasons of simplification) and to a unit through om:hasUnit.

3. Common Greenhouse Ontology

The CGO contains 382 classes, 99 properties (both data and object properties), and 12 individuals. This includes the classes and properties from the SOSA ontology. For an overview of all classes and their hierarchy in the ontology, consult the ontology itself (https://ontology.tno.nl/ddings/Greenhouse_ontology_v1.3.ttl, accessed on 22 November 2021). We can divide the content into four main categories: the greenhouse, which is the center concept of the ontology; features, which are the set properties of the greenhouse such as its dimensions; parts, which are the objects that can be found in greenhouses; and finally, measurements. The measurements in the greenhouse are modeled using SOSA [2] and OM [3] as can be seen in Figure 1. SOSA classes are connected to CGO classes by being superclasses of possible Features of Interest and Observable Properties, OM classes are connected to data properties of CGO and as a subtype of sosa:Result.

Each of the categories contain a variety of classes. The parts category contains over 150 classes and provides an elaborated, yet not complete, overview of parts of a high-tech greenhouse. Important subsets are systems and construction hierarchies. The systems subset describes a wide set of systems in a greenhouse ranging from broader ventilation systems to specific geothermal heat pumps. The construction subset includes classes such as screens and ventilation vents. These classes are all connected to the center of CGO, the greenhouse class, by the object property "part of". As mentioned above, measurements are modeled through SOSA, but features of the greenhouse, such as its orientation and location, are expressed with data properties.

4. Evaluation

For testing the completeness of the CGO, we created competency questions (CQ) in collaboration with greenhouse experts. Competency questions are for an ontology what the requirements are for a software. Competency questions can be questions about the data, the relations between concepts, and broader analytical questions. In Table 1, five examples of CQs are shown. The complete set of CQs was created through several practical use cases and can be found in the CGO. For querying, SPARQL queries were made from these CQ's, and restrictions were expressed using SHACL constraints.

Table 1. This table shows five examples of CGO CQs.

What is the location of sensor X? What is the size of the ridge foils?	Which sensors hang next to the ridge foils? For which crops are ridge foils used?	What is the delivered heating power based on pipe X at time Y and location Z?

Three sets of restrictions were defined via SHACL: cardinality restrictions constrain the properties of greenhouse parts, the observation time is validated to have a normalized time interval, and we constrain which features and observable properties can co-occur. The domain and range restrictions were modeled via OWL constraints. All CQs included in the CGO can be answered using its classes and properties together with sensor data (for more information about the data see [10]), thereby providing a complete ontology for expressing greenhouses and their components.

5. Conclusions

Modern high-tech greenhouses have systems that continuously make measurements which cannot be queried together without linking the relevant data. In this paper, we introduced the Common Greenhouse Ontology, a standard for sharing data on greenhouses and their measurable components. The ontology was created with domain experts and incorporates existing ontologies, SOSA and OM. It includes subsets of systems, construction parts, and features such as the location of the greenhouse. It was evaluated using competency questions, and SPARQL queries. The results of the evaluation show that the Common Greenhouse Ontology can answer competency questions from different use cases. For future work, the ontology can be expanded by forming competency questions from new use cases. Another interesting direction would be to connect it to other standards and platforms such as AGROVOC [4]. Expanding the CGO will further increase its value as a data interoperability and standardization solution.

Supplementary Materials: The CGO can be downloaded online at https://ontology.tno.nl/ddings/Greenhouse_ontology_v1.3.ttl (accessed on 22 November 2021).

Author Contributions: Conceptualization, R.B., R.v.D. and C.B.; methodology, R.B. and C.B.; software, R.B. and R.v.D.; validation, R.B., R.v.D., C.B., S.v.L., L.v.R. and J.T.; formal analysis, C.B. and S.v.L.; investigation, R.v.D. and R.B.; data curation, C.B.; writing—original draft preparation, R.B.;

writing—review and editing, R.v.D., C.B., S.v.L., L.v.R. and R.B.; visualization, R.B., R.v.D. and C.B.; supervision, J.T. All authors have read and agreed to the published version of the manuscript.

Funding: This research was funded by a public-private collaboration (PPS) with funding number HT17228. The public funding was provided by Topsector Tuinbouw en Uitgangsmaterialen and Topsector High-Tech Systemen en Materialen, and the private funding through Stichting Hortivation.

Informed Consent Statement: Not applicable.

Data Availability Statement: Data available in a publicly accessible repository. The data presented in this study are openly available in DDINGS-CGO at https://gitlab.com/ddings/common-greenhouse-ontology (accessed on 22 November 2021).

Acknowledgments: We would like to thank Jack Verhoosel for initiating the idea for a CGO, Charlotte Lelieveld for leading the DDINGS project, Athanasios Sapounas and Bart Slager for bringing in their expert knowledge, and Barry Nouwt for guiding the creative process of the early architecture.

Conflicts of Interest: The authors declare no conflict of interest. The funders had no role in the design of the study; in the collection, analyses, or interpretation of data; in the writing of the manuscript, or in the decision to publish the results.

References

1. Verhoosel, J.P.C.; Nouwt, B.; Bakker, R.M.; Sapounas, A.; Slager, B. A Datahub for Semantic Interoperability in Data-Driven Integrated Greenhouse Systems. In Proceedings of the EFITA Conference, Rhodes Island, Griekenland, 7–29 June 2019.
2. Neuhaus, H.; Compton, M. The semantic sensor network ontology. In Proceedings of the AGILE Workshop on Challenges in Geospatial Data Harmonisation, Hannover, Germany, 2 June 2009.
3. Rijgersberg, H.; Wigham, M.L.I.; Top, J.L. How semantics can improve engineering processes: A case of units of measure and quantities. *Adv. Eng. Inform.* **2011**, *25*, 276–287. [CrossRef]
4. Caracciolo, C.; Stellato, A.; Morshed, A.; Johannsen, G.; Rajbhandari, S.; Jaques, Y.; Keizer, J. The AGROVOC linked dataset. *Semant. Web* **2013**, *4*, 341–348. [CrossRef]
5. Jonquet, C.; Toulet, A.; Arnaud, E.; Aubin, S.; Yeumo, E.D.; Emonet, V.; Graybeal, J.; Laporte, M.A.; Musen, M.A.; Pesce, V.; et al. AgroPortal: A vocabulary and ontology repository for agronomy. *Comput. Electron. Agric.* **2018**, *144*, 126–143. [CrossRef]
6. Stucky, B.J.; Guralnick, R.; Deck, J.; Denny, E.G.; Bolmgren, K.; Walls, R. The plant phenology ontology: A new informatics resource for large-scale integration of plant phenology data. *Front. Plant Sci.* **2018**, *9*, 517. [CrossRef] [PubMed]
7. Guarino, N.; Oberle, D.; Staab, S. What Is an Ontology. In *Handbook on Ontologies*; Staab, S., Ruder, R., Eds.; Springer: Berlin/Heidelberg, Germany, 2009; pp. 154–196.
8. de Almeida Falbo, R. Experiences in using a method for building domain ontologies. In Proceedings of the 16th International Conference on Software Engineering and Knowledge Engineering, Banff, AB, Canada, 20–24 June 2004.
9. Noy, N.; McGuinness, D. Ontology development 101: A guide to creating your first ontology. *Development* **2001**, *32*, 1–25.
10. Bouter, C.; Kruiger, H.; Verhoosel, J. Domain-Independent Data Processing in an Ontology Based Data Access Environment using the SOSA. In Proceedings of the Joint Ontology Workshops 2021, Episode VII: The Bolzano Summer of Knowledge, Co-located with the 12th International Conference on Formal Ontology in Information Systems (FOIS 2021), and the 12th International Conference on Biomedical Ontologies (ICBO 2021), Bolzano, Italy, 11–18 September 2021.

Proceeding Paper

Depth Image Selection Based on Posture for Calf Body Weight Estimation [†]

Yuki Yamamoto [1,*], Takenao Ohkawa [1], Chikara Ohta [2], Kenji Oyama [3] and Ryo Nishide [4]

1   Graduate School of System Informatics, Kobe University, Kobe 657-0013, Japan; ohkawa@kobe-u.ac.jp
2   Graduate School of Science, Technology and Innovation, Kobe University, Kobe 657-8501, Japan;
    ohta@port.kobe-u.ac.jp
3   Food Resources Education and Research Center, Kobe University, Kasai 675-2103, Japan; oyama@kobe-u.ac.jp
4   The Center for Data Science Education and Research, Shiga University, Hikone 522-8522, Japan;
    ryo-nishide@biwako.shiga-u.ac.jp
*   Correspondence: y_yuuki@cs25.scitec.kobe-u.ac.jp
†   Presented at the 13th EFITA International Conference, online, 25–26 May 2021.

Abstract: We are developing a system to estimate body weight using calf depth images taken in a loose barn. For this purpose, depth images should be taken from the side, without calves overlapping and without their backs bent. However, most of the depth images that are taken successively and automatically do not satisfy these conditions. Therefore, we need to select only the depth images that match these conditions, as to take many images as possible. The existing method assumes that a calf standing sideways and upright in front of cameras is in a suitable pose. However, since such cases rarely occur, not many images were selected. This paper proposes a new depth image-selection method, focusing on whether a calf is sideways, and the back is not bent, regardless of whether the calf is still or walking. First, depth images including only a single calf are extracted. The calf was identified using radio frequency identification (RFID) when its depth image was taken. Then, the calf area was extracted by background subtraction and contour detection with a depth image. Finally, to judge the usable depth images, we detected and evaluated the calf's posture, such as the angle of the calf to the camera and the slope of the dorsal line. We used the mean absolute percentage error (MAPE) to assess the efficiency of our method. As two times the number of depth images were extracted, our method achieved an MAPE of 12.45%, while the existing method achieved an MAPE of 13.87%. From this result, we have confirmed that our method makes body weight estimation more accurate.

Keywords: cattle; calf; weight estimation; image processing; depth camera

Citation: Yamamoto, Y.; Ohkawa, T.; Ohta, C.; Oyama, K.; Nishide, R. Depth Image Selection Based on Posture for Calf Body Weight Estimation. *Eng. Proc.* **2021**, *9*, 20. https://doi.org/10.3390/engproc2021009020

Academic Editors: Charisios Achillas and Lefteris Benos

Published: 25 November 2021

1. Introduction

There is a high demand to select the breeding stock with a high reproductive ability in cattle breeding farms. The reproductive ability can be measured as the maternal ability as well as the calving interval. The weight changes in calves are often used as an indicator of maternal ability [1]. However, manually measuring the calf's weight burdens farmers and stresses calves. Hence, several methods using computer vision were studied to automate this process [2]. We are developing a system to estimate body weight using calf depth images taken in a loose barn, where the calves can move without stress. Depth images appropriate for body weight estimations are ones taken from the side, without calves overlapping and without their backs bent. In [3], a method to select such depth images from all those that were taken has been proposed, under the consideration that standing sideways and still upright in front of a camera is a suitable pose. However, such cases rarely occurred, since the calves were walking around freely. As a result, few images were selected.

Therefore, this paper proposes a method to select appropriate depth images regardless of whether calves are still or walking, as some depth images taken when a calf was walking across in front of the camera can still be useful. This method does not utilize any constraint condition to judge whether the calf is still and upright but focuses on whether the calf is sideways and the back is not bent. When detecting the calf's posture, such as the angle of the calf to the camera and the slope of the dorsal line, our method is more accurate than the previous method, as it corrects the calf's contour extracted from a depth image and considers the local depth of body parts.

2. Methodology

2.1. Overview of Calf Body Weight Estimation System

Figure S1 shows the process flow of our automatic calf body weight estimation system in a loose barn. First, depth images including only a single calf are selected according to an individual identification, using radio frequency identification (RFID). Next, the calf's area is cut out from the depth image. If it is judged that the calf's posture is appropriate for body weight estimation, its body part is modeled, because chest girth and waist girth have a high correlation with body weight, and its volume is calculated. Finally, the body weight is estimated based on the volume [4,5].

2.2. Individual Identification by Using RFID

The RFID system can read multiple electronic tags at once. The tag reader that we used is RWLU1002, manufactured by MASPRO DENCO Corp., using a specific low-power 920 MHz radio with a maximum transmission power of 24 dBm. The reader antenna is RAF4031, also manufactured by MASPRO DENCO Corp., with a narrow half-value angle of 22 degrees. Two RFID tags hang down each side from a belt around a calf's neck, which means that the calf can be identified from its right and left sides. In addition, the RFID reader antenna and the camera are aligned in the same direction. Therefore, by comparing the shooting time with the identification time, how many calves are taken in the image and what each calf is can be known, which enables the depth image including only a single calf to be selected. However, this selection does not always succeed, especially when other calves' RFID tags are hidden.

2.3. Selecting Depth Images Based on Calf's Posture

The calf's body part is modeled using a three-dimensional successive cylindrical model vertical to the direction of the camera [4,5]. When the axis of the body is inclined, i.e., the angle of the calf to the camera is not vertical or the calf's dorsal line is not horizontal, the body part cannot accurately be modeled (Figures S2 and S3). Therefore, these images need to be excluded. For this posture judgment, the calf's area on the depth image is extracted using the background subtraction method [6].

The angle between the calf and the camera is estimated based on the average depth values in each part: head, shoulder top, and rump (Figure 1a). Since calves move their heads freely, even if they are not inclined from rump to head, their body parts may be bent, which leads to a less accurate weight estimation. For this reason, not only the entire body, but also each part of the body, is checked to ensure it is straight and sideways. It is critical to accurately record the angle of the calf's axis using the depth values near the backbone, which is less affected by bulges on the sides of the body. The slope of the calf's dorsal line is estimated to be that of the line from the root of the neck to the rump (Figure 1b). The issue in this process is that depth data near the contour are likely to lose the depth image of a moving calf, so that the slope of the dorsal line may not be accurately detected. Therefore, the dorsal contour is complementarily smoothed before dorsal slope detection.

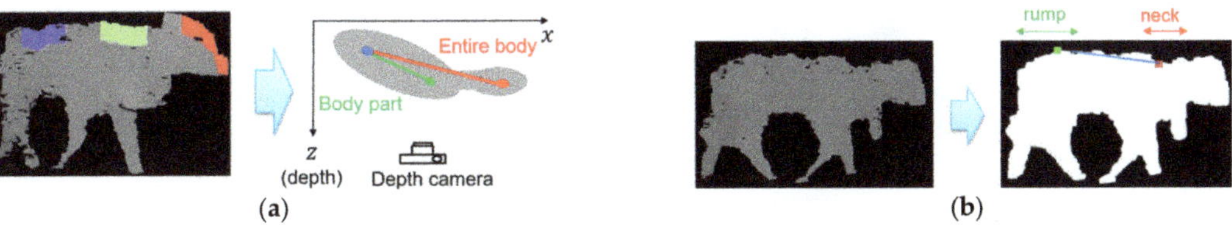

Figure 1. Calf's posture judgment: (**a**) The angle to the camera; (**b**) The slope of dorsal line.

3. Experiment

3.1. Data Set

To assess the effectiveness of the proposed method, we conducted experiments to compare it with an existing image-selection method [3]. In the experiment, we targeted calves, from newborns to calves weighing under 100 kg, at the Food Resources Education and Research Center of Kobe University. The depth camera we used is RealSense Depth Camera D415 manufactured by Intel. Using RFID, 65,000 images were judged to include only one calf among those taken from 6 December 2020 to 25 January 2021. Note that 463 images included multiple calves rather than a single calf. Among these, 45,000 images taken from 6 December 2020 to 7 January 2021, and 20,000 images taken from 8 January 2020 to 25 January 2021, were used for training and testing, respectively, for body weight estimation [3]. Two image-selection methods, i.e., the existing method [3] and the proposed method, were applied for those images. In the training phase, first, body volumes were estimated using the images chosen by each image-selection method. Then, a linear regression equation between the calculated volumes and the manually measured body weights, and a coefficient of determination, were determined. In the testing phase, the body volumes were estimated using the images chosen by each image-selection method. Then, the body weights were calculated using the regression equation obtained in the training phase. Finally, we calculated mean absolute percentage error (MAPE) between estimated body weight $W_i^{\text{estimated}}$ and actual body weight W_i^{actual}:

$$MAPE = \frac{\left| W_i^{\text{actual}} - W_i^{\text{estimated}} \right|}{W_i^{\text{actual}}} \times 100. \tag{1}$$

Note that only circle fitting was used to model the calf's body part to calculate the volume [4,5]. Moreover, when obtaining multiple images of one calf on the same day, we used the average of the volumes, because this improves the accuracy of body weight estimation.

3.2. Result

Table 1 shows the experimental results for each of the two methods: the number of images selected, the total number of calf-days, the coefficient of determination, and MAPE.

Table 1. The experimental results for each of the two methods.

Method	Number of Images/Calves (Training Data)	Number of Images/Calves (Test Data)	Coefficient of Determination	MAPE (%)
Existing method [3]	375/60	69/19	0.6303	13.87
Proposed method	650/78	226/27	0.6548	12.45

4. Discussion

Thanks to the posture judgment, which focuses on the angle to the camera and the slope of the dorsal line, we could acquire almost twice as many depth images by using the proposed method as by using the previous method. On the other hand, the existing method had a larger MAPE, even though depth images of an upright calf accounted for a larger percentage of the selected images. From these results, we consider it crucial to acquire more depth images, instead of selecting only high-quality ones. Thus, we have confirmed that our method makes body weight estimation more accurate. The proposed method and the existing method have selected four depth images and two depth images with multiple calves overlapping, respectively (Figure S4). Therefore, such images should be eliminated using outlier detection.

5. Conclusions

This paper proposed an image selection method to choose appropriate depth images, regardless of whether calves are still or walking, for body weight estimation. We have confirmed that our method acquires more depth images from the experimental results and makes the body weight estimation more accurate. In future work, we should establish a method to eliminate inappropriate depth images. It also needs to be tested on other farms to verify its versatility. For this to operate on a farm, we plan to expand the research on calf body weight estimation.

Supplementary Materials: The following are available online at https://www.mdpi.com/article/10.3390/engproc2021009020/s1, Figure S1: Process flow of calf body weight estimation, Figure S2: Example of a bad model of the body when the angle to the camera is inclined, Figure S3: Example of a bad model of the body when the dorsal line slopes, Figure S4: Examples of images with multiple calves overlapping.

Author Contributions: Conceptualization, T.O.; methodology, Y.Y. and C.O.; software, Y.Y.; validation, Y.Y. and K.O.; formal analysis, Y.Y.; investigation, Y.Y., T.O., C.O., K.O. and R.N.; resources, K.O.; data curation, Y.Y.; writing—original draft preparation, Y.Y.; writing—review and editing, T.O., C.O., K.O. and R.N.; visualization, Y.Y.; supervision, T.O.; project administration, K.O.; funding acquisition, K.O. All authors have read and agreed to the published version of the manuscript.

Funding: This research was funded by the subsidized projects (2016–2019) of Japan Livestock Technology Association, the Council for Information Utilization of Beef Cattle Improvement, and the Ministry of Agriculture, Forestry and Fisheries. Note that any implications in this work are not their official opinions.

Data Availability Statement: The data presented in this study are available on request from the corresponding author. The data are not publicly available due to privacy.

Conflicts of Interest: The authors declare no conflict of interest.

References

1. Meyer, K. Variance components due to direct and maternal effects for growth traits of Australian beef cattle. *Livest. Prod. Sci.* **1992**, *31*, 179–204. [CrossRef]
2. Wang, Z.; Shadpour, S.; Chan, E.; Rotondo, V.; Wood, M.K.; Tulpan, D. ASAS-NANP SYMPOSIUM: Applications of machine learning for livestock body weight prediction from digital images. *J. Anim. Sci.* **2021**, *99*, skab022. [CrossRef] [PubMed]
3. Fukuda, N.; Ohkawa, T.; Ohta, C.; Oyama, K.; Takaki, Y.; Nishide, R. Image Extraction Based on Depth Information for Calf Body Weight Estimation. In Proceedings of the 12th EFITA-HAICTA-WCCA Congress, Rhodes, Greece, 27–29 June 2019.
4. Yamashita, A.; Ohkawa, T.; Oyama, K.; Ohta, C.; Nishide, R.; Honda, T. Calf Weight Estimation with Stereo Camera Using Three-Dimensional Successive Cylindrical Model. *J. Inst. Ind. Appl. Eng.* **2018**, *6*, 39–46. [CrossRef]
5. Nishide, R.; Yamashita, A.; Takaki, Y.; Ohta, C.; Oyama, K.; Ohkawa, T. Calf Robust Weight Estimation Using 3D Contiguous Cylindrical Model and Directional Orientation from Stereo Images. In Proceedings of the Ninth International Symposium on information and Communication Technology, Danang City, Vietnam, 6 December 2018; pp. 208–215.
6. opencv_contrib/bgfg_gsoc.cpp at 6520dbaa224a661ca8105b1ab0b71451fd715f4c · opencv/opencv_contrib · GitHub. Available online: https://github.com/opencv/opencv_contrib/blob/6520dbaa224a661ca8105b1ab0b71451fd715f4c/modules/bgsegm/src/bgfg_gsoc.cpp (accessed on 14 June 2021).

Proceeding Paper

Capturing and Evaluating the Effects of the Expansive Species *Ailanthus altissima* on Agro-Ecosystems on the Ionian Islands [†]

Yorghos Voutos [1,*], Nicole Godsil [2], Anna Sotiropoulou [1], Phivos Mylonas [1], Pavlos Bouchagier [3], Themis Exarchos [1], Aristotelis Martinis [2] and Katerina Kabassi [2]

[1] Department of Informatics, Ionian University, 49100 Corfu, Greece; annasoti@ionio.gr (A.S.); fmylonas@ionio.gr (P.M.); exarchos@ionio.gr (T.E.)

[2] Department of Environment, Ionian University, 29100 Zakynthos, Greece; n.godsil@ionio.gr (N.G.); amartinis@ionio.gr (A.M.); kkabassi@ionio.gr (K.K.)

[3] Department of Food Science and Technology, Ionian University, 28100 Argostolion, Greece; pavlosbouchagier@yahoo.com

* Correspondence: c16vout@ionio.gr

† Presented at the 13th EFITA International Conference, online, 25–26 May 2021.

Abstract: There is a significant number of agricultural systems with rich and special biodiversity, characterized as high-nature-value farming systems (HNV) in the Ionian Islands region. These agro-ecosystems cover a significant area in this region and are divided in olive groves and vineyards which, in some cases, cover a significant part of the protected areas (Natura 2000 and SPA). There are solid olive groves but also a large number of scattered trees or clusters, as well as vineyards, which are largely identified as high-quality wine producers. Finally, there are smaller but extremely important examples of HNV, such as the Englouvi plateau in Lefkada. In this study, we propose a method to survey the spread of *Ailanthus altissima* in olive groves and vineyards (HNV areas) with the scope of evaluating the considered agro-ecosystems, based on the importance of ecosystems and ecosystem services they provide, and preparing a management plan for HNV areas.

Keywords: remote sensing; geographic information systems; decision making; AHP

Citation: Voutos, Y.; Godsil, N.; Sotiropoulou, A.; Mylonas, P.; Bouchagier, P.; Exarchos, T.; Martinis, A.; Kabassi, K. Capturing and Evaluating the Effects of the Expansive Species *Ailanthus altissima* on Agro-Ecosystems on the Ionian Islands. *Eng. Proc.* **2021**, *9*, 19. https://doi.org/10.3390/engproc2021009019

Academic Editors: Dimitrios Kateris and Maria Lampridi

Published: 25 November 2021

Publisher's Note: MDPI stays neutral with regard to jurisdictional claims in published maps and institutional affiliations.

1. Introduction

Approximately half the area of Greece used for agricultural purposes has a high nature value (HNV) [1]. Total agricultural and forest land covers 51% of the total area, of which 18% is forests, 15% is forest land used for grazing, and 18% is cultivated land. HNV are the oldest and most biodiversity-rich farming and forestry systems [2]. They are also characterized as semi-natural habitats, and wild species have been interdependent with low-intensity management by local communities [1].

Alianthus altissima is an opportunistic plant that thrives in full sun and disturbed areas; it is a deciduous, very-fast-growing, invasive tree weed, versatile at different lighting intensities, adapting to almost all soil types (from the most fertile to barren, rocky); it does not thrive in soils that drain freely; it adjusts to a wide range of temperatures (such as high temperatures), and it adapts to saline soils [3]. There are direct dangers from its spread, such as: its effect on the soil fertility, fast decomposition rate of plant residues, reduction of photosynthesis, increase of microbium activity, and change in pH [3].

According to the literature [4], ecosystem services, i.e., the benefits that humans reap from nature, have been the subject of different classifications, according to generally accepted definitions in terms of their content. More specifically, they take into consideration many aspects of living organisms and biomes, such as biodiversity, production/supply services (production of food, water, biomass), climate regulation (rainfall, groundwater, and waste services), cultural/intangible services (aesthetic value of rural landscape), leisure services, spiritual uplift and inspiration, and support services (soil formation, soil retention, photosynthesis, concentration, and utilization of nutrients).

In this study, we propose a method to survey the spread of *Ailanthus altissima* in olive groves and vineyards (HNV areas) with the scope of evaluating the considered agro-ecosystems based on the importance of ecosystems and ecosystem services, and prepare a management plan for HNV areas. We used the geographic information system (GIS), remote sensing (RS), information technology (IT), and communication technology to capture the level of threat and develop effective response plans, through a fuzzy synthetic evaluation system [5].

2. Study Area

The area of interest is the HNV area of the four largest islands of the Ionian Islands (Table 1). The study area is located at the Ionian region and consisted of 372,000 acres of crops. Furthermore, the area is covered by vines and olive trees, 35,000 acres (9.4%) and 337,000 (90.6%), respectively (Figure 1). The Ionian landscape consists mainly of solid olive groves, scattered trees and vines that shape a complex geography, with points of retreat and/or abandonment, tourism and access to the coastline and connection with rural history. These historic crops offer many of the rural landscape characteristics of insular ecosystems, which combine rich habitats and local identity [6]. Despite their highly acclaimed importance, regarding biodiversity and ecosystem services, vineyards and olive groves are tormented by abandonment, deforestation/desertification and slow development with shrinking trends.

Table 1. Total cover of HNV areas.

HNV Areas	Total Cover
Corfu	46.2%
Zakynthos	26.1%
Lefkada	16.1%
Cephalonia	11.6%

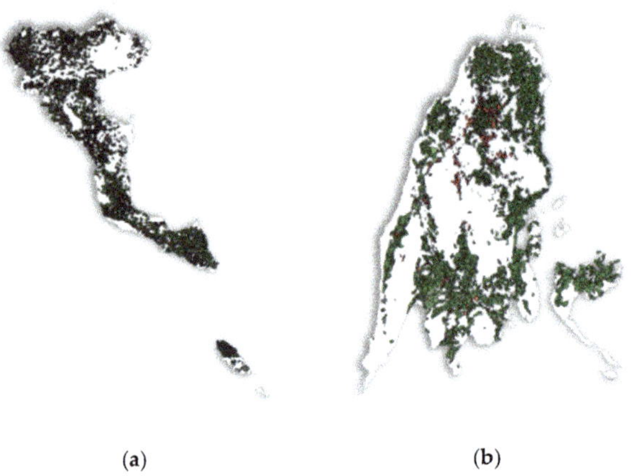
(a)   (b)

Figure 1. Affected HNV sites form Corfu (**a**) and Lefkada (**b**).

3. Methodology

Expansive alien species are a serious threat to these systems, producing the danger of spreading unwanted vegetation on agricultural land, while their risk is also identified and recognized in community legislation on biodiversity protection (Annex II of Article 6 of the CFP Regulation) [7]. A very dangerous area-expanding type of weed, *Ailanthus altissima*, which now threatens the natural environment, including crops (olive and viticulture) in the Ionian region, is the research subject that this work focuses on.

3.1. Mapping

More specifically, we created land cover/land use maps based on orthophotos of 0.5 resolution (SENTINEL-2) [8], which where overlayed by geospatial data (polygon type vectors) from Corine 2018 [9] and ILOT [10]. All of the study area was analyzed and produced mapping products, indicating the affected sites in the Ionian region (Figure 1). Furthermore, we created land cover/land use (LCLU) maps based on the aforementioned vector data, combined with field observations, in order to determine the affected sites in the Ionian Islands region [11].

3.2. Policy Recommendations

The initial findings of the study are crucial to identify the possible interactions between the parameters of alien species' spread, which lead to the construction of specific models and corresponding evaluation scenarios [12]. In the case of the Ionian region, the lack of data has created difficulties in the past during the process of the issue. We produced continuous data and comparable information on the subject, which is expected to enhance the on-site research outcomes and further support public evaluation of the studied agroecosystems.

4. Results

The proposed decision support system was built by applying AHP (Tables S2–S4) that is then supported by SWOT analysis processes (Figure S2). Through our research, we proposed a decision support system based on an applicable evaluation framework for assessing project alternatives by employing multi-criteria assessment. More specifically, we employed the analytical hierarchy process (AHP), through a model that is based on criteria including factors related to *A. altissima*, ecological impacts, and LCLU on the farming system [13].

5. Conclusions

This work is the first major study of invasive species in the Ionian region. It proposes a management plan for areas of high natural value against the invasion of *A. Alissima*. Furthermore, it introduces a novel method for managing HNV farming areas and a tool for the further alignment of National policies to EU policies, concerning sustainability, environmental protection, and green growth. It may work as a tool to highlight the designation of origin for olive and vine products in the region. The results will be publicly available through a web platform (Figure S3) at a future point of our project.

Supplementary Materials: The following are available online at https://www.mdpi.com/article/10.3390/engproc2021009019/s1 https://tinyurl.com/efita241: Figure S1: Coverage in acres for Olives (a) and Vines (b) within the area of interest. Figure S2: Results from the SWOT analysis. Figure S3: Proposed Web map. Table S1: The applied criteria for the AH process. Table S2: The results of the AH process. Table S3: AHP scoring rate.

Author Contributions: Conceptualization, Y.V. and P.B.; methodology, Y.V. and N.G.; software, A.S., P.M. and T.E.; validation, A.M. and K.K., data curation Y.V., A.S. and T.E.; supervision P.B. and P.M. All authors have read and agreed to the published version of the manuscript.

Funding: This research was funded by the European Union and Greece (Partnership Agreement for the Development Framework 2014–2020) under the Regional Operational Programme Ionian Islands 2014–2020 for the project "Assessment of the impact on biodiversity of High Nature Value Areas in the Region of Ionian Islands due to the invasion of the allocthonus plant species *Ailanthus altissima* (HNV-Threat)" (MIS code 5034911).

Institutional Review Board Statement: Not applicable.

Informed Consent Statement: Not applicable.

Data Availability Statement: The data presented in this study are available on request from the corresponding author. The data are not publicly available due to the reason that this is an ongoing study.

Conflicts of Interest: The authors declare no conflict of interest.

References

1. Dimalexis, T.; Markopoulou, D.; Kourakli, P.; Manolopoulos, A.; Vitaliotou, M.; Chouvardas, D. *Identification of High Nature Value Agricultural and Forestry Land*; Hellenic Ornithological Society: Athens, Greece, 2008; p. 5.
2. Keenleyside, C.; Beaufoy, G.; Tucker, G.; Jones, G. *High Nature Value Farming throughout EU-27 and Its Financial Support under the CAP*; Report Prepared for DG Environment, Contract ENV B.1/ETU/2012/0035; Institute for European Environmental Policy: London, UK, 2014.
3. Miller, J.H. Ailanthus altissima (Mill.) Swingle Ailanthus. In *Silvics of North America*; US Department of Agriculture: Washington, DC, USA, 1990; pp. 101–104.
4. Fisher, B.; Turner, R.K. Ecosystem services: Classification for valuation. *Biol. Conserv.* **2008**, *141*, 1167–1169. [CrossRef]
5. Koschke, L.; Fürst, C.; Frank, S.; Makeschin, F. A multi-criteria approach for an integrated land-cover-based assessment of ecosystem services provision to support landscape planning. *Ecol. Indic.* **2012**, *21*, 54–66. [CrossRef]
6. Kefalas, G.; Kalogirou, S.; Poirazidis, K.; Lorilla, R.S. Landscape transition in Mediterranean islands: The case of Ionian islands, Greece 1985–2015. *Landsc. Urban Plan.* **2019**, *191*, 103641. [CrossRef]
7. Commission of the European Communities. Communication from the commission to the council, the European parliament, the European economic and social committee, and the committee of the regions: A mid-term assessment of implementing the EC biodiversity action plan. *J. Int. Wildl. Law Policy* **2009**, *12*, 108–120. [CrossRef]
8. Sentinel Online-ESA-Sentinel Online. Available online: https://earth.esa.int/web/sentinel/home (accessed on 13 April 2021).
9. CLC 2018—Copernicus Land Monitoring Service. Available online: https://land.copernicus.eu/pan-european/corine-land-cover/clc2018 (accessed on 30 July 2021).
10. OPEKEPE.-Home Page. Available online: https://www.opekepe.gr/en/ (accessed on 30 July 2021).
11. Lorilla, R.S.; Poirazidis, K.; Kalogirou, S.; Detsis, V.; Martinis, A. Assessment of the spatial dynamics and interactions among multiple ecosystem services to promote effective policy making across mediterranean island landscapes. *Sustainability* **2018**, *10*, 3285. [CrossRef]
12. Fernández Martínez, P.; de Castro-Pardo, M.; Barroso, V.M.; Azevedo, J.C. Assessing Sustainable Rural Development Based on Ecosystem Services Vulnerability. *Land* **2020**, *9*, 222. [CrossRef]
13. Benke, K.K.; Steel, J.L.; Weiss, J.E. Risk assessment models for invasive species: Uncertainty in rankings from multi-criteria analysis. *Biol. Invasions* **2011**, *13*, 239–253. [CrossRef]

Proceeding Paper

An Easily Installed Method of the Estimation of Soybean Yield Based on Meteorological Environments with Regression Analysis [†]

Akihiro Nitta [1,*], Yuya Chonan [2], Satoshi Hayashi [2], Takuji Nakamura [2], Hiroyuki Tsuji [2], Noriyuki Murakami [2], Ryo Nishide [3], Takenao Ohkawa [1] and Seiichi Ozawa [4]

[1] Graduate School of System Informatics, Kobe University, Kobe 657-0013, Japan; ohkawa@kobe-u.ac.jp
[2] Hokkaido Agricultural Research Center, NARO, Sapporo 062-0045, Japan; y.chonan@affrc.go.jp (Y.C.); hayashis@affrc.go.jp (S.H.); takuwan@affrc.go.jp (T.N.); tuzihiro@affrc.go.jp (H.T.); noriyuki@affrc.go.jp (N.M.)
[3] The Center for Data Science Education and Research, Shiga University, Hikone 522-8522, Japan; ryo-nishide@biwako.shiga-u.ac.jp
[4] Graduate School of Engineering, Kobe University, Kobe 657-08501, Japan; ozawasei@kobe-u.ac.jp
[*] Correspondence: Aki_0811@cs25.scitec.kobe-u.ac.jp; Tel.: +81-78-803-6333
[†] Presented at the 13th EFITA International Conference, online, 25–26 May 2021.

Abstract: A simple method for estimating soybean yield under ideal environments in Japan is proposed. Several models that simulate soybean yield have been proposed in other countries; however, direct adaptation to Japanese species is difficult in terms of climatic and regional characteristics. In addition, they often require variety-specific information or various environmental information, which is sometimes hard to simulate. Therefore, we attempted to create a simple simulation model with meteorological data as the main input to the model. The proposed model ignores the features that need setting for each cultivated field and is composed of a statistical model instead of a physiological analysis for the sake of brevity. Although the prediction accuracy of the model needs to be improved, we can use it as a decision support system for soybean cultivation because it requires only location information and can be easily introduced by many farmers.

Keywords: regression analysis; data mining; smart farming

Citation: Nitta, A.; Chonan, Y.; Hayashi, S.; Nakamura, T.; Tsuji, H.; Murakami, N.; Nishide, R.; Ohkawa, T.; Ozawa, S. An Easily Installed Method of the Estimation of Soybean Yield Based on Meteorological Environments with Regression Analysis. *Eng. Proc.* **2021**, *9*, 26. https://doi.org/10.3390/engproc2021009026

Academic Editors: Dimitrios Aidonis and Aristotelis Christos Tagarakis

Published: 30 November 2021

Publisher's Note: MDPI stays neutral with regard to jurisdictional claims in published maps and institutional affiliations.

1. Introduction

In recent years, due to the decrease in number and aging of farmers, there is a need to improve and renovate agricultural work in Japan. Accurate predictions of the growth of crops lead to appropriate cultivation management and improve the production yield. Crop growth tends to be affected by the surrounding environment. In particular, the yield is the result cumulatively affected by the growing environment over the whole cultivation period [1]. Collecting and recording various detailed data such as plant heights or leaves are indexes which may pave the way to generate a highly accurate model. However, the application of such models is limited only to fields with special equipment or requiring additional work to collect these detailed data. Therefore, predicting growth using information available without special measurement equipment or additional work is important. Meteorological conditions can be obtained from climate satellites and are the dominant environmental factors in cultivation [2].

Soybean is a principal crop in Japan and more sensitive to climatic conditions than rice. To obtain high yields, cultivating the appropriate species in a suitable season is important. Applying growth models is an effective way to improve plans. In other countries, work on developing models for soybean has been conducted, but these models are difficult to apply in Japan due to differences in the varieties and the climate. In this respect, our approach enables regional flexibility by readjusting models. In this study, we aim to quantitatively evaluate the cultivation techniques of soybean farmers in Japan as a part of cultivation management support. We estimate the climatic productivity [1] through a growth model using meteorological data as inputs. By comparing the climatic productivity with the actual yield, we can evaluate the conditions of our field. In this study, the growth model consists of two sub-models. One is the prediction of the developmental progress, and the other is the prediction of seedling formation. By integrating the results of the analysis in each sub-model, climatic productivity is finally determined. The proposed growth model in this study requires only the longitude and latitude for obtaining meteorological data, which makes it easy to install.

2. Materials and Methods

Crop growth is often described as a system that integrates several sub-processes such as developmental progress or dry matter production. In this study, the growth model has two sub-models: the vegetative process, which describes the growth of the plant body; and the maturation process, which describes the formation of the seedlings.

2.1. Vegetative Process

The growth of soybean is the cumulative outcome of quantitative changes in the plant body due to the growing environment. The rate of this quantitative change is the developmental Rate (DVR), and its integration is the developmental index (DVI). In detail, DVI has a relationship with DVR, as detailed in Equation (1), which is called the DVR model [3]. To express the continuous vegetative process of soybeans, Table S1 shows the correspondence of DVI to the growth stage of soybeans defined by Fehr et al. [4].

$$\text{DVR}_i = \sum_{j=1}^{i} \text{DVR}_j \tag{1}$$

Temperature and day length have a significant effect on the developmental phenomena of crops [1]. Temperature and day length can be used to calculate DVR, as shown in Equation (2). Equation (2) is based on the model for rice [5], which divides the periods before and after the flowering stage due to differences in the reactivity of soybean.

$$\text{DVR}_i = \begin{cases} \dfrac{1}{G_v} \cdot \dfrac{1 - \exp\left(B_L\left(L_j - L_c\right)\right)}{1 - \exp\left(-A_T\left(T_j - T_k\right)\right)} & \text{for } 0 \leq \text{DVI}_i < 1 \\[3mm] \dfrac{\alpha}{G_v} \cdot \dfrac{1 - \exp\left(B_L\left(L_j - L_c\right)\right)}{1 - \exp\left(-A_T\left(T_j - T_k\right)\right)} + \varepsilon & \text{for } 1 \leq \text{DVI}_i \leq 2 \end{cases} \tag{2}$$

A_T and B_L are coefficients representing the sensitivity to temperature and day length, respectively. T_i and L_i are the respective temperature and day length on day i. T_k is the temperature at which the DVR halves under a certain day length, L_c is the critical day length for growth to proceed, and G_v is the minimum number of days required for the VE-R2 stage. δ is a fixed DVR value that is added over days, and α is the ratio of the effect of weather to the effect of δ. Some parameters need to be optimized. In this study, we used Bayesian optimization [6] with each parameter as an input variable and the total number of growing days as an objective variable.

2.2. Reproductive Process

We describe the estimation of climatic productivity as a prediction of the maturation process. The climatic productivity is calculated in three steps: division of the growing period, characterization of the meteorological data, and construction of a regression model. First, DVI divides the whole cultivation period into three periods: $0 \leq \text{DVI} < 1.0$, $1.0 \leq \text{DVI} < 1.2$ and $1.2 \leq \text{DVI} \leq 2.0$. Soybeans respond differently to the environment at specific growth stages, such as around the flowering stage, and we chose this partitioning method based on previous studies [7]. Then, the following six meteorological elements are obtained as features for each divided period: average temperature, average solar radiation, average precipitation, percentage of days with precipitation, average wind speed, and maximum wind speed. We examined a wide variety of meteorological elements as features. The elements mentioned above were selected after determining multicollinearity by calculating variance inflation factors (VIFs) [8]. There are six types of meteorological elements acquired for three sections; therefore, the total number of input features is eighteen. Finally, these features are used to construct a regression model to obtain the climatic productivity. To suppress overlearning and to reduce dimensionality, LASSO regression is applied as a method to construct regression model. Moreover, we also examined the prediction accuracy in comparison with the conventional method without dividing the growing period. Climatic productivity was estimated in this study; therefore, the fields were assumed to be under ideal conditions, without any growth-inhibiting factors such as insect damage or disease, and with sufficiently fertile soil.

3. Results

The dataset used for the analysis was the cultivation records of 293 fields collected from different locations in Japan. The fields in this dataset show high yield potential, making them ideal for the estimation of climatic productivity. Therefore, these fields were kept free from factors inhibiting growth such as insect damage or disease, and the soil conditions sufficiently were fertile. The weather information was obtained from NARO's Agro-Meteorological Grid Square Data System [9,10], which was developed and operated by the National Agriculture and Food Research Organization (NARO).

In terms of the sub-model for predicting the vegetative process, the RMSE between the estimated and measured growing days was 3.91 days for the interval $0 \leq \text{DVI} < 1$ and 5.45 days for the interval $1 \leq \text{DVI} \leq 2$. This result indicates that the sub-model can sufficiently estimate the growth process of soybean. Furthermore, the results imply the capability for evaluating the cultivation environment based on the growth stage in the sub-model for predicting the maturation process. Focusing on individual plots, heavy rain caused growth inhibition in some fields, which resulted in large errors. The lack of accounting for precipitation in the sub-model led to a large error. Regarding the sub-model for predicting the maturation process, the RMSE between the estimated and measured values was 67.2 g m^{-2}, and the RMSE without dividing the growing period was 76.9 g m^{-2}. There was a significant difference in the results between the divided and undivided growing periods. Thus, dividing the growing period by growth stages is beneficial to evaluate the effects of climatic factors.

Figure S1 is a scatter plot of the estimated and measured values for the applied growth model. Figure S1 shows that the higher (or lower) the yield is, the larger the error. There is normality in the yield of this dataset, with values biased around the mean. The tendency of the regressor to especially reduce the error of data near the mean probably accounts for this result. For this, accurate estimation of off-center values is important. On the other hand, the results were inadequate because the expected error was less than 50 g m^{-2}. The limitation of the input information for the convenience of operation caused this issue. Furthermore, we believe that the analysis ignored the differences among species, which is the reason for the small amount of data in this study, also contributing to the issue. In the future, we will improve the input features and investigate the analysis method, while guaranteeing the simplicity of the operation.

4. Conclusions

In this study, we modeled the growth of soybean, taking into account the effect of meteorological factors, and estimated the climatic productivity. This model requires no special measurement equipment or additional work because we anticipated a practical operating situation. The information necessary for the introduction of the model is meteorological data, which can be obtained if the longitude, latitude, and the dates of the growing period are available. Hence, the proposed model has high installation applicability. The model has two sub-models, one for the vegetative process and the other for the maturation process.

This model accurately predicts the growth of the plant body and estimates the formation of the seedlings. In predicting the vegetative process, the model succeeded in achieving sufficiently plausible results. Although the estimation of the maturation process needs continuous improvement based on the simplicity of operation, the climatic productivity assesses the productivity by comparison with the actual yield.

Supplementary Materials: The following are available online at https://github.com/Aki0811/EFITA2021_ID130, Table S1: Stage of Soybean Development and Corresponding DVI; Figure S1: Scatter plot of estimated and measured yields.

Author Contributions: Conceptualization, T.O.; methodology, A.N.; software, A.N.; validation, A.N., Y.C., S.H., T.N., H.T. and N.M.; formal analysis, A.N.; investigation, A.N., Y.C., S.H., T.N., H.T. and N.M.; resources, Y.C., S.H., T.N., H.T., N.M. and S.O.; data curation, A.N.; writing—original draft preparation, A.N.; writing—review and editing, Y.C., S.H., T.N., H.T., N.M., R.N., T.O. and S.O.; visualization, A.N.; supervision, T.O.; project administration, T.O.; funding acquisition, T.N. and T.O. All authors have read and agreed to the published version of the manuscript.

Funding: This research was partly funded by Ministry of Agriculture, Forestry and Fisheries: Development of diagnostic methods and countermeasure techniques for overcoming high-yield inhibitory factors.

Data Availability Statement: The data presented in this study are available on request from the corresponding author. The data are not publicly available due to privacy.

Conflicts of Interest: The authors declare no conflict of interest.

References

1. Horie, T. A Model for Evaluating Climatic Productivity and Water Balance of Irrigated Rice and Its Application. *Southeast Asian Stud.* **1987**, *25*, 62–74.
2. Garner, W.W.; Allard, H.A. Effect of the relative length of day and night and other factors of the environment on growth and reproduction in plant. *J. Agric. Res.* **1920**, *18*, 553–606. [CrossRef]
3. Wit, C.T.; Brouwer, R.; Penning, F.W.T. The simulation of photosynthetic systems. In Proceedings of the International Biological Program/Production Process, Technical Meeting, Trebon, Czech Republic, 14–21 September 1969; pp. 44–70.
4. Fehr, W.R.; Caviness, C.R. "Stage of soybean development", Agriculture and Home Economic Experiment Station. *Spec. Rep.* **1977**, *80*, 1–12.
5. Horie, T.; Nakagawa, H. Modelling and Prediction of Developmental Process in Rice: 1. Structure and method of parameter estimation of a model for simulating developmental process toward heading. *Jpn. J. Crop. Sci.* **1990**, *59*, 687–695. [CrossRef]
6. Shahriari, B.; Swersky, K.; Wang, Z.; Adams, R.P.; Freitas, N.D. Taking the Human out of the Loop: A Review of Bayesian Optimization. *Proc. IEEE* **2016**, *104*, 148–175. [CrossRef]
7. Nakano, S.; Kumagai, E.; Shimada, S.; Sameshima, R.; Ohno, H.; Homma, K.; Shiraiwa, T. Modeling of Phenological Development Stage and Impact of Elevated Air Temperature on the Phenological Development of Soybean Cultivars in Japan. *Jpn. J. Crop Sci.* **2015**, *84*, 408–417. [CrossRef]
8. Marquardt, D.W. Generalized Inverses, Ridge Regression, Biased Linear Estimation, and Nonlinear Estimation. *Technometrics* **1970**, *12*, 605–607. [CrossRef]
9. Ohno, H.; Sasaki, K.; Ohara, G.; Nakazono, K. Development of grid square air temperature and precipitation data compiled from observed, forecasted, and climatic normal data. *Clim. Biosph.* **2016**, *16*, 71–79. [CrossRef]
10. Kominami, Y.; Sasaki, K.; Ohno, H. *User's Manual for the Agro-Meteorological Grid Square Data*, 4th ed.; NARO: Sapporo, Japan, 2019; p. 67.

engineering proceedings

Proceeding Paper

Sensitivity of Greek Organisations in Sustainability Issues [†]

Alexandros Chrysos-Anestis [1], Charisios Achillas [1,*], Dimitrios Folinas [1], Dimitrios Aidonis [1] and Michael Chrissos Anestis [2]

[1] Department of Supply Chain Management, International Hellenic University, 60100 Katerini, Greece; alexanderchrissos@gmail.com (A.C.-A.); folinasd@ihu.gr (D.F.); daidonis@ihu.gr (D.A.)

[2] Department of Business Administration, Marketing and Tourism, International Hellenic University, 57400 Thessaloniki, Greece; michail.chrysos@gmail.com

* Correspondence: c.achillas@ihu.edu.gr; Tel.: +30-2351-047-864

† Presented at the 13th EFITA International Conference, online, 25–26 May 2021.

Abstract: Recently, the world has been faced with a variety of environmental, social, and economic problems. The effects of climate change and the lack of resources are constantly intensifying, while at the same time the impact of industrial production has become an international issue. Undoubtedly, this global paradigm and these relevant social and economic challenges require joint efforts at an international level. During the past few decades, a number of companies in Greece have undertaken initiatives towards sustainable development (SD) by adopting "green" practices. This work presents the findings of a survey that has been conducted in 2020, which investigates the extent of business contribution to the United Nations' 17 Sustainable Development Goals (SDGs). Research was based on the analysis of sustainability reporting published by Greek companies. The key findings of the survey show that issues related to environmental protection, societal well-being, and citizens' quality of life have attracted an increasing level of awareness in the Greek industry sectors. Issues such as climate change, as well as sustainable production and consumption, are becoming topics within companies' day-to-day agenda.

Keywords: sustainability reporting; climate change; sustainable production and consumption; sustainable development goals

Citation: Chrysos-Anestis, A.; Achillas, C.; Folinas, D.; Aidonis, D.; Anestis, M.C. Sensitivity of Greek Organisations in Sustainability Issues. *Eng. Proc.* **2021**, *9*, 4. https://doi.org/10.3390/engproc2021009004

Academic Editors: Dimitrios Kateris and Maria Lampridi

Published: 19 November 2021

Publisher's Note: MDPI stays neutral with regard to jurisdictional claims in published maps and institutional affiliations.

1. Introduction

In the last few decades, more and more organisations have engaged in sustainability reporting (SR) activities due to the uncertain situation of various crises in Greek society. Such crises were the economic crisis (2008), environmental crisis (e.g., Chernobyl, 1986; Desert Storm, 1991), and now the social crisis (e.g., COVID-19 pandemic). For example, during the economic crisis, Greek companies made an effort to be more active, especially regarding environmental issues, human resources, health and safety, and contributions to local communities. However, public concerns about controlling, monitoring, and measuring the impact of their activities on the environment and community life in business areas lead many businesses to publish sustainability reporting (SR). For that purpose, the United Nations (UN) enriched 17 Sustainable Developments Goals (SDG) with 169 targets related to economic, social, and environmental outcomes, observing the effects caused by the day-to-day operations of firms. By the 2030 Agenda, all 183 UN Member States have committed to accomplishing the SDG [1]. However, there is a gap in the literature review as previous studies have not completely enveloped the Greek reality. Thus, the aim of this study is to identify SR practices in an emerging field in addition to business contributions to the UN' 17 SDGs. To achieve that, data mining techniques are applied to utilize the statistics available in the Greek community and provide contributions to various audiences including marketers, for-profit and non-profit stakeholders, and the academic community generally.

2. Conceptual Framework

Stakeholders are increasingly expecting more data about an organisation's environmental and social activities, in addition to its economic performance. Thus, firms adopt SR practices (which are multidimensional in orientation) to enhance the communication with stakeholders [2]. Corporate reports usually focus on disclosing environmental issues [3], on the social dimension (corporate social responsibility [CSR] reports) [4], or on the triple-bottom-line approach [5]. Particularly in Greece, firms engage in CSR activities as they appear to be developing [6] and offer a number of benefits [7]. Although Greek companies are truly concerned about the environment, resources, quality of life, and social matters, they are unaware of CSR principles [8].

Interestingly, the evolution of corporate SR seems to have several similarities with the concept of sustainable development (SD) [9]. Specifically, SD defined primarily by Brundtland Report [10], the most widely accepted suggestion, as *"the development that meets the needs of the present generation without compromising the ability of future generations to meet their own needs"*. In the wake of the "Brundtland report", however, numerous international organisations proposed similar sustainable claims. These practices can make progress in the concept of SR by introducing the UN' SDGs [11]. The UN' SDGs could assist multinational companies in improving their CSR performance and contribution to SD [12]. Therefore, CSR reports can support such solutions in business activities that are socially responsible, environmentally friendly and, at the same time, economically valuable [13] representing the business's overall strategy and objectives, covering issues and topics, and offering information about the company's results.

3. Materials and Methods

Due to the volume of information about corporate SR, an empirical analysis was applied. More precisely, sustainability-in-data-mining techniques are used to discover new information hidden within large databases [14] and distinguish between huge quantities of irrelevant data with interesting connections. The research subject in this study were large-sized organisations (European Commission: No 70/2001, EN 2001 L 10/33) that published sustainability reports in Greece. The case of Greece is a unique case study due to the difficulties that the country faces (see introduction). In total, fifty (50) CSR reports were collected and analyzed with regard to the 17 SDGs set by the UN and the annual reports, providing readers with the opportunity to study the latest data of the organisation. In particular, CSR reports were grouped into the companies' industry: (a) production organisations (13/50), (b) service organisations (10/50), (c) food production (8/50), (d) product merchandising (6/50), (e) financial service (5/50), and (f) energy (8/50).

4. Results

As shown in Table 1, all 50 organisations contribute to the above SDGs. Especially indicators such as "SDG No. 8—decent work and economic growth", "SDG No. 12—climate action", and "SDG No. 13—sustainable consumption and production" gathered the highest demand. To expand the contribution level of the companies to the SDGs, some subcategories were divided. Taking the SDG No. 3 first, it was divided into the following primary categories: improving health benefits (41/50) and health and safety training activities (15/50). Turning to the SDG No. 4, it concerned staff training and information (46/50) and youth support (33/50). Regarding the SDG No.8, it was divided into supporting economic growth (47/50), establishing decent work (50/50), respect for human rights (46/50), and reduction of occupational accidents (36/50). In the SDG No. 12, 48/50 organisations reported in their most recent CSR that they reduce waste generation each year through the recycling and reuse of certain materials. While the 49 companies stated that they operate responsibly in terms of energy consumption. Finally, all organisations seem to contribute to the SDG No. 13 categorized into environmentally friendly raw materials (50/50) and renewable energy sources (40/50).

Table 1. Corporate sustainability reports in relation to the UN' SDGs.

Sustainable Development Goals (SDG)'s			Organisations	
No.	**Goals**	**Outline Description**	**n = 50**	**%**
1.	No Poverty	End poverty in all its forms everywhere	20	40
2.	Zero Hunger	End hunger, achieve food security and improved nutrition and promote sustainable agriculture	20	40
3.	Good Health and Well-being	Ensure healthy lives and promote well-being for all at ages	45	90
4.	Quality Education	Ensure inclusive and equitable quality education and promote lifelong learning opportunities for all	46	92
5.	Gender Equality	Achieve gender equality and empower all women and girls	37	74
6.	Clean Water and Sanitation	Ensure availability and sustainable management of water and sanitation for all	37	74
7.	Affordable and Clean Energy	Ensure access to affordable, reliable, sustainable and modern energy for all	26	52
8.	Decent Work and Economic Growth	Promote sustained, inclusive and sustainable economic growth, full and productive employment and decent work for all	50	100
9.	Industry, Innovation, and Infrastructure	Build resilient infrastructure, promote inclusive and sustainable industrialization and foster innovation	35	70
10.	Reduced Inequalities	Reduce income inequality within and among countries	35	70
11.	Sustainable Cities and Communities	Make cities and human settlements inclusive, safe, resilient and sustainable	35	70
12.	Sustainable Consumption and Production	Ensure sustainable consumption and production patterns	50	100
13.	Climate Action	Take urgent action to combat climate change and its impacts by regulating emissions and promoting developments in renewable energy	50	100
14	Life Below Water	Conserve and sustainably use the oceans, seas, marine resources for sustainable development	7	14
15.	Life on Land	Protect, restore and promote sustainable use of terrestrial ecosystems, sustainably manage forests, combat diserfivitation and halt and reversing land degradation and halt biodiversity loss	23	46
16.	Peace, Justice, and Strong Institutions	Promote peaceful and inclusive societies for sustainable development, provide access to justice for all and build effective, accountable and inclusive institutions at all levels	35	70
17.	Partnerships for the Goals	Strengthen the means of implementation and revitalize the global partnership for sustainable development	21	42

5. Discussion and Conclusions

This paper contributes to the literature concerning CSR practices and UN SDGs. Although there are many researchers that focus on corporate reports, examining the content and quality of the information disclosed in sustainability reports, there is a limited number of studies that have carried out an analysis of the current status of corporate reporting practices in relation to the sustainability directions emerged from the Greek reality. In the light of this limitation, this paper examines CSR's activities in relation to the 17 UN' SDGs using an empirical analysis carried out in a sample of 50 Greek organisations that operate in six different industry sectors and publish sustainability reports.

The results of the empirical analysis indicate the CSR practices in relation to UN SDGs from various sectors, which is in line with Tsalis et al. [15] and Avrampou et al. [16], who highlight the need for further research on the content of sustainability reports from other industries. In the Greek context especially, outcomes of the empirical analysis are relevant since the investigated CSR practices were published by a leading group of Greek companies in corporate sustainability. Thus, this paper provides some evidence for the initial reactions of the Greek business community to the introduction of the 2030 Agenda for Sustainable Development.

To be more specific, our findings suggest that all companies declared their contributions to at least one SDG. Issues related to environmental protection, societal well-being, and citizens' quality of life attracted an increasing level of awareness in the Greek industry sectors. Obviously, the current paper could also be a fruitful environment for designing future research in sustainability science. Scrutinizing a larger sample of companies operating in different countries with different institutional regimes could be extremely conducive to uncover that research in more detail. Except for that, further studies should identify factors, such as listing status of organisations, size, industry type, and CSR certification.

Author Contributions: Conceptualization, C.A. and A.C.-A.; methodology, C.A. and A.C.-A.; investigation, A.C.-A. and M.C.A.; writing—original draft preparation, A.C.-A. and M.C.A.; writing—review and editing, C.A., D.F., and D.A.; supervision, C.A.; project administration, C.A.; funding acquisition. All authors have read and agreed to the published version of the manuscript.

Funding: This research was co-funded by the European Regional Development Fund of the European Union and Greek national funds through the Operational Program Competitiveness, Entrepreneurship and Innovation 2014–2020, under the call "European R&T Cooperation—Action of Granting Greek Bodies that Successfully Participated in Joint Calls for Proposals of the European ERA-NET Networks 2019b" (MELS project, project code: T11EPA4-00076).

Institutional Review Board Statement: Not applicable.

Informed Consent Statement: Not applicable.

Data Availability Statement: The study did not report any additional data.

Acknowledgments: The authors would like to acknowledge the contribution of the partners of the MELS project (www.mels-project.eu, accessed on 16 November 2021) in collecting and proposing country-specific GHG-based DS tools to be evaluated.

Conflicts of Interest: The authors declare no conflict of interest. The funders had no role in the design of the study; in the collection, analyses, or interpretation of data; in the writing of the manuscript; or in the decision to publish the results.

References

1. United Nations. The 17 Goals. 2015. Available online: https://sdgs.un.org/goals (accessed on 4 July 2021).
2. Sotorrío, L.L.; Sánchez, J.L.F. Corporate social reporting for different audiences: The case of multinational corporations in Spain. *Corp. Soc. Responsib. Environ. Manag.* **2010**, *17*, 272–283. [CrossRef]
3. Aerts, W.; Cormier, D. Media legitimacy and corporate environmental communication. *Account. Organ. Soc.* **2009**, *34*, 1–27. [CrossRef]
4. Dhaliwal, D.S.; Radhakrishnan, S.; Tsang, A.; Yang, Y.G. Nonfinancial Disclosure and Analyst Forecast Accuracy: International Evidence on Corporate Social Responsibility Disclosure. *Account. Rev.* **2012**, *87*, 723–759. [CrossRef]
5. Gray, M.R.; Milne, M. Towards reporting on the triple bottom line: Mirages, methods and myths. In *The Triple Bottom Line*; Routledge: London, UK, 2013; pp. 92–102.
6. Skouloudis, A.; Evangelinos, K.; Nikolaou, I.; Filho, W.L. An overview of corporate social responsibility in Greece: Perceptions, developments and barriers to overcome. *Bus. Ethic A Eur. Rev.* **2011**, *20*, 205–226. [CrossRef]

7. Leonidas, P.; Mary, G.; Theofilos, P.; Amalia, T. Managers' Perceptions and Opinions towards Corporate Social Responsibility (CSR) in Greece. *Procedia Econ. Finance* **2012**, *1*, 311–320. [CrossRef]
8. Panayiotou, N.A.; Aravossis, K.G.; Moschou, P. Greece: A Comparative Study of CSR Reports. In *Global Practices of Corporate Social Responsibility*; Springer: Berlin/Heidelberg, Germany, 2008; pp. 149–164.
9. Hąbek, P.; Wolniak, R. Assessing the quality of corporate social responsibility reports: The case of reporting practices in selected European Union member states. *Qual. Quant.* **2016**, *50*, 399–420. [CrossRef] [PubMed]
10. WCED, S.W.S. World commission on environment and development. *Our Common Future* **1987**, *17*, 1–91.
11. Jones, P.; Hillier, D.; Comfort, D. The Sustainable Development Goals and business. *Int. J. Sales Retail. Mark.* **2016**, *5*, 38–48.
12. Redman, A. Harnessing the Sustainable Development Goals for businesses: A progressive framework for action. *Bus. Strat. Dev.* **2018**, *1*, 230–243. [CrossRef]
13. Gaweł, E.; Jałoszyńska, A.; Orłowski, M.; Ratajczak, E.; Ratajczak, J.; Riera, B. Corporate social responsibility as an instrument of sustainable development of production enterprises. *Manag. Syst. Prod. Eng.* **2015**, *19*, 152–155.
14. Natek, S.; Zwilling, M. Student data mining solution–knowledge management system related to higher education institutions. *Expert Syst. Appl.* **2014**, *41*, 6400–6407. [CrossRef]
15. Tsalis, T.A.; Malamateniou, K.E.; Koulouriotis, D.; Nikolaou, I.E. New challenges for corporate sustainability reporting: United Nations' 2030 Agenda for sustainable development and the sustainable development goals. *Corp. Soc. Responsib. Environ. Manag.* **2020**, *27*, 1617–1629. [CrossRef]
16. Avrampou, A.; Skouloudis, A.; Iliopoulos, G.; Khan, N. Advancing the Sustainable Development Goals: Evidence from leading European banks. *Sustain. Dev.* **2019**, *27*, 743–757. [CrossRef]

Proceeding Paper

Conceptual Framework to Integrate Economic Drivers of Decision Making for Technology Adoption in Agriculture [†]

Thiago L. Romanelli [1,*], Francisco Muñoz-Arriola [2,3] and Andre F. Colaço [4]

[1] College of Agriculture, University of São Paulo, Av. Pádua Dias, 11, Piracicaba/SP, São Paulo CEP 13418-900, Brazil
[2] Department of Biological Systems Engineering, University of Nebraska-Lincoln, 246 Chase Hall, Lincoln, NE 68583, USA; fmunoz@unl.edu
[3] School of Natural Resources, University of Nebraska-Lincoln, 620 Hardin Hall, Lincoln, NE 68583, USA
[4] CSIRO, Waite Campus, Locked Bag 2, Glen Osmond, SA 5064, Australia; andre.colaco@csiro.au
* Correspondence: romanelli@usp.br; Tel.: +55-19-3447-8523
† Presented at the 13th EFITA International Conference, online, 25–26 May 2021.

Abstract: This study evaluates how much technology adoption could cost in a variety of crop-production scenarios. Cost-reduction simulations consider scenarios of higher input use efficiency such as reducing the usage of diesel, labor, irrigation, fertilizer, herbicide, and seed, among others. The scenarios aim to increase yields by integrating the effect of each input-reduction on the total operating costs. Agricultural production estimates for Nebraska in the US indicates that a technology that saves 1% of diesel is cost-effective, costing between USD 0.15/ha and USD 0.32/ha (for corn). Improvements on input use efficiency should be prioritized to incentivize technology development and adoption. This study balances input costs and crop production, allowing the identification of adoption cost thresholds tailored to specific farming scenarios. It also enabled interpretations regarding optimal scenarios for technology adoption. In addition, this study indicates that irrigated systems foster the adoption of technologies more than in dryland cropping systems.

Keywords: profitability; agricultural management; sustainability; cost; efficiency

Citation: Romanelli, T.L.; Muñoz-Arriola, F.; Colaço, A.F. Conceptual Framework to Integrate Economic Drivers of Decision Making for Technology Adoption in Agriculture. *Eng. Proc.* **2022**, *9*, 43. https://doi.org/10.3390/engproc2021009043

Academic Editors: Dimitrios Aidonis and Aristotelis Christos Tagarakis

Published: 23 January 2022

Publisher's Note: MDPI stays neutral with regard to jurisdictional claims in published maps and institutional affiliations.

1. Introduction

Food production is expected to intensify in the next 50 years to sustain the increasing demand of food supplies. Agricultural practices will determine the level of food production and, to a great extent, the state of the global environment [1]. Historically, agriculture has been associated with technological improvements such as those during the Green Revolution, and more recently in information technologies and robotics [2]. The integration of such improvements in agricultural practices have led to the maximation of production and economies and, sometimes, increasing environmental degradation [3]. Sustainable agriculture can be seen as a broad term that merges elements of sustainable development in the form of resource conservation, technologic suitability, social compliance, and economic viability with agroecosystem resilience, human livelihoods, and agricultural productivity [4,5]. A sustainable intensification of agriculture might be possible if the use of external inputs, the improvements of management techniques and practices, and the efficient use of natural resources and purchased inputs are balanced [3,6]. However, the operationalization of novel technologies, designed to minimize resource loss or maximize production, are also limited by triggers of innovation, growth, and prosperity [2,7].

In agriculture, technology enables labor efficiency, increases in revenues, and food security; yet, technological innovations are not adopted immediately or completely throughout a population [8–10]. The diffusion of new technologies through users and markets depends on understanding of the economics, innovation, and cross-sectional patterns of technology adoption [11]. In food production, precision agriculture (PA) exemplifies the

Eng. Proc. **2022**, *9*, 43. https://doi.org/10.3390/engproc2021009043   https://www.mdpi.com/journal/engproc

introduction of technology, widely benefiting from constant innovations such as optimizing soil sampling, mapping yield variability, variable rate application of inputs, non-invasive sensing of plant status, soil conductivity, and auto guidance systems [2,9,12,13].

This study estimates the cost thresholds for technological adoption based on distinct scenarios of crop budgets, assuming improvements in production costs effect (less input consumption maintaining yield or more yield maintaining input levels). A cost threshold is the investment that would tie the profit improvement, leading to no profit change. These production scenarios will support a framework aimed to help producers or technology suppliers to determine a cost threshold based on their own production costs driven. The scenarios evaluated in this study are representative of Nebraska's crops which account for USD 11.7 billion (5th place in USA) being ranked as the main state producing corn (3.7 million ha), soybean (2.0 million ha), and wheat (1.0 million ha) [14].

2. Material and Methods

Technological adoption for Nebraska's farming is assessed through the development of production scenarios based on the input demands and yields observed. The proposed scenarios for corn production are built by a panel of experts and reported in the Nebraska Crop Budget [15], Table S1. We considered two basic effects of technology adoption to determine the economic threshold: (1) cost reduction (less inputs applied or lower prices) and (2) income increase (through higher yield or better price due to higher quality). The thresholds for technologies involve an increase in yield, data of the expected yields and market prices to determine the cost limits (Equation (1)). The variation of income (yield $\times$ market price) represents how much could be paid for a given technological option and is calculated as follows:

$$\mathrm{ET} = \mathrm{Y} \times \mathrm{d\%} \times \mathrm{P}, \tag{1}$$

where ET (economic threshold) is the cost limit for adopting an increasing yield technology (USD ha^{-1}); Y, is yield (t ha^{-1}); d %, is the expected change on yield (%); and P, is the market price of the product (USD t^{-1}). The effect on the production cost was estimated through sensitivity analyses of input variables such as diesel, labor, irrigation (when applicable), fertilizer, herbicide, interest rate, repair cost, land ownership, and seed, changing individually +10% of their original cost value. For example, the expected increase in input use efficiency (i.e., fertilizer, herbicide, and seed) could be achieved by the site-specific application, high assimilation of nutrients by plant roots, and improved quality of seeds.

Improvements on the usage/cost of diesel could be achieved by variation in the price and/or machine efficiency. More efficiency in labor could be achieved by lowering wages, improving efficiency, or even the cost one could pay for unmanned vehicles (when a 100% improvement and no labor costs of field activities). Production costs are composed by the input requirement (i.e., fertilizer, seed, pesticide, etc.) and their respective prices as in Equation (2):

$$\mathrm{OCIi} = \mathrm{ADi} \times \mathrm{IPi} \times \mathrm{ASi}, \tag{2}$$

where OCIi is the operational costs of the ith input (USD ha^{-1}), AD is the doses applied of the ith input (kg ha^{-1}, L ha^{-1}, unit ha^{-1}), IPi (Table S2) is the input prices of the ith applied input (USD kg^{-1}, USD L^{-1}, USD unit^{-1}), and ASi is the share of areas in which the ith inputs are applied (%). The total production cost reduction was divided by the percentage change applied, resulting in an economic improvement of 10% (USD ha^{-1}) associated with the cost of the technological adoption. Assumptions about the size and age of the equipment were made according to [15]. The labor wage was USD 20 h^{-1}.

3. Results and Discussion

For profit to be achieved, the cost of technology should be below the values stated in Table 1. For instance, if there is a tractor equipped with a more efficient diesel engine, providing 10% less fuel consumption, a corn producer in a situation of system 17#15 will benefit if he or she pays less than USD 3.16 ha^{-1} for this asset to operate. This value is the threshold of the additional cost of this tractor to represent a benefit. The same solution for situation 17#17 will be worthy if it costs less than USD 1.53 ha^{-1}. Knowing the cost limit across different scenarios may require R&D companies to just know how much area the distinct systems represent, avoiding unviable solutions.

Table 1. Cost threshold for 10% higher efficiency on input use for studied scenarios.

Scenario	Diesel	Fertilizer	Labor	Herbicide	Irrigation	Seed	Production System
				USD ha^{-1}			
17#15	3.16	9.71	4.53	16.52	0.00	6.99	Corn, Conventional
17#16	2.97	14.31	3.97	15.62	0.00	6.40	Corn, Conventional
17#17	1.53	15.99	2.46	11.73	0.00	3.61	Corn, No-till
17#23	1.63	15.91	2.84	18.92	0.00	4.05	Corn, Ecofallow
17#24	18.73	18.95	12.66	7.26	30.07	28.54	Corn, Ridge
17#25	18.88	19.64	12.95	8.23	30.07	28.89	Corn, Ridge
17#26	3.16	28.32	9.11	13.78	4.23	11.19	Corn
17#27	12.60	27.65	4.85	20.37	21.23	15.9	Corn, No-till
17#29	18.78	26.59	7.47	13.55	30.74	23.87	Corn

In terms of cost reduction, seed, fertilizer, irrigation (when applicable), and herbicide should be the focus for improvement in corn scenarios. In these scenarios, seed was the highest or the second highest in priority for 14 scenarios (all, except 17#15, 17#23, 17#24, and 17#29). Fertilizer was of priority for 12 of the scenarios. Irrigation of priority was for three out of nine irrigated scenarios. Herbicide application was one of the main variables for five scenarios (all dryland). The exceptions were for the less intensified scenario (17#15), for which ownership was second, and for scenario 17#24, in which field efficiency was second.

The effect of labor reduction can be used to estimate how much the user could pay for unmanned vehicles in their production systems, which would be considered a 100% improvement (i.e., no labor cost), keeping all other input requirements the same. This consideration is aligned with [16], who concluded that spraying using UAV, agricultural income, and hours worked in agricultural production contribute positively to adopting technologies. For instance, for a corn scenario (17#17), a full unmanned mechanized operation would be worthwhile below USD 24.60 ha^{-1}, while for other scenarios (17#23 and 17#24), it would be viable around USD 125 ha^{-1}. In a 5-year period for corn production (2012–2016), the variations in commodity prices and crop yields show lower average yields (7.7 t ha^{-1} in 2012 and 9.9 t ha^{-1} in 2013) and better prices (USD 263 and USD 242 t^{-1}, respectively) than those high average-yield years (10.7 to 11.0 t ha^{-1}, 2014–16), with USD 137 to USD 162 t^{-1}.

Technology adoptions may be more economically suitable when the environmental conditions are not favorable for high yields. Consequently, the investment in technologies in years with higher yields may lead to unwanted consequences due to the decrease in prices and greater effort to raise revenues. This result suggests that international-market trends should be monitored by producers to manage risks of adopting high-value technologies. For a more universal technology adoption framework that includes the adoption of complex technological arrays such as in PA or the shift of large-scale farming management practices, long term analyses can be more beneficial. A farm manager would be more confident to adopt the technology based on scenarios involving long-term economic cycles.

4. Conclusions

The priorities for technology development and adoption in Nebraska were the efficient use of seeds, fertilizer, irrigation, and herbicide application. Thus, an efficient, productive system led to reductions in the total production costs. The proposed approach estimates how much producers can pay for technological adoption considering the specific characteristics of their production systems. In addition, we identified that the years with high yields might be the least suitable for a profitable technology adoption due to the lower market prices and the consequent lower incomes.

Supplementary Materials: The following supporting information can be downloaded at: https://www.mdpi.com/article/10.3390/engproc2021009043/s1, Table S1: Characteristics of production systems evaluated for 2017 [15], Table S2: Cost of some agricultural inputs [15].

Author Contributions: Conceptualization, T.L.R. and F.M.-A.; methodology, T.L.R., F.M.-A. and A.F.C.; validation, T.L.R., F.M.-A. and A.F.C.; formal analysis, T.L.R., F.M.-A. and A.F.C.; investigation, T.L.R., F.M.-A. and A.F.C.; resources, T.L.R., F.M.-A. and A.F.C.; data curation, T.L.R.; writing—original draft preparation, T.L.R., F.M.-A. and A.F.C.; writing—review and editing, T.L.R., F.M.-A. and A.F.C.; funding acquisition, T.L.R. All authors have read and agreed to the published version of the manuscript.

Funding: This research was funded by the CNPq—National Council for Scientific and Technological Development, grant number 307114/2019.

Institutional Review Board Statement: Not applicable.

Informed Consent Statement: Not applicable.

Data Availability Statement: We just cited the multiple sources of data used and reported in the paper as results emerged from the proposed hypothesis testing.

Acknowledgments: The first author acknowledges Fulbright Brazil for the support on his position as Fulbright Chair in Agricultural Sciences at the University of Nebraska, Lincoln, in 2017. Some research ideas were part of University of Nebraska-Lincoln ARD Irrigation Sustainability project.

Conflicts of Interest: The authors declare no conflict of interest.

References

1. Tilman, D.; Cassman, K.G.; Matson, P.A.; Naylor, R.; Polasky, S. Agricultural sustainability and intensive production practices. *Nature* **2002**, *418*, 671–677. [CrossRef] [PubMed]
2. Shekhar, S.; Colleti, J.; Munoz-Arriola, F.; Ramaswamy, L.; Krinz, C.; Varshney, L.; Richardson, D. Intelligent Infrastructure for Smart Agriculture: An Integrated Food, Energy and Water System. *arXiv* **2017**, arXiv:1705.01993.
3. Bongiovanni, R.; Lowenberg-Deboer, J. Precision Agriculture and Sustainability. *Precis. Agric.* **2004**, *5*, 359–387. [CrossRef]
4. Lee, D.R. Agricultural sustainability and technology adoption: Issues and policies for developing countries. *Am. J. Agric. Econ.* **2005**, *87*, 1325–1334. [CrossRef]
5. Uden, D.R.; Allen, C.R.; Munoz-Arriola, F.; Ou, G.; Shank, N. A Framework for Tracing Social–Ecological Trajectories and Traps in Intensive Agricultural Landscapes. *Sustainability* **2018**, *10*, 1646. [CrossRef]
6. Romanelli, T.L.; Milan, M. Machinery management as an environmental tool-material embodiment in agriculture. *CIGR J.* **2012**, *14*, 1–16.
7. Fiksel, J. Designing Resilient, Sustainable Systems. *Environ. Sci. Technol.* **2003**, *37*, 5330–5339. [CrossRef] [PubMed]
8. Pannell, D.J.; Bennett, A.L. Economic feasibility of precision weed management: Is it worth the investment. In *Precision Weed Management in Crops and Pastures*; Medd, R.W., Pratley, J.E., Eds.; CRC for Weed Management Systems: Adelaide, Australia, 1999; pp. 138–148.
9. Hassall, J. Future Trends in Precision Agriculture—A Look into the Future of Agricultural Equipment. 2010. Available online: https://www.nuffieldscholar.org/sites/default/files/reports/2009_AU_James-Hassall_Future-Trends-In-Precision-Agriculture-A-Look-Into-The-Future-Of-Agricultural-Equipment.pdf (accessed on 24 December 2021).
10. Higgins, V.; Bryant, M.; Howell, A.; Battersby, J. Ordering adoption: Materiality, knowledge and farmer engagement with precision agriculture technologies. *J. Rural Stud.* **2017**, *55*, 193–202. [CrossRef]
11. Maertens, M.; Barrett, C.B. Measuring social networks' effects on agricultural technology adoption. *Am. J. Agric. Econ.* **2012**, *95*, 353–359. [CrossRef]
12. Spekken, M.; Molin, J.P.; Romanelli, T.L. Cost of boundary maneuvers in sugarcane production. *Biosyst. Eng.* **2015**, *129*, 112–126. [CrossRef]
13. Colaço, A.F.; Pagluca, L.G.; Romanelli, T.L.; Molin, J.P. Economic viability, energy and nutrient balances of site-specific fertilization for citrus. *Biosyst. Eng.* **2020**, *200*, 138–1565. [CrossRef]
14. USDA-United States Department of Agriculture. National Agricultural Statistics Service. 2016. Available online: https://www.nass.usda.gov/Statistics_by_State/Nebraska/index.php (accessed on 24 December 2021).

15. Klein, R.; Wilson, R.; Johnson, J.; Jansen, J.A. 2017 Nebraska Crop Budgets. Nebraska Cooperative Extension EC04-872. 2017. Available online: https://cropwatch.unl.edu/budgets (accessed on 15 August 2017).
16. Chen, Q.; Wachenheim, C.; Zheng, S. Land scale, cooperative membership and benefits information: Unmanned aerial vehicle adoption in China. *Sustain. Futures* **2020**, *2*, 100025. [CrossRef]

engineering proceedings

Proceeding Paper

Investment in Information and Communication Technology in Agriculture and Soybean Production Stability: The Case of China [†]

Vesna Jablanovic

Faculty of Agriculture, University of Belgrade, 11080 Belgrade, Serbia; vesnajab@ptt.rs
† Presented at the 13th EFITA International Conference, online, 25–26 May 2021.

Abstract: The paper creates the chaotic soybeans production growth model. This model includes investment in ICTs in agriculture. Further, the paper discovers a sequence of Elliot waves in soyabeans market in China in the period 1991–2015. Investment in information and communication technologies (ICTs) in agriculture improved both the productivity and soybean production stability. The stable convergent fluctuations of soybeans production existed in China in the observed period.

Keywords: information and communication technology; agriculture; chaos; Elliot waves

Citation: Jablanovic, V. Investment in Information and Communication Technology in Agriculture and Soybean Production Stability: The Case of China. *Eng. Proc.* **2021**, *9*, 34. https://doi.org/10.3390/engproc 2021009034

Academic Editors: Charisios Achillas and Lefteris Benos

Published: 13 December 2021

Publisher's Note: MDPI stays neutral with regard to jurisdictional claims in published maps and institutional affiliations.

1. Introduction

E-agriculture involves designing, developing, and applying new ways to use information and communication technologies (ICTs) with a primary focus on agriculture. Soybean finds a principal place in the agricultural production systems of many countries, including the United States, Brazil, Argentina, China, and India.

It is important to improve access to valuable information that can help agricultural producers and consumers to make the best possible economic decisions and use the resources available in the most productive manner [1].

ICTs that can be used for e-agriculture may include: telephone (interactive voice response); computers and websites (agriculture information and markets); broadcasting (expertise sharing advisory, community); satellite (weather, universal accessibility, remote sensing); mobile (advisory, sales, banking, networking); internet and broadband (knowledge sharing, social media, e-community, banking, market platform, trading, …); sensor networks (real time information, better data quality and quantity, decision making), data storage and analytics (precision agriculture) [1].

Using e-agriculture can lead to greater efficiencies in agricultural extension, disaster risk management and early warning systems, enhanced market access and financial inclusion, etc. Using e-agriculture can improve market information. Lower transaction costs, improved market coordination and more transparent rural markets are results of this type of agriculture [1].

ICTs, GIS, remote sensing, precision farming and many other technologies or processes hold great promises for improved agricultural productivity. Achieving improved and sustainable agricultural production and productivity growth largely depends on the advancement of agricultural research and its effective applications at farmer's fields through the transfer of technology and innovation. FAO estimates that 91 percent of the global food production increase towards the year 2050 should come from yield increases of current arable lands based on the advancement of agricultural research [2].

China's digital economy has expanded rapidly in recent years. While average digitalization of the economy remains lower than in advanced economies, digitalization is already high in certain regions and sectors, in particular e-commerce. Such transformation has boosted productivity growth, with varying impact on employment across sectors.

Digitalization will continue to reshape the Chinese economy. The government should play an important role in maximizing the benefits of digitalization while minimizing related risks, such as potential labor disruption, privacy infringement, emerging oligopolies, and financial risks [3].

Soybean finds a principal place in the agricultural production systems of many countries including the United States, Brazil, Argentina, China, and India.

2. The Model

There are four components of the gross domestic product (real output), (Y): (i) consumption, (C); (ii) investment, (I); (iii) government purchases, (G); and (iv) net export (N_x), as follows:

$$Y_{d,t} = C_t + I_t + G_t + N_{xt} \tag{1}$$

The consumption function is explained by (Equation (2)), i.e.,

$$C_t = \alpha\, Y^2{}_t - 1 \quad 0 < \alpha < 1 \tag{2}$$

The investment in ICTs that can be used for e-agriculture and soybean production is explained by (Equation (3)), i.e.,

$$I_{i,t} = \beta\, Y_t - 1 \quad 0 < \beta < 1 \tag{3}$$

where β stands for the investment in ICTs ratio. The relation between an investment in ICTs that can be used for e-agriculture and soybean production and investment (I) is:

$$I_{i,t} = \omega\, I_t \quad 0 < \omega < 1 \tag{4}$$

where ω is the ratio showing the amount of investment in ICTs in relation to the total investment. Further, government spending function (Equation (5)) and net export function (Equation (6)) are given as:

$$G_t = \gamma\, Y_t \quad 0 < \gamma < 1 \tag{5}$$

$$N_{x,t} = \eta\, Y_t \quad 0 < \eta < 1 \tag{6}$$

where γ is the government spending ratio and η is the net exports ratio. The relation between the real output, Y_t, and soybean production, $Y_{s,t}$, is:

$$Y_{s,t} = \delta\, Y_t \quad 0 < \delta < 1 \tag{7}$$

Now, putting Equations (1)–(7) together, we obtain:

$$Y_{s,t} = \left[\frac{\beta}{\omega(1 - \gamma - \eta)} \right] Y_{s,t-1} - \left[\frac{\alpha}{\delta(\gamma + \eta - 1)} \right] Y_{s,t-1}{}^2 \tag{8}$$

It is important to introduce $y\,s$ as $Y_s = Y_s / Y_s{}^m$, where $0 < Y_s < 1$. In this sense, it is obtained

$$y_{s,t} = \left[\frac{\beta}{\omega(1 - \gamma - \eta)} \right] y_{s,t-1} - \left[\frac{\alpha}{\delta(\gamma + \eta - 1)} \right] y_{s,t-1}{}^2 \tag{9}$$

The model given by Equation (9) is called the logistic model.

3. The Logistic Map

The iteration process for the logistic map (Equation (10))

$$z_t = \pi\, z_{t-1} (1 - z_{t-1}), \quad \pi \in [0, 4], \; z_t \in [0, 1] \tag{10}$$

is equivalent to the iteration of the growth model (Equation (9)) when we use the identification:

$$z_t = \left[\frac{\alpha\,\omega\,(1-\gamma-\eta)}{\beta\,\delta\,(\eta+\gamma-1)}\right] y_{s,t} \text{ and } \pi = \left[\frac{\beta}{\omega\,(1-\gamma-\eta)}\right] \tag{11}$$

In accordance with Equations (9) and (11), we obtain:

$$z_t = \left[\frac{\alpha\,\omega\,(1-\gamma-\eta)}{\beta\,\delta\,(\eta+\gamma-1)}\right] y_{s,t} = \left[\frac{\alpha\,\omega\,(1-\gamma-\eta)}{\beta\,\delta\,(\eta+\gamma-1)}\right] \left\{ \left[\frac{\beta}{\omega\,(1-\gamma-\eta)}\right] y_{s,t-1} - \left[\frac{\alpha}{\delta\,(\eta+\gamma-1)}\right] y_{s,t-1}^2 \right\} =$$

$$= \left[\frac{\alpha}{\delta\,(\eta+\gamma-1)}\right] y_{s,t-1} - \left[\frac{\alpha^2\omega\,(1-\gamma-\eta)}{\beta\,\delta^2\,(\eta+\gamma-1)^2}\right] y_{s,t-1}^2$$

In accordance with Equations (10) and (11), we obtain:

$$z_t = \left[\frac{\beta}{\omega\,(\eta+\gamma-1)}\right] \left[\frac{\alpha\,\omega\,(1-\gamma-\eta)}{\beta\,\delta\,(\eta+\gamma-1)}\right] y_{s,t-1} \left\{ 1 - \left[\frac{\alpha\,\omega\,(1-\gamma-\eta)}{\beta\,\delta\,(\eta+\gamma-1)}\right] y_{s,t-1} \right\} =$$

$$= \left[\frac{\alpha}{\delta\,(\eta+\gamma-1)}\right] y_{s,t-1} - \left[\frac{\alpha^2\,\omega\,(1-\gamma-\eta)}{\beta\,\delta^2\,(\eta+\gamma-1)^2}\right] y_{s,t-1}^2$$

The dynamic properties of the logistic map (Equation (10)) have been widely analyzed [4,5]. It is obtained that: (i) For parameter values $0 < \pi < 1$, all solutions will converge to $z = 0$; (ii) For $1 < \pi < 3.57$, there exist fixed points, the number of which depends on π; (iii) For $1 < \pi < 2$, all solutions monotonically increase to $z = (\pi - 1)/\pi$; (iv) For $2 < \pi < 3$, fluctuations will converge to $z = (\pi - 1)/\pi$; (v) For $3 < \pi < 4$, all solutions will continuously fluctuate; (vi) For $3.57 < \pi < 4$, the solution becomes "chaotic".

4. Empirical Evidence

The soybean production growth stability in the period 1991–2015 can be explained by(Equation (12)):

$$y_{s,t} = \pi\,Y_{s,t-1} - \sigma\,Y_{s,t-1}^2 \tag{12}$$

where $Y_{s,t}$—soybean production, $\pi = \left[\frac{\beta}{\omega\,(1-\gamma-\eta)}\right]$, $\sigma = \left[\frac{\alpha}{\delta\,(\eta+\gamma-1)}\right]$.

Now, it is important to obtain the wave in soybean production movements. Elliot Wave [6] can be identified if we choose the following years in soybean production movements; 1991, 1994, 1996, 2002, 2004, 2007, 2008, 2015. In this case the soybean production movements are divided into: (i) trends (five waves in the direction of the main trend); and (ii) corrections (three corrective waves) (www.fao.org).

Now, the model (Equation (12)) is estimated (see Table 1). According to the logistic map, for $2 < \pi < 3$, fluctuations of the soybean production converge to $z = (\pi - 1)/\pi$. In this case, the equilibrium value is $z = (2.729108 - 1)/2.729108$, or $z = 0.6336$, or 11,025,767.808 tonnes. The golden ratio is $z = 0.618$.

Table 1. The estimated model (Equation (12)) $R = 0.54780$, Variance explained 30.009%.

$N = 8$	π	σ
Estimate	2.729108	1.992962
Std. Err.	0.340378	0.385135
t(6)	8.017877	5.174711
p-level	0.000201	0.002065

5. Conclusions

A key hypothesis of this work is based on the idea that the coefficient $\pi = \left[\frac{\beta}{\omega\,(1-\gamma-\eta)}\right]$ plays a crucial role in explaining the local growth stability of soybean production, where β is the investment in ICTs ratio, γ is the share of government spending in the real gross domestic product, η is the share of net export in the real gross domestic product, and ω is the ratio showing the amount of investment in ICTs in relation to the total investment. An estimated value of the coefficient π (2.729108) confirms the existence of stable fluctuations of soybean production in China in the period under consideration. The equilibrium value of soybean production in China was 11,025,767.808 tonnes in the period 1991–2015. This equilibrium value ($z = 0.6336$) approaches the golden ratio ($z = 0.618$).

Funding: This research received no external funding.

Institutional Review Board Statement: Not applicable.

Informed Consent Statement: Not applicable.

Data Availability Statement: The data presented in this study are openly available in http://www.fao.org/faostat/en/#data/QC (accessed on 13 February 2020).

Conflicts of Interest: The author declares no conflict of interest.

References

1. Food and Agriculture Organization of the United Nations, International Telecommunication Union. E-Agriculture Strategy Guide—A Summary. Available online: https://www.fao.org/3/i6909e/i6909e.pdf (accessed on 13 February 2020).
2. Food and Agriculture Organization of the United Nations. *Information and Communication Technologies for Sustainable Agriculture—Indicators from Asia and Pacific*; Sylvester, G., Ed.; Food and Agriculture Organization of the United Nations: Bangkok, Thailand, 2013; ISBN 978-92-5-108107-5.
3. Zhang, L.; Chen, S. *China's Digital Economy: Opportunities and Risks*; Working Paper No. 19/16; International Monetary Fund: Tokyo, Japan, 2019.
4. Li, T.; Yorke, J. Period Three Implies Chaos. *Am. Math. Mon.* **1975**, *8*, 985–992. [CrossRef]
5. May, R.M. Mathematical Models with Very Complicated Dynamics. *Nature* **1976**, *261*, 459–467. [CrossRef] [PubMed]
6. Frost, A.J.; Prechter, R.P. *Elliott Wave Principle: A Key to Market Behavior*, 10th ed.; New Classics Library: Gainesville, GA, USA, 2005.

engineering proceedings

Proceeding Paper

Rice Contract Farming in Vietnam: Insights from a Qualitative Study [†]

Mai Chiem Tuyen [1,2,3,*], Prapinwadee Sirisupluxana [2], Isriya Bunyasiri [2] and Pham Xuan Hung [3]

[1] Graduate School, Kasetsart University, Bangkok 10900, Thailand
[2] Department of Agricultural and Resource Economics, Faculty of Economics, Kasetsart University, Bangkok 10900, Thailand; fecopds@ku.th or fecopds@gmail.com (P.S.); isriya.b@ku.th (I.B.)
[3] Faculty of Economics and Development Studies, University of Economics, Hue University, Hue City 530000, Vietnam; pxhung@hueuni.edu.vn
* Correspondence: maichiemtuyen.t@ku.th or mctuyen@hueuni.edu.vn; Tel.: +84-94-929-6668
† Presented at the 13th EFITA International Conference, online, 25–26 May 2021.

Abstract: Since 2002, Vietnam has implemented rice contract-farming policies to develop the linkage among stakeholders in the agricultural sector; however, there is very low participation of farmers. Therefore, this study aims to determine the perception on both advantages and disadvantages of rice contract farming (RCF); identify the reasons for non-participation and drop-out of rice contract farming; and indicate the typology of contract by using data from documentation, key informant interviews, and focus group discussions. The results indicate that farmers considered the guaranteed output price and stable income as the most advantages of RCF while the main disadvantages were reducing the household's freedom or losing flexibility in making decisions on-farm production, management, and selling product; possible delays in payments, in input delivery, in harvesting, and output delivery. In addition, farmers did not want to participate in RCF because of reducing the household's freedom in making decisions, not complying with RCF, not trusting cooperatives as well as enterprises, and because selling paddy to middlemen is easier and simpler. Farmers dropped out of RCF because the contracting companies breached the contract provisions. Farmers mentioned many provisions of the contract but the most important to them were payment, price options, and delivery arrangement.

Keywords: contract farming; perception; participation; typology; rice; Vietnam

Citation: Tuyen, M.C.; Sirisupluxana, P.; Bunyasiri, I.; Hung, P.X. Rice Contract Farming in Vietnam: Insights from a Qualitative Study. *Eng. Proc.* **2021**, *9*, 6. https://doi.org/10.3390/engproc2021009006

Academic Editors: Charisios Achillas and Lefteris Benos

Published: 19 November 2021

Publisher's Note: MDPI stays neutral with regard to jurisdictional claims in published maps and institutional affiliations.

1. Introduction

Currently, with climate change being intense, the demand for high-quality food has been increasing. In this light, improving linkages between smallholders and companies to achieve efficient and sustainable production becomes an urgent issue [1]. In addition, small households are facing many production and marketing constraints [2]. Therefore, contract farming (CF) has been proposed as a relevant measure for small-scale farmers in developing countries [3].

In Vietnam, rice has been the main crop with a proportion of over 50% of the total planted area, but 85.13% of households using paddy lands are small households [4]. Therefore, since 2002, the Vietnam government has implemented CF policies to develop stable and sustainable production by the decision 80/2002/QD-TTg, and lately by decree 98/2018/ND-CP. Regarding these policies, paddy would be sold through contract; at least 30% of the total output by 2005 and over 50% by 2010 of the rice industry was directed toward producing high-quality rice to improve the value and competitiveness. However, to date, rice output sold via contract is only 9.77% [5]; planted area under CF accounts for only 1.78% with the participation of 5.33% of total paddy cultivation households, while CF agreements are unsuccessful up to 70–80% [6].

The literature review indicates that the main motivation of farmers toward the rice contract farming (RCF) schemes are better output market, supportive benefits for the production, and better performance of rice farming [7], while the factors affecting farmers' participation are commitment, sharing of benefits and risks, household awareness about the benefits of contracts, policy environment, management skills and enterprises capacity of link firms, and price issues [8]. Other studies revealed the perception of specific factors affecting participation such as risk perception and perceived credit uncertainty [9,10]. However, there is a lack of research on farmers' perception on both the advantages and disadvantages of RCF to generate a representative indicator or component factors, and then add this indicator or these factors in the participation model as the independent variable(s) to investigate the factors affecting farmers' participation; a choice experiment on contracts' attributes to meet farmers' preference to encourage them to participate in RCF. Therefore, this study aims to determine the perception on both advantages and disadvantages of RCF; identify factors affecting farmers' participation and the reasons for non-participation and drop-out of RCF; indicate the typology of RCF in Vietnam. This study also creates the foundation and direction for quantitative research through a household survey.

2. Materials and Methods

This study utilized multi-stage sampling for study site selection. The Mekong River Delta (MRD) was selected as it represents the largest region for rice production in Vietnam. In this area, An Giang, Can Tho, and Kien Giang provinces were selected for investigating RCF because they were the first province applying RCF, the main rice exporter, and the largest cultivated rice area, respectively. Finally, in each province, we selected two representative districts, then two representative communes in each selected district.

Data for this study were collected from the General Statistic Office of Vietnam (GSO); the Department of Cooperatives and Rural Development (DCRD); previous empirical studies; key informant interviews (KII); and focus group discussions (FGD). Key informants were selected by the purposive sampling method since it allowed for the researchers to select experienced respondents. The participants of FGD were contracted farmers and non-contracted farmers, and or directors of agri-cooperatives. KII and FGD were performed by using a semi-structured questionnaire. Collected data were analyzed by typology [11,12], constant comparison [13], and content analysis methods [14].

3. Results and Discussion

3.1. Perception on Advantages and Disadvantages of RCF

The results indicate that RCF gave farmers many advantages, such as secure output markets or access to new markets; stable price; more stable income; easier access to marketing support services; reduction of pre and post-harvest losses; more profit; reliable supply of inputs; easier access to inputs and services; easier access to credit and better credit benefits (reducing the cash pressure on households due to priority in receiving credit or inputs); better technical assistance/production support; improved production/technical efficiency and management skills in rice production; and easier introduction of new techniques, new varieties, and practices. However, farmers were still faced with some disadvantages during doing RCF including reducing the household's freedom or losing flexibility in making decisions on-farm production, management, and selling of product; possible delays in payments and late delivery of input; possible delays in harvesting and delivery of output; possible high price of inputs provided by the contractor; increasing dependency and vulnerability of farmers if buyers are unreliable or exploit monopoly; increasing environmental risks from monoculture; unequal bargaining power between farmers and buyers; and getting indebted from loans provided by buyers. These are nearly the same as the government officials' perceptions.

3.2. Farmers' Participation in RCF

According to the FGD and KII, the factors affecting farmers' participation in RCF were from both farmers and contracting companies as well as the contract provisions. On the farmer side, these factors were their trust in agri-cooperatives and companies, farmers' perception on RCF, and parcel location, while the factors from the companies were commitments and prestige. In addition, farmers pointed out that the payment schedule, price options, delivery arrangement, and the complicated level of production methods affected their participation.

The research also indicated that farmers did not participate in RCF because of reducing their freedom in making the decision; being used to old production practices; being afraid to change; not trusting cooperatives or enterprises; not accepting that parcels were not in the large field; the perception that farmers' egos were large; failure to agree to simultaneously sow the same seed. In addition, selling rice to traders had some obvious benefits such as flexible price offers, fast determination to price, fast weighing, and immediate payment. Farmers did not participate in RCF due to an unacceptable price of the complicated production requirement; difficulty in price negotiation when the market price changed; strict requirement of input materials and output quality; high price of input provided by contractors; lacking contracting companies. Moreover, farmers witnessed the discredit of the contracting companies in the previous crops or the surrounding areas when they breached the contracts' terms such as delay in payment; delay in delivery; delay in harvesting, or extending the harvest time. This discredit was also the reason for dropping out of RCF of farmers.

3.3. Typology of RCF

The study results indicate that most contracts are in written form. These contracts were signed before starting of the crop season, after sowing, or before harvesting depending on the types of contracts by different stakeholders. Except for market-specifying contractors, other contractors always supply inputs to farmers, but it was quite flexible. In case farmers had received inputs in advance but did not want to sell paddy to contractors because of an unacceptable price offered, farmers had to pay the value of received input to contractors after harvesting. Some contractors required farmers to use at least 500 thousand VND of the input value to be bought paddy (output). Some contractors did not force farmers to use their inputs, but they inform farmers to avoid using the banned active ingredients. In addition to inputs, farmers received services as well as technical assistance. To create prestige and trust, contractors normally deposited 1500–2500 thousand VND/ha, sometimes about 4000 thousand VND/ha when paddy price was estimated to increase.

In production, some contractors required specified methods to produce rice following the GAP, organic or high-quality standard, while most contractors accepted the production methods consulted by the relevant offices of the local authorities such as "1 must, 5 decrease" and "3 decreases, 3 increases". Then, the contracting companies accepted to buy all paddy quantities, as long as they met the minimum standard requirements. This quantity was specified in the contract by estimating the cultivated area and yield of the crop season.

The most important provisions of the contracts discussed enthusiastically by the farmers were payment, price options, and delivery arrangements. Farmers preferred immediate payment or 50% before harvesting and the rest 3–5 days after delivery, however, sometimes they had to accept the delayed payment from contractors. For the price, there were four options of price including fixed price, market price, adjusted price, and premium price. About delivery arrangement, most paddies were delivered immediately after harvesting; however, companies sometimes delayed because they did not have enough transporting boats at the peak of harvesting time, as well as having limited storing and processing capacity.

4. Conclusions

There were many advantages of RCF pointed out by the farmers and the officials in Vietnam mainly relating to the output while the main disadvantages were linked with production. The factors affecting farmers' participation and the reasons for their non-participation in RCF were from not only both farmers and companies but also the contract provisions. Therefore, the solutions to encourage farmers' participation involves determining a reference price to set up reasonable price options, while enhancing the fulfillment of commitments, thereby increasing trust among the stakeholders.

Author Contributions: Conceptualization, M.C.T., P.S., I.B. and P.X.H.; Formal analysis, M.C.T.; Investigation, M.C.T.; Methodology, M.C.T.; Supervision, P.S. and I.B.; Writing—original draft, M.C.T.; Writing—review and editing, M.C.T. All authors have read and agreed to the published version of the manuscript.

Funding: This research was funded by the Deutscher Akademischer Austauschdienst (DAAD), Germany and Southeast Asian Regional Center for Graduate Study and Research in Agriculture (SEARCA), Philippines through the In-Country/In-Region Scholarship Programme (57433468), award number 91715475. The APC was funded by the University of Economics, Hue University.

Institutional Review Board Statement: Not applicable.

Informed Consent Statement: Informed consent was obtained from all subjects involved in the study.

Data Availability Statement: The data presented in this study are available on request from the corresponding author. However, the data are not publicly available due to confidentiality requirements for the thesis.

Acknowledgments: This study is one part of the Ph.D. thesis in the Philosophy Program in Agricultural and Resource Economics, Graduate School, Kasetsart University.

Conflicts of Interest: The authors declare no conflict of interest. The funders had no role in the design of the study; in the collection, analyses, or interpretation of data; in the writing of the manuscript, or in the decision to publish the results.

References

1. Calicioglu, O.; Flammini, A.; Bracco, S.; Bellù, L.; Sims, R. The future challenges of food and agriculture: An integrated analysis of trends and solutions. *Sustainability* **2019**, *11*, 222. [CrossRef]
2. Barrett, C.B.; Bachke, M.E.; Bellemare, M.F.; Michelson, H.C.; Narayanan, S.; Walker, T. Smallholder participation in contract farming: Comparative evidence from five countries. *World Dev.* **2012**, *40*, 715–730. [CrossRef]
3. Da Silva, C.A.; Rankin, M. *Contract Farming for Inclusive Market Access*; Food and Agriculture Organization of the United Nations (FAO): Rome, Italy, 2013; p. 227.
4. General Statistics Office of Vietnam. *Results of the Rural, Agricultural and Fishery Census 2016*; Statistical Publishing House: Hanoi, Vietnam, 2018; p. 686.
5. DCRD. *Results of Implementing Consumption Linkage of Key Products*; Department of Cooperatives and Rural Development (DCRD): Hanoi, Vietnam, 2020.
6. Hieu, D. Promoting Cooperation, Develop Large Field Models in Mekong River Delta 2016. Available online: https://dangcongsan.vn/kinh-te/thuc-day-lien-ket-xay-dung-canh-dong-lon-o-dong-bang-song-cuu-long-417866.html (accessed on 20 March 2020).
7. Nhan, T.; Yutaka, T. Analysis of contract farming between paddy farmers and an agribusiness firm in the Mekong Delta of Vietnam. *Agric. Mark. J. Jpn.* **2018**, *27*, 60–67.
8. Van Song, N.; Huyen, V.N.; Thuy, N.T.; Van Tien, D.; Van Ha, T.; Diep, N.X.; Huyen, V.T.K. A Way Forward To Promote The Farming Contracts Between Firms And Farmers In Cultivation Productions: A case study of vietnam. *Rev. Argent. Clínica Psicológica* **2020**, *XXIX*, 731–742.
9. Abdul-Rahman, M. *Contract Farming and Adoption of Improved Technologies in Maize Production in the Northern Region of Ghana*; Agricultural Economics, University for Development Studies: Tamale, Ghana, 2017.
10. Soullier, G.; Moustier, P. Impacts of contract farming in domestic grain chains on farmer income and food insecurity. Contrasted evidence from Senegal. *Food Policy* **2018**, *79*, 179–198. [CrossRef]
11. Lofland, J.; Lofland, L.H. *Analyzing Social Settings*; Wadsworth Publishing Company: Belmont, CA, USA, 1971; p. 136.
12. Smalley, R. *Plantations, Contract Farming and Commercial Farming Areas in Africa: A Comparative Review*; Future Agricultures Consortium: Brighton, UK, 2013; p. 73.
13. Strauss, A.L. *Qualitative Analysis for Social Scientists*; Cambridge University Press: Cambridge, UK, 1987; p. 319.
14. Weber, R.P. *Basic Content Analysis*; Sage: Riverside County, CA, USA, 1990; p. 96.

Proceeding Paper

A Review of the State-of-Art, Limitations, and Perspectives of Machine Vision for Grape Ripening Estimation [†]

Eleni Vrochidou *, Christos Bazinas, George A. Papakostas , Theodore Pachidis and Vassilis G. Kaburlasos

Human-Machines Interaction Laboratory (HUMAIN-Lab), Department of Computer Science, International Hellenic University (IHU), 65404 Kavala, Greece; chrbazi@cs.ihu.gr (C.B.); gpapak@cs.ihu.gr (G.A.P.); pated@cs.ihu.gr (T.P.); vgkabs@cs.ihu.gr (V.G.K.)
* Correspondence: evrochid@cs.ihu.gr
† Presented at the 13th EFITA International Conference, online, 25–26 May 2021.

Abstract: This work highlights the most recent machine vision methodologies and algorithms proposed for estimating the ripening stage of grapes. Destructive and non-destructive methods are overviewed for in-field and in-lab applications. Integration principles of innovative technologies and algorithms to agricultural agrobots, namely, *Agrobots*, are investigated. Critical aspects and limitations, in terms of hardware and software, are also discussed. This work is meant to be a complete guide of the state-of-the-art machine vision algorithms for grape ripening estimation, pointing out the advantages and barriers for the adaptation of machine vision towards robotic automation of the grape and wine industry.

Keywords: grape ripeness estimation; machine vision; harvesting robot; image analysis; sustainable agriculture

Citation: Vrochidou, E.; Bazinas, C.; Papakostas, G.A.; Pachidis, T.; Kaburlasos, V.G. A Review of the State-of-Art, Limitations, and Perspectives of Machine Vision for Grape Ripening Estimation. *Eng. Proc.* **2021**, *9*, 2. https://doi.org/10.3390/engproc2021009002

Academic Editors: Charisios Achillas and Lefteris Benos

Published: 19 November 2021

Publisher's Note: MDPI stays neutral with regard to jurisdictional claims in published maps and institutional affiliations.

1. Introduction

Grape harvesting is a great challenge that requires cost-effective solutions. The choice of harvest time determines the quality of the yield. Especially in wine production, deciding on the optimal harvest time concerning certain compounds, such as anthocyanins, is crucial in order to obtain optimal wine quality. Additionally, different vineyards ripen at different rates. Identifying maturity zones could enhance the efficiency of the harvesting operation. Grape harvesting based on grape ripening stage estimation increases the sustainable production of grape fruit products in the following three main directions: (1) improvement of the quality of harvested grapes due to homogenously ripened and equally fresh fruit, (2) reducing the post-harvest waste along the supply chain, and (3) reducing the production costs and human labor due to sustainable resource management.

In this context, an automated approach for grape ripeness estimation is considered necessary. Research focuses on the development of non-destructive, cost-effective, and environmentally friendly techniques. Machine vision has recently gained considerable acceptance for agricultural-related tasks. To this end, machine vision has been employed in in-field practical applications for the assessment of grape ripeness. The reported results reveal that image processing has good potential to be used for grape ripeness estimation, not as a substitute, but rather as an attractive alternative to chemical analyses, due to its simplicity, flexibility, and low cost. The aim of this work is to increase the insight into the application of machine vision techniques for grape ripening estimation. The techniques reported here are recent, and could be utilized as effective and rapid tools for decision making regarding the desirable harvest time.

2. Machine Vision Methods for Grape Ripeness Estimation

Different grape cultivars display a different refractometric index, which is related to maturity. For instance, table grapes must have at least 16° Brix to be considered ripened,

Eng. Proc. **2021**, *9*, 2. https://doi.org/10.3390/engproc2021009002

while Sauvignon Blanc may ripen at 20–22° Brix, etc. However, even this is relevant, since the proper harvesting time is not necessarily related to a standard ripening value, but to a desired ripening value depending on the harvested variety. In the wine industry, the health condition and maturity of harvested grapes need to be evaluated in order to determine the exact procedure, diffusional, enzymatic, or biochemical, to be applied. Objective criteria to judge the ripeness stage of grapes are those related to chemical attributes, such as titratable acidity (TA), soluble solid content (SSC), etc. Subjective maturation criteria are just as important, and include the sensory characteristics chosen to be discriminative among samples.

During ripening, many physical and biochemical changes occur that affect grape characteristics. Usually, red–green–blue (RGB) imaging is the most cost-effective way to determine color channel values and other characteristics, such as shape and texture. The RGB color space depends on the device on which it is displayed; therefore, it is not adequate when it comes to absolute measures. For this reason, the CIELAB color space is used as an international standard for color measurements. However, alternative color spaces have also been investigated, such as hue–saturation–intensity (HIS), hue-saturation value (HSV), etc. Classic near-infrared (NIR) spectroscopy has been proven as a powerful analytical tool to designate bioactive compounds in grapes. The main advantage of NIR spectroscopy over other methods is the chemical-free sample preparation and the ability to determine efficiently optical properties of the fruit that are strongly related to the chemical and physical properties, and, thus, to maturity. Hyperspectral imaging is also considered to be an evolving tool, since it combines conventional imaging with spectroscopy, and, thus, one can obtain both the spectral and spatial information of an object. Table 1 summarizes the recent applications of hyperspectral imaging, color imaging, and NIR spectroscopy for grape ripeness estimation.

A limitation of the use of machine vision algorithms in grape-related applications is attributed to the lack of large-scale public datasets of grape bunch images for testing innovative methodologies and performing comparative evaluation reports. Researchers must rely on the data collected by themselves, which are neither universal nor comparable. The complexity of agricultural environments in terms of diversity and dynamically changing conditions, such as illumination and vegetation, are the main obstacles that machine vision technology attempts to overcome. Robust algorithms, able to operate in heterogeneous environments, need to be developed. The application of visual technology in the field involves the integration of multiple parameters and principles that can only be encountered in natural setups, in practical applications, and not in laboratory settings. In recent years, the development of graphical peripheral units (GPUs) helped to increase the computational power and allowed powerful real-time pattern recognition techniques to be used on-site. In the future, machine vision algorithms are expected to play a vital role in sustainable agricultural automation, towards improving local economies and promoting ecology.

Table 1. Applications of hyperspectral imaging for grape ripeness estimation.

Cultivar	Maturity Index	Spectral Range/ Color Space	Number of Images	Prediction Model	Evaluation	Ref.
Touriga Franca (TF), Touriga Nacional (TN), Tinta Barroca (TB)	pH, anthocyanin	380–1028 nm	225	Neural Network (NN)	pH R^2: 0.723 (TF), 0.661 (TN), 0.710 (TB) Anthocyanin R^2: 0.906 (TF), 0.751 (TN), 0.697 (TB)	[1]
cv. Malagousia	TSS, pH	510–790 nm	12	Multiple linear regressor (MLR), support vector machine (SVM)	Classification rates: TSS 83.33% (MLR), TSS 83.33% (SVM), pH 75% (MLR), pH 75% (SVM)	[2]
Syrah, Tempranillo	Flavanols, total phenols	900–1700 nm	200	principal component analysis, modified partial least squares, K-means, linear discriminant	Classification rates: 83.3% (leave-one-out cross-validation), 76.9% (external validation)	[3]
Vitis vinifera L. cv., Tempranillo, Syrah	Anthocyanins	900–1700 nm	99	Principal component analysis	0.86 R^2	[4]
Kyoto grapes	SSC, pH	HIS, NTSC, YCbCr, HSV, CMY	180	Multiple linear regressor (MLR), partial least-squares regressor (PLSR)	Mean square error: pH 0.0987 (MLR), pH 0.1257 (PLSR), SSC 0.7982 (MLR), SSC 0.9252 (PLSR)	[5]
<Not defined>	Visual assessment	RGB, HSV	289	Dirichlet mixture model (DMM)	125.04 perplexity, 0.29 average perplexity per color	[6]
Italia, Victoria	Visual assessment	HSV, CIELAB	800	Random forest (RF)	Cross-validation classification accuracy: up to 100% (Italia), up to 0.92 (Victoria)	[7]
White grape Sonaka	Visual assessment	RGB HSV	4000	Convolutional NN (CNN), support vector machine (SVM)	Classification rates: 79% (CNN), 69% (SVM)	[8]
Cabernet Sauvignon	Visual assessment	RGB	13	Neural network (NN)	Average error: 5.36%	[9]
Syrah, Cabernet Sauvignon	TSS, flavonoid, anthocyanin	RGB	1008	Convolutional neural networks	Classification rates: 93.41% (Syrah), 72.66% (Cabernet)	[10]

3. Conclusions

Nowadays, fruits are still harvested manually by fruit pickers, based on subjective criteria and extensive sampling, followed by chemical analyses. Maturity is critical for the storage life of harvested fruits, and the quality of table grapes and produced wines. The degree of maturity defines the way grapes are further processed. Machine vision algorithms are employed for grape ripeness estimation, as a non-destructive, labor-saving, cost-effective, and eco-friendly alternative. This work explores the possible applications of machine vision technology in grape ripeness estimation, and discusses the limitations and challenges hindering their successful adoption for sustainable agricultural automation.

Author Contributions: Conceptualization, G.A.P. and E.V.; methodology, E.V.; investigation, E.V. and C.B.; resources, E.V. and C.B.; writing—original draft preparation, E.V. and C.B.; writing—review and editing, E.V, G.A.P., T.P. and V.G.K.; visualization, C.B.; supervision, G.A.P.; project administration, G.A.P., T.P. and V.G.K.; funding acquisition, V.G.K. and G.A.P. All authors have read and agreed to the published version of the manuscript.

Funding: This research has been co-financed by the European Regional Development Fund of the European Union and Greek national funds through the Operational Program Competitiveness, Entrepreneurship and Innovation, under the call RESEARCH—CREATE—INNOVATE (project code: T1EDK-00300).

Institutional Review Board Statement: Not applicable.

Informed Consent Statement: Not applicable.

Data Availability Statement: Not applicable.

Conflicts of Interest: The authors declare no conflict of interest.

References

1. Gomes, V.; Fernandes, A.; Martins-Lopes, P.; Pereira, L.; Mendes Faia, A.; Melo-Pinto, P. Characterization of Neural Network Generalization in the Determination of PH and Anthocyanin Content of Wine Grape in New Vintages and Varieties. *Food Chem.* **2017**, *218*, 40–46. [CrossRef] [PubMed]
2. Iatrou, G.; Mourelatos, S.; Gewehr, S.; Kalaitzopoulou, S.; Iatrou, M.; Zartaloudis, Z. Using Multispectral Imaging to Improve Berry Harvest for Wine Making Grapes. *Ciência e Técnica Vitivinícola* **2017**, *32*, 33–41. [CrossRef]
3. Baca-Bocanegra, B.; Nogales-Bueno, J.; Heredia, F.; Hernández-Hierro, J. Estimation of Total Phenols, Flavanols and Extractability of Phenolic Compounds in Grape Seeds Using Vibrational Spectroscopy and Chemometric Tools. *Sensors* **2018**, *18*, 2426. [CrossRef] [PubMed]
4. Hernández-Hierro, J.M.; Nogales-Bueno, J.; Rodríguez-Pulido, F.J.; Heredia, F.J. Feasibility Study on the Use of Near-Infrared Hyperspectral Imaging for the Screening of Anthocyanins in Intact Grapes during Ripening. *J. Agric. Food Chem.* **2013**, *61*, 9804–9809. [CrossRef] [PubMed]
5. Xia, Z.; Wu, D.; Nie, P.; He, Y. Non-Invasive Measurement of Soluble Solid Content and PH in Kyoho Grapes Using a Computer Vision Technique. *Anal. Methods* **2016**, *8*, 3242–3248. [CrossRef]
6. Hernández, S.; Morales, L.; Urrutia, A. Unsupervised Learning for Ripeness Estimation from Grape Seeds Images. *Int. J. Smart Sens. Intell. Syst.* **2017**, *10*, 594–612. [CrossRef]
7. Cavallo, D.P.; Cefola, M.; Pace, B.; Logrieco, A.F.; Attolico, G. Non-Destructive and Contactless Quality Evaluation of Table Grapes by a Computer Vision System. *Comput. Electron. Agric.* **2019**, *156*, 558–564. [CrossRef]
8. Kangune, K.; Kulkarni, V.; Kosamkar, P. Grapes Ripeness Estimation Using Convolutional Neural Network and Support Vector Machine. In Proceedings of the 2019 Global Conference for Advancement in Technology, GCAT 2019, Bangalore, India, 12–18 October 2019. [CrossRef]
9. Kaburlasos, V.G.; Vrochidou, E.; Lytridis, C.; Papakostas, G.A.; Pachidis, T.; Manios, M.; Mamalis, S.; Merou, T.; Koundouras, S.; Theocharis, S.; et al. Toward Big Data Manipulation for Grape Harvest Time Prediction by Intervals' Numbers Techniques. In Proceedings of the 2020 International Joint Conference on Neural Networks (IJCNN), Glasgow, UK, 19–24 July 2020; pp. 1–6. [CrossRef]
10. Ramos, R.P.; Gomes, J.S.; Prates, R.M.; Simas Filho, E.F.; Teruel, B.J.; Dos Santos Costa, D. Non-Invasive Setup for Grape Maturation Classification Using Deep Learning. *J. Sci. Food Agric.* **2021**, *101*, 2042–2051. [CrossRef] [PubMed]

Proceeding Paper

Prediction of Corn and Sugar Prices Using Machine Learning, Econometrics, and Ensemble Models [†]

Roberto F. Silva [1,*], Bruna L. Barreira [2] and Carlos E. Cugnasca [2]

[1] Institute of Mathematics and Computer Sciences (ICMC), University of São Paulo (USP), São Carlos 13566-590, Brazil

[2] Polytechnic School, University of São Paulo (USP), São Paulo 05508-010, Brazil; brunalyuger@gmail.com (B.L.B.); carlos.cugnasca@usp.br (C.E.C.)

* Correspondence: roberto.fray.silva@gmail.com

† Presented at the 13th EFITA International Conference, online, 25–26 May 2021.

Abstract: This paper explores the use of several state-of-the-art machine learning models for predicting the daily prices of corn and sugar in Brazil in relation to the use of traditional econometrics models. The following models were implemented and compared: ARIMA, SARIMA, support vector regression (SVR), AdaBoost, and long short-term memory networks (LSTM). It was observed that, even though the prices time series for both products differ considerably, the models that presented the best results were obtained by: SVR, an ensemble of the SVR and LSTM models, an ensemble of the AdaBoost and SVR models, and an ensemble of the AdaBoost and LSTM models. The econometrics models presented the worst results for both products for all metrics considered. All models presented better results for predicting corn prices in relation to the sugar prices, which can be related mainly to its lower variation during the training and test sets. The methodology used can be implemented for other products.

Keywords: agricultural prices; econometrics; ensembles; machine learning; price prediction

Citation: Silva, R.F.; Barreira, B.L.; Cugnasca, C.E. Prediction of Corn and Sugar Prices Using Machine Learning, Econometrics, and Ensemble Models. *Eng. Proc.* **2021**, *9*, 31. https://doi.org/10.3390/engproc2021009031

Academic Editors: Charisios Achillas and Lefteris Benos

Published: 2 December 2021

Publisher's Note: MDPI stays neutral with regard to jurisdictional claims in published maps and institutional affiliations.

1. Introduction

The different agricultural products value chains are essential for producing and distributing food, medicines, clothes, among many other products [1,2]. Two of the most important agricultural value chains worldwide are the sugar and corn chains [3]. One essential activity for all agents in those value chains is to correctly predict the agricultural products' prices [1,4–7]. The quality of this prediction impacts decision-making and revenue generation for all agents in the value chains. However, most of the literature on time series analysis for price prediction focuses on prediction stock market prices, such as the works by [8–10]. Traditionally, econometrics models such as autoregressive integrated moving average (ARIMA), seasonal ARIMA (SARIMA), and SARIMA with exogenous factors (SARIMAX) are used [1,4–10]. In the last decade, several machine learning (ML) models showed better performance, with lower mean absolute error (MAE), mean squared error (MSE), root mean squared error (RMSE), and higher R2 score [1,4–7]. In addition, the long short-term memory network (LSTM) is considered a state-of-the-art model for price prediction [6].

Ouyang et al. [11] compared the use of ARIMA, LSTNet (an LSTM-based network), and different configurations of artificial neural networks to predict the prices of twelve agricultural commodities. They have concluded that the LSTNet presented the best results. Kanchymalay et al. [4] compared, using a multi-layer perceptron, a support vector regression (SVR), and a Holt Winter exponential smoothing method to predict crude palm oil prices. The authors concluded that the SVR model presented the best results. Ref. [6] evaluated using a backpropagation neural network and an LSTM to predict high and low prices of soybean futures, concluding that the LSTM presented better results.

The main objective of this work is to evaluate the use of state-of-the-art ML models to predict the daily prices of corn and sugar in Brazil in relation to traditional econometrics models. The following models were implemented: ARIMA, SARIMA, SVR, AdaBoost, LSTM, and ML ensembles with different configurations. All the models were evaluated on the test subset, composed of the whole year of 2019, considering three metrics: MAE, MSE, and R2 score.

2. Methodology

The methodology used in this paper was composed of five main steps:

i. Data gathering, collecting daily prices for sugar (from 2004 to 2019) and corn (from 2003 to 2019) from the CEPEA agricultural prices database [12]; ii. data preprocessing, encompassing the following tasks: identifying and handling missing data and outliers, and dividing the datasets into subsets for both products. The subsets used were: (i) training: beginning of the dataset until the validation subset; (ii) validation: cross-validation using the blocking time series method; and (iii) testing: 2019; iii. exploratory data analysis, considering an analysis of each price time series and an autocorrelation analysis with the implementation of the augmented Dickey–Fuller test (ADF) test, autocorrelation (ACF) and partial autocorrelation functions (PACF), and their respective plots; iv. models implementation and hyperparameters analysis, considering the following models: (i) econometrics: ARIMA and SARIMA; (ii) ML: SVR and AdaBoost; (iii) DL: LSTM; and (iv) ensembles: E1 (AdaBoost + SVR), E2 (SVR + LSTM, and E3 (AdaBoost + LSTM); and v. models comparison for both products considering three error metrics: MAE, MSE, and R2.

The implementation was realized using Python on a Google Collaboratory CPU (https://colab.research.google.com/, accessed on: 20 April 2021), with the following technical specifications: Intel(R) Xeon(R) CPU @ 2.30GHz CPU, 12GB of RAM. The libraries used were: NumPy (https://numpy.org/, accessed on: 20 April 2021), Pandas (https://pandas.pydata.org/, accessed on: 20 April 2021), Matplotlib (https://www.tensorflow.org/, accessed on: 20 April 2021), TensorFlow (https://www.tensorflow.org/, accessed on: 20 April 2021), Scikit-Learn (https://scikit-learn.org/), and Statsmodels (https://www.statsmodels.org/, accessed on: 20 April 2021).

3. Results and Discussions

The result of the ADF test showed that both agricultural products' price series present autocorrelation. The analysis of the ACF and PACF plots also pointed out that there is seasonality in the data. Figure 1 illustrates the prices in R$/ton for both sugar and corn. It is important to observe that: (i) for sugar, the highest price peaks were observed between 2016 and 2017, between 2010 and 2012, and between 2006 and 2007; (ii) for sugar, the lowest prices were observed in 2004, 2008, and between 2013 and 2015; (iii) for corn, the highest price peaks were observed between 2016 and 2018, and between 2007 and 2008; (iv) for corn, the lowest prices were observed in 2006, 2010, 2015, and 2018; and (v) for both products, the were increasing significantly during 2020. Those points reflect several factors, such as the fluctuations in product demand worldwide, the occurrence of financial crises, the fluctuation in product supply (influenced mainly by the occurrence of droughts), the variation of exchange rates, and the impacts of the COVID-19 pandemics. Table 1 presents a comparison of the final models implemented on the test subset (the year of 2019), considering the best hyperparameters values identified during the cross-validation procedure. It is essential to observe that, even though the prices time series for both products differed considerably, the models that presented the best results were: (i) SVR, (MAE: 0.287 for corn and 0.430 for sugar); (ii) ensemble of the SVR and LSTM models (MAE: 0.335 for corn and 0.458 for sugar); (iii) ensemble of the AdaBoost and SVR models (MAE: 0.395 for corn and 0.476 for sugar); and (iv) ensemble of the AdaBoost and LSTM models (MAE: 0.425 for corn and 0.500 for sugar). One of the reasons that may explain the superior performance for the SVR is the small dataset size.

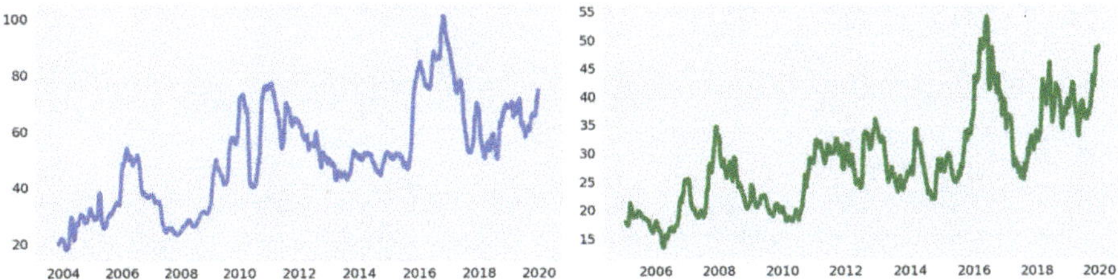

Figure 1. Prices in R$/ton for sugar (left, in blue) and corn (right, in green) from 2004 to 2020. Source: CEPEA, 2021 [12].

Table 1. Final models comparison for predicting prices of sugar and corn in Brazil.

Model	Sugar				Corn			
	MAE	MSE	R2	Diff. Best MAE	MAE	MSE	R2	Diff. Best MAE
ARIMA	8.496	109.422	−6.218	2593%	2.846	12.923	0.091	890%
SARIMA	3.406	19.058	−0.257	980%	2.922	13.044	0.083	917%
SVR	0.430	0.315	0.979	X	0.287	0.145	0.990	X
AdaBoost	0.560	0.519	0.966	77%	0.616	0.538	0.962	114%
LSTM	0.575	0.585	0.961	82%	0.432	0.324	0.977	50%
E1 (AdaBoost+SVR)	0.476	0.374	0.975	19%	0.395	0.239	0.983	38%
E2 (SVR+LSTM)	0.458	0.369	0.976	17%	0.335	0.195	0.986	17%
E3 (AdaBoost+LSTM)	0.500	0.421	0.972	58%	0.425	0.278	0.980	48%

Legend: in green: best model for the commodity; in red: worst model for the commodity.

The econometrics models presented the worst results for both products for the three metrics considered, implying that those models did not capture the trends in those datasets. Two of the main reasons that may explain this observation are: (i) both products' prices were significantly volatile during the period; and (ii) both datasets presented non-stationary data and varying trends. It is also important to observe that all models presented better results for predicting corn prices in relation to the sugar prices, which can be related mainly to its lower price variation. Additionally, the observation that ML models and ensembles presented better results indicates that ML models and ensembles could improve agricultural products prices prediction results. This observation has important implications for researchers and practitioners, as it can help improve the quality of the price predictions for a given agricultural product. Furthermore, practitioners could use it to implement the ML models used in this work and backtest their results for different agricultural products and periods. This could provide valuable information for decision-making. It is essential to note that the methodology used in this work can be implemented for other products. Lastly, the main limitations observed were: (i) the small datasets used, which could have impacted on the LSTM results; (ii) the unknown market dynamics, making it challenging to generate new features; and (iii) the lack of standard datasets and model implementations in the literature for comparing the results obtained for the different agricultural products.

4. Conclusions and Future Works

Agricultural products value chains are essential for producing and distributing food, medicines, and clothes, among many other products. Therefore, improving product prices prediction is essential to improve decision-making by the different value chain agents. However, most works in the literature focus on predicting stock market prices. In this work, the use of traditional econometrics (ARIMA and SARIMA), ML (SVR and LSTM), and ML ensembles models (with different configurations) was evaluated for predicting daily prices for corn and sugar in Brazil.

It was observed that: (i) the SVR model presented the best results for both products, followed by the SVR and LSTM ensemble; (ii) the econometrics models presented the worst results for both products; and (iii) all models presented better results for predicting corn prices in relation to the sugar prices. Future work is related to: implementing other ML models, using unsupervised learning to improve pattern detection, implementing deep reinforcement learning models to allow for autonomous decision making, and evaluating other datasets and periods.

Author Contributions: Conceptualization, R.F.S.; methodology, R.F.S., B.L.B.; investigation, R.F.S., B.L.B.; writing—original draft preparation, R.F.S.; writing—review and editing, B.L.B., C.E.C.; supervision, C.E.C. All authors have read and agreed to the published version of the manuscript.

Funding: This work was carried out with the support from Itaú Unibanco S.A., linked to the Centro de Ciência de Dados (C2D) of the Polytechnic School of the University of São Paulo (USP).

Conflicts of Interest: The authors declare no conflict of interest.

References

1. Kamilaris, A.; Kartakoullis, A.; Prenafeta-Boldu, F.X. A review on the practice of big data analysis in agriculture. *Comput. Electron. Agric.* **2017**, *143*, 23–37. [CrossRef]
2. Verdouw, C.N.; Beulens, A.J.M.; Trienekens, J.H.; Wolfert, J. Process modelling in demand-driven supply chains: A reference model for the fruit industry. *Comput. Electron. Agric.* **2010**, *73*, 174–187. [CrossRef]
3. FAO. *Food and Agriculture Organization of the United Nations. FAOSTAT Database*; FAO: Rome, Italy, 2021; Available online: http://www.fao.org/faostat/en/#data/QC (accessed on 1 March 2021).
4. Kanchymalay, K.; Salim, N.; Sukprasert, A.; Krishnan, R.; Hashim, U.R.A. Multivariate time series forecasting of crude palm oil price using machine learning techniques. In *IOP Conference Series: Materials Science and Engineering*; IOP Publishing: Bristol, UK, 2017; Volume 226, p. 012117.
5. Khamis, A.; Abdullah, S.N.S.B. Forecasting wheat price using backpropagation and NARX neural network. *Int. J. Eng. Sci.* **2014**, *3*, 19–26.
6. Wang, C.; Gao, Q. High and low prices prediction of soybean futures with LSTM neural network. In Proceedings of the 2018 IEEE 9th International Conference on Software Engineering and Service Science (ICSESS), Beijing, China, 23–25 November 2018; pp. 140–143.
7. Xiong, T.; Li, C.; Bao, Y. Seasonal forecasting of agricultural commodity price using a hybrid STL and ELM method: Evidence from the vegetable market in China. *Neurocomputing* **2018**, *275*, 2831–2844. [CrossRef]
8. Kara, Y.; Boyacioglu, M.A.; Baykan, O.K. Predicting direction of stock price index movement using artificial neural networks and support vector machines: The sample of the Istanbul Stock Exchange. *Expert Syst. Appl.* **2011**, *38*, 5311–5319. [CrossRef]
9. Di Persio, L.; Honchar, O. Artificial neural networks architectures for stock price prediction: Comparisons and applications. *Int. J. Circuits Syst. Signal Process.* **2016**, *10*, 403–413.
10. Weng, B.; Lu, L.; Wang, X.; Megahed, F.M.; Martinez, W. Predicting short-term stock prices using ensemble methods and online data sources. *Expert Syst. Appl.* **2018**, *112*, 258–273. [CrossRef]
11. Ouyang, H.; Wei, X.; Wu, Q. Agricultural commodity futures prices prediction via long-and short-term time series network. *J. Appl. Econ.* **2019**, *22*, 468–483. [CrossRef]
12. CEPEA. Preços Agropecuários. 2021. Available online: https://www.cepea.esalq.usp.br/br (accessed on 5 January 2021).

Proceeding Paper

Crop Water Availability Mapping in the Danube Basin Based on Deep Learning, Hydrological and Crop Growth Modelling [†]

Silke Migdall [1,*], Sandra Dotzler [1], Eva Gleisberg [1], Florian Appel [1], Markus Muerth [1], Heike Bach [1], Giulio Weikmann [2], Claudia Paris [2], Daniele Marinelli [2] and Lorenzo Bruzzone [2]

1 VISTA GmbH, Gabelsbergerstr, 51, 80333 Munich, Germany; dotzler@vista-geo.de (S.D.); gleisberg@vista-geo.de (E.G.); appel@vista-geo.de (F.A.); muerth@vista-geo.de (M.M.); bach@vista-geo.de (H.B.)
2 Department of Information Engineering and Computer Science, University of Trento, Povo, I-38123 Trento, Italy; giulio.weikmann@unitn.it (G.W.); claudia.paris@unitn.it (C.P.); daniele.marinelli@unitn.it (D.M.); lorenzo.bruzzone@unitn.it (L.B.)
* Correspondence: migdall@vista-geo.de; Tel.: +49-89-4521-614-21
† Presented at the 13th EFITA International Conference, online, 25–26 May 2021.

Abstract: The Danube Basin has been hit by several droughts in the last few years. As climate change makes weather extremes and temperature records in late winter and early spring more likely, water availability and irrigation possibilities become more important. In this paper, the crop water demand at field and national scale within the Danube Basin is presented using a dense time series of multispectral Sentinel-2 data, for crop type maps derived with deep learning techniques and physically-based models for crop parameter retrieval and crop growth modelling.

Keywords: food security; Thematic Exploitation Platform; cloud processing; crop water demand; crop type classification; biophysical parameter retrieval; water stress; water use efficiency

Citation: Migdall, S.; Dotzler, S.; Gleisberg, E.; Appel, F.; Muerth, M.; Bach, H.; Weikmann, G.; Paris, C.; Marinelli, D.; Bruzzone, L. Crop Water Availability Mapping in the Danube Basin Based on Deep Learning, Hydrological and Crop Growth Modelling. *Eng. Proc.* **2021**, *9*, 42. https://doi.org/10.3390/engproc2021009042

Academic Editors: Dimitrios Aidonis and Aristotelis Christos Tagarakis

Published: 24 January 2022

Publisher's Note: MDPI stays neutral with regard to jurisdictional claims in published maps and institutional affiliations.

1. Introduction

More droughts, stronger storms, more unpredictable winters—climate change brings more volatile conditions to European agriculture, along with challenges of managing water distribution to fulfil irrigation needs and to support all other ecosystem functions equally. Moreover, growing population and increased food consumption call for measures to stabilise and sustainably heighten the efficiency of agricultural production. To counter some of these trends, data-driven agriculture can provide a new level of information quality for agricultural processes. This is where this study has its target: set within the Horizon 2020 project ExtremeEarth, the issues of water availability and water demand for agricultural production are addressed. The goal is to support irrigation management decisions by combining big data Earth Observation (EO) analysis with water balance and crop growth modelling within European pilot catchments.

In this paper, the first focus is on a wide-scale analysis of the whole Austrian territory as part of the Danube catchment. Its location in the centre of Europe and its different geographic regions are perfect examples of the challenges faced by agriculture and the way big data EO analysis can offer a unique insight into large scale processes and challenges. The second focus is on site-specific analysis for fields in Southern Germany. Cloud Computing and European platform infrastructures were used to pre-process a dense time series of multispectral Sentinel-2 data that were then analysed with a multitemporal deep learning architecture to derive crop type maps. Finally, the water stress of different crops was modelled and the irrigation water demands of various crops were calculated using physically-based models for crop parameter retrieval and crop growth modelling. The results are a big step towards understanding the challenges of water management in the Danube.

2. Methods

2.1. Satellite Data Pre-Processing

Figure 1 shows the block-scheme of the proposed method. First, pre-processing of the Sentinel-2 images was conducted on the Food Security TEP, the ESA-supported Thematic Exploitation Platform for agriculture and aquaculture (https://foodsecurity-tep.net/, accessed date: 5 January 2022). Its infrastructure allows direct access to all Sentinel-2 satellite imagery and many other datasets. Scalable resources for processing, data handling and storage help to overcome the challenges of big data for large-area EO applications. Within the ExtremeEarth project, a federation with the deep learning platform Hopsworks (https://www.hopsworks.ai, accessed date: 5 January 2022) is being established that allows the exploitation of deep learning approaches for agriculture using the Food Security TEP datasets and resources.

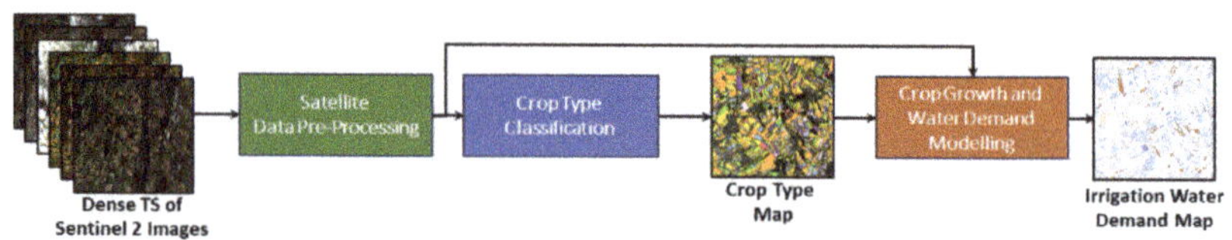

Figure 1. Block-scheme of the system architecture for automatic irrigation water demand mapping.

A dense time series of multi-spectral Sentinel-2 data were the input needed for crop type classification, water stress and water demand. Particularly for the deep learning model, the highest quality atmospheric corrections, including high accuracy cloud and cloud shadow masking in addition to cirrus correction, were necessary. This was offered by VISTA's image processing chains (VIAs), which were implemented as processors on the Food Security TEP. They comprised sophisticated methods for all necessary pre-processing steps in addition to a crop-type independent derivation of plant physiological parameters, so that a time series of high-quality atmospherically corrected multispectral Sentinel-2 data could be delivered for deep learning analysis. Finally, the time series (TSs) were harmonised from the temporal viewpoint by generating a TS of 12 monthly composites for each tile [1]. This regularised the TSs of images acquired on different tiles and mitigated the impact of cloud coverage.

2.2. Crop Type Classification

The goal of this step was to automatically produce crop type and crop boundary maps at country scale. To this end, the multitemporal long short-term memory (LSTM) deep learning model was trained by considering the TimeSen2Crop dataset reporting 16 crop types [2]. This publicly available dataset is made up of one million samples collected in Austria for the agronomic year (i.e., period from one year's harvest to the next one) ranging from September 2017 to August 2018. The LSTM deep learning model was trained to generate the crop type map of the whole Austrian cultivated area (~36,000 km^2). This multitemporal deep learning model was designed to handle sequential data by classifying the current samples while also exploiting previous ones. The hyper-parameters of the LSTM were selected according to the standard grid-search approach by testing all combinations of the number of network layers (2, 3, 4) and the number of cells per layer (100, 125, 200, 225, 300). The cross-entropy loss was evaluated at each training step, while the optimization problem was addressed by stochastic gradient descent (RMSProp optimiser). The initial learning rate and the weight decay were equal to 10^{-3} to 0.4, respectively. The distributed LSTM was implemented in Hopsworks.

2.3. Crop Growth and Water Demand Modelling with PROMET

Based on the crop type classification, a representative sample of pixels (355,000) was distributed over the study area with an overall area of 83.879 km^2. Samples were located at a certain distance from field boundaries or roads to guarantee pure crop-specific remote sensing information. For each crop, the development of the leaf area, represented by the LAI, was retrieved by inversion of Sentinel-2 time series, for 15 tiles covering 100×100 km each, using the radiative transfer model SLC [3]. The leaf area information was then assimilated into the crop growth model PROMET [4]. The physically based agro-hydrological model allowed the simulation of, for example, photosynthesis, evapotranspiration, soil moisture, biomass increase, phenological development, and crop water stress, for every sample in hourly temporal resolution.

Crop water stress was indicated as soon as photosynthesis was limited by water shortage and was displayed as a normalised index ranging between 0 (max. stress) and 1 (no stress). Yield in t/ha and water use efficiency (WUE) linked yield with water loss through the plant (total evapotranspiration) derived for each season [5]. The model was forced by meteorological parameters (precipitation, air temperature, humidity, radiation and wind speed) extracted from ERA5 reanalysis data (e.g., available from https://cds.climate.copernicus.eu/, accessed date: 5 January 2022).

3. Results

3.1. Simulation Results at Field Scale

The crop water demand for some fields along the Bavarian–Austrian borders with different crop types such as soybean, corn, winter barley and winter wheat was simulated with the crop growth model PROMET for the year 2018 (Figure 2). In 2018, the soybean fields had the largest crop water demand which was not satisfied by precipitation, while the other crop types had significantly less crop water demand. Crop type assignment is very important for correct quantification of crop-specific irrigation water demand, and it differs from year to year, thus, it is important to simulate it with meteorological data and Earth Observation information.

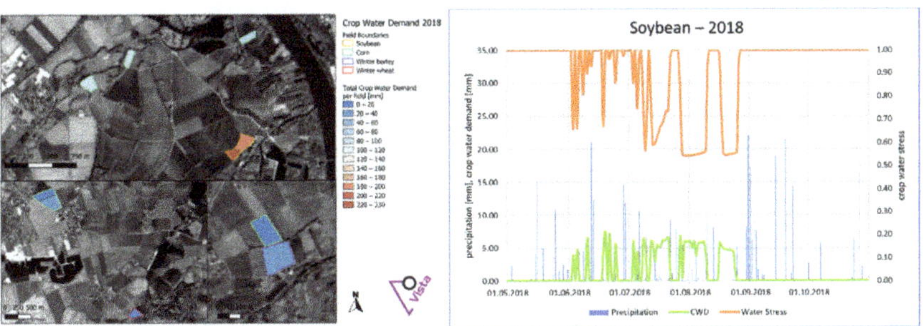

Figure 2. Water demand from different crops of a farm in Southern Bavaria and respective soil moisture and crop water stress time series of a soybean field in the depicted area in 2018.

3.2. Simulation Results on a National Scale for Austria

To upscale the approach to derive crop water demand for different crop types on a national scale, a validated crop type map and weather data from 2018 were used for NUTS 2 regions in Austria. Regional differences due to anomalies in spring and summer precipitation affected soil moisture development and crop water stress in the eastern parts of Austria. As a result, severe crop water stress occurred in May and June increasing the simulated irrigation water demand for various crops, especially soybean (Figure 3). Comparison of winter wheat and soybean clearly showed that differences in growing periods and phenology led to differing irrigation water requirements for each crop type, highlighting the need for accurate, high resolution crop type maps for this analysis.

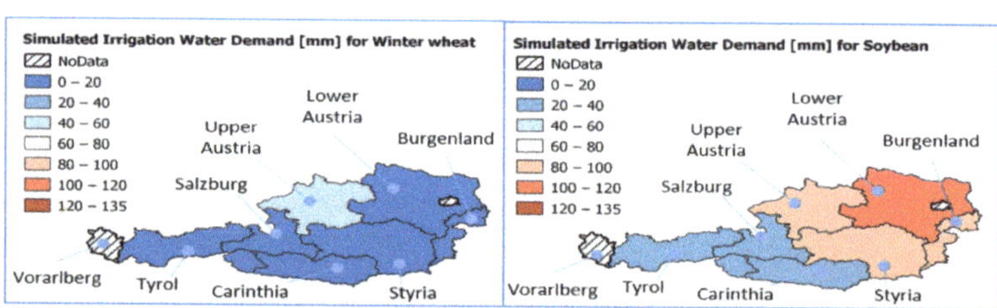

Figure 3. Irrigation water demand in mm for winter wheat and soybean for Austria in the year 2018.

4. Conclusions and Outlook

This work provides a small example of the capabilities of solution-driven methods for water demand calculations making use of Sentinel-2, crop type mapping and crop growth modelling. Using EO (big) data and DL to retrieve detailed crop type information and local phenological and biomass development, and feeding the information into the physical model, enabled the qualification and quantification of the water stress of a variety of crops at farm and regional level in Austria for the extreme year 2018. During ExtremeEarth, the presented methods will be extended to more seasons and regions. In combination with water availability calculations (integrating the hydrological cycle of the catchments) in addition to predictions of water availability in the summer months, these water demand calculations can lead to irrigation policy advice, suitable from local to regional level. EO-derived crop information has proven its potential to support and improve crop and water related challenges in future farming and food security management.

Supplementary Materials: The following are available online at https://www.mdpi.com/article/10.3390/engproc2021009042/s1.

Author Contributions: Conceptualization, S.M., C.P., F.A. and G.W.; methodology, S.M., F.A., M.M., C.P. and G.W.; software, F.A., M.M., G.W. and D.M.; validation, E.G., S.D., C.P., G.W., D.M.: visualisation, E.G. and S.D., C.P. and D.M.; formal analysis, E.G., S.D. investigation, E.G., S.D., C.P., G.W.; resources, C.P., G.W. and M.M.; data curation, C.P., G.W., D.M.; writing, S.M., M.M., C.P., D.M.; supervision, H.B., L.B., S.M. and C.P.; project administration, F.A. and L.B.; funding acquisition, S.M., F.A., H.B. and L.B. All authors have read and agreed to the published version of the manuscript.

Funding: This research was funded by the European Union's Horizon 2020 research and innovation program under grant agreement No. 825258.

Institutional Review Board Statement: Not applicable.

Informed Consent Statement: Not applicable.

Data Availability Statement: The crop classification data is available in a publicly accessible repository. It can be found here: https://rslab.disi.unitn.it/timesen2crop/, accessed date: 5 January 2022.

Acknowledgments: Thank you to Wolfgang Angermair, whose fields have been analysed in this study.

Conflicts of Interest: The authors declare no conflict of interest.

References

1. Paris, C.; Weikmann, G.; Bruzzone, L. Monitoring of agricultural areas by using Sentinel 2 image time series and deep learning techniques. In Proceedings of the SPIE 2020, Image and Signal Processing for Remote Sensing XXVI, Online Only, 21–25 September 2020; p. 115330K. [CrossRef]
2. Weikmann, G.; Paris, C.; Bruzzone, L. TimeSen2Crop: A Million Labeled Samples Dataset of Sentinel 2 Image Time Series for Crop-Type Classification. *IEEE J. Sel. Top. Appl. Earth Obs. Remote Sens.* **2021**, *14*, 4699–4708. [CrossRef]
3. Verhoef, W.; Bach, H. Simulation of hyperspectral and directional radiance images using coupled biophysical and atmospheric radiative transfer models. *Remote Sens. Environ.* **2003**, *87*, 23–41. [CrossRef]

4. Mauser, W.; Bach, H. PROMET—Large scale distributed hydrological modelling to study the impact of climate change on the water flows of mountain watersheds. *J. Hydrol.* **2009**, *376*, 362–377. [CrossRef]
5. Probst, E.; Klug, P.; Mauser, W.; Dogaru, D.; Hank, T. Water Use Efficiency of selected crops in the Romanian Plain—Model studies using Sentinel-2 Satellite Images. *Sci. Pap. Ser. E Land Reclam. Earth Obs. Surv. Environ. Eng.* **2018**, *7*, 198–208.

Proceeding Paper

A Smart Farming System for Circular Agriculture [†]

Aristotelis C. Tagarakis [1,*], Christos Dordas [2], Maria Lampridi [1], Dimitrios Kateris [1] and Dionysis Bochtis [1]

[1]   Centre for Research and Technology-Hellas (CERTH), Institute for Bio-Economy and Agri-Technology (IBO), 6th km Charilaou-Thermi Rd., GR 57001 Thessaloniki, Greece; m.lampridi@certh.gr (M.L.); d.kateris@certh.gr (D.K.); d.bochtis@certh.gr (D.B.)

[2]   School of Agriculture, Aristotle University of Thessaloniki, 54124 Thessaloniki, Greece; chdordas@agro.auth.gr

*   Correspondence: a.tagarakis@certh.gr

†   Presented at the 13th EFITA International Conference, online, 25–26 May 2021.

Abstract: Agricultural activities such as fertilization and other crop management techniques contribute to greenhouse gas emissions and pollution; the cost of such activities is considerably high, with nitrogen pollution costing the EU up to 320 billion euros annually. In the present study, an integrated system has been proposed which utilizes smart farming tools and smart processing methodologies following the concept of a circular economy to reduce the impact of agricultural activity on climate change. Circular agriculture and precision farming together with the use of appropriate crop management tools may contribute to better resource use efficiency and sustainable agriculture.

Keywords: bioeconomy; circular agriculture; precision farming; smart farming; fertilization; resource use efficiency; sustainability

Citation: Tagarakis, A.C.; Dordas, C.; Lampridi, M.; Kateris, D.; Bochtis, D. A Smart Farming System for Circular Agriculture. *Eng. Proc.* **2021**, *9*, 10. https://doi.org/10.3390/engproc 2021009010

Academic Editors: Charisios Achillas and Dimitios Aidonis

Published: 24 November 2021

1. Introduction

Climate change is a major threat for the natural environment, and there is an imperative need to reduce greenhouse gas emissions. Agriculture contributes for about 10% of the greenhouse gas emissions and up to 95% of the emissions of ammonia in the European Union [1]. In order to reduce greenhouse gas emissions, new management approaches, such as circular agriculture, should be adapted, and new technologies that can minimize inputs for crop production must be developed and utilized. Circular agriculture is a modern agricultural management concept that promotes the reuse of all resources that can be used by the production system itself [2–4]. The utilization of such systems directly leads to the minimization of agricultural waste, while their application can reduce nutrient and water losses and emissions such as greenhouse gasses [4].

The latest advances in sensing and communication systems in agriculture have paved the way towards developing and following more efficient management schemes that account for the temporal and spatial variability of field and crop properties for timely and accurate field applications [5]. To be able to take advantage of these advanced sensing and management tools, the development of integrated systems with user-friendly interfaces is a prerequisite. As such, precision farming technologies can be utilized in conjunction with Internet of Things (IoT) sensing systems, in both crop and animal production processes, to feed the integrated system with real-time data [4,6].

In this study, a holistic system for circular agriculture is proposed (BIOCIRCULAR) to be applied at small–medium vertical production dairy farms. Part of the system involves integrated soil fertility management which combines the use of appropriate amounts of organic and inorganic fertilizers together with green manure and good agronomic practices [4,6,7]. Preliminary results of the crop fertilization study for crops grown for feed (*Zea mays* L.), as part of the crop production sub-system, are also discussed.

2. System Requirements and Design

The objective of the present study was to develop a precision field management and smart farming tool applied at every stage of the production chain. In order to properly design and develop such a tool, the system and user requirements were initially set. To that end, the procedures of the typical production dairy farms were divided into three main processes and analyzed, defining the inputs and outputs for each process step (Figure 1).

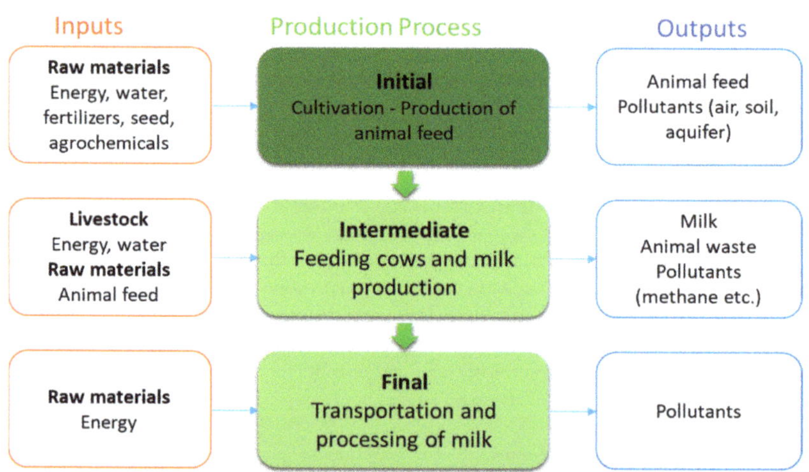

Figure 1. Production chain processes of small–medium vertical production dairy farms.

The system includes all procedures throughout the production chain and was therefore broken down into the basic sub-systems that are connected and interacting with a central circular precision management information system, which is the heart of the circular farm ecosystem (Figure 2). The sub-systems include: (a) crop production, (b) feed production, (c) animal production, and (d) animal waste management. Within the animal waste sub-system procedures, energy and organic fertilizer production are also included as biproducts to be used by the system itself (circular management). IoT and precision farming systems are involved in each sub-system to monitor and record the parameters needed for informed management.

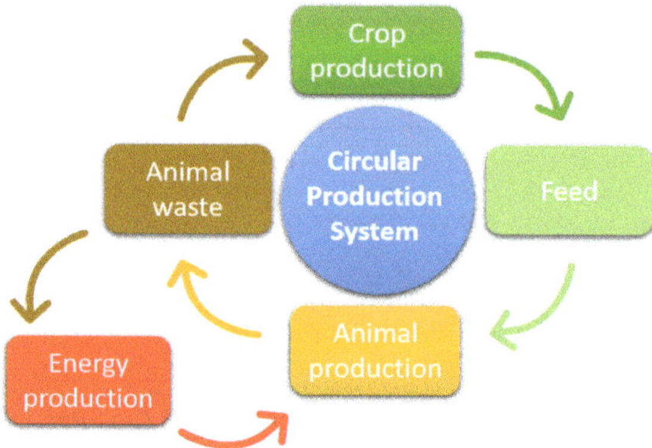

Figure 2. Circular production system.

3. The Platform

The circular production system can receive and process all the information from each subsystem installed at every production level in the farm and provides the results in a simple and manageable form, supporting users to make appropriate decisions (Figure 3). Examples of input data include: (a) for the crop production sub-system—weather and soil moisture data from IoT, satellite and drone images, and crop and soil properties mapping proximal sensor data; (b) for the feed production sub-system—silage moisture and temperature data from IoT; (c) for the animal production sub-system—animal housing environmental conditions (temperature, humidity, CO_2 and methane concentrations, etc.) monitored using IoT sensing systems, animal health status (temperature, activity, and behavior) using wireless IoT-enabled collars; (d) for the animal waste treatment sub-system—waste volume, waste water content and temperature, produced volume of biogas (if equipped with biogas production facilities), using IoT systems.

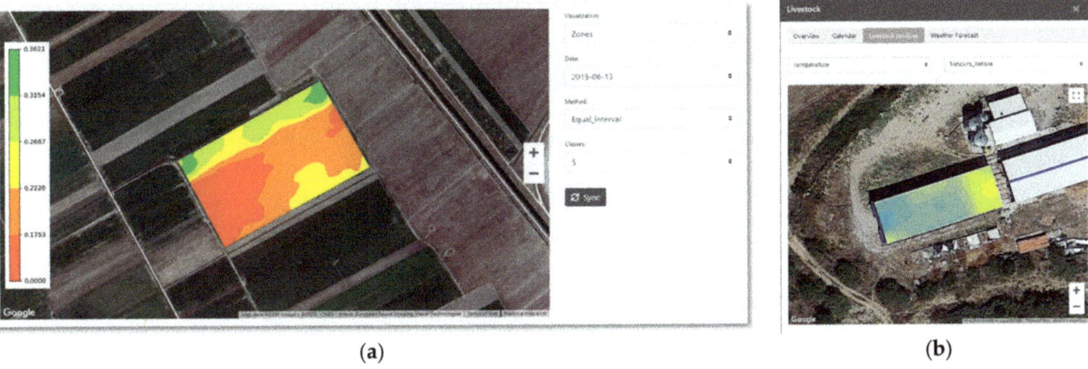

(a)   (b)

Figure 3. Some of the basic functions of the platform; (**a**) automated mapping of within-field crop properties and (**b**) mapping of environmental conditions inside the animal housing facilities.

4. Crop Production Experiments

Crop production is the initial production stage in a circular dairy system and consists of the first sub-system in BIOCIRCULAR. Within the scope of the project, field experiments were setup to study and develop management practices that may lead to increased production efficiency by minimizing the environmental footprint. In this work, the yield response of maize crop (*Zea mays* L.), grown for forage, to the application of different types of fertilizer (organic—manure and green manure; chemical fertilizers) was also investigated. Two maize hybrids were included in the study, namely Dekalb6777 and Pioneer1291. Five different fertilization treatments were tested: green manure (organic), manure (organic), conventional fertilizer (synthetic), slow-release fertilizer (synthetic), and control (no fertilizer applied).

The results showed that cattle manure can provide the nutrients in an available form that are essential for plant growth, while enhancing soil water-retention properties, leading to increased productivity of maize (Figure 4). Additionally, green manure with common vetch can provide adequate amounts of nitrogen, which is very important for maize. Therefore, efficient management of natural resources together with integrative crop management approaches can reduce inputs in crop production systems.

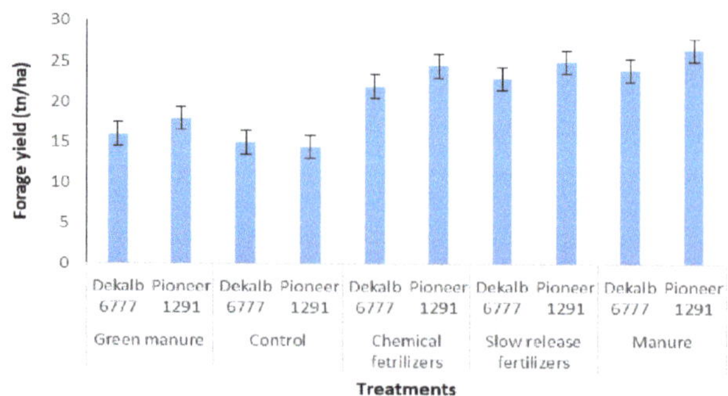

Figure 4. Forage yield of the different treatments that were tested with two maize hybrids during the 2019 growing season.

5. Conclusions

The benefits of using circular farming systems and digital technology are numerous. Implementing circular precision farming approaches may lead to sustainable development in dairy production, increasing efficiency, reducing inputs, and minimizing waste. More specifically, these systems enable monitoring of processes in the production stages, facilitating traceability towards producing certified products. In addition, the use of precision farming can increase the efficiency of the resources' utilization, by appropriately managing the spatial variability of fields. Furthermore, by supporting farmers in making optimal management decisions, it can increase the use efficiency of inputs, leading to their reduction, thus providing a financial benefit for the producers. Moreover, the circular characteristics of these systems minimize the production of animal waste and diminishes losses of nutrients and agrochemicals, thus reducing the environmental footprint.

The system presented in this study aims to constitute a dynamic, intelligent decision support system for circular agriculture that combines information from each available source throughout the whole production chain and supports farmers and farm managers in taking informed and timely decisions based on real-time data. Furthermore, the preliminary results from the crop fertilization experiment showed that using cattle manure as the sole source of fertilizer may lead to increased yields as compared to synthetic fertilizers. This result confirms the perspective of using animal waste in circular dairy production systems for increased sustainability.

Author Contributions: Conceptualization, A.C.T. and C.D.; methodology, A.C.T., C.D. and M.L.; validation, A.C.T., D.K. and D.B.; formal analysis, A.C.T. and C.D.; investigation, A.C.T., C.D. and M.L.; resources, A.C.T., C.D. and D.B.; data curation, A.C.T. and C.D.; writing—original draft preparation, A.C.T., C.D. and M.L.; writing—review and editing, A.C.T., D.K. and D.B.; visualization, A.C.T.; supervision, A.C.T., C.D. and D.B.; project administration, A.C.T., C.D. and D.B.; funding acquisition, A.C.T., C.D. and D.B. All authors have read and agreed to the published version of the manuscript.

Acknowledgments: This research has been supported by the European Union and Greek national funds through the Operational Programme Competitiveness, Entrepreneurship and Innovation, under the call RESEARCH–CREATE–INNOVATE (project code: T1EDK-03987) "BioCircular: Bioproduction System for Circular Precision Farming".

Conflicts of Interest: The authors declare no conflict of interest.

References

1. *Eurostat Pocketbooks Agriculture, Forestry and Fishery Statistics*; European Union: Luxembourg, 2013; ISBN 9789279330056.
2. von Braun, J. Bioeconomy—The global trend and its implications for sustainability and food security. *Glob. Food Sec.* **2018**, *19*, 81–83. [CrossRef]

3. Kircher, M. Bioeconomy—Present status and future needs of industrial value chains. *New Biotechnol.* **2021**, *60*, 96–104. [CrossRef] [PubMed]
4. Rodias, E.; Aivazidou, E.; Achillas, C.; Aidonis, D.; Bochtis, D. Water-Energy-Nutrients Synergies in the Agrifood Sector: A Circular Economy Framework. *Energies* **2020**, *14*, 159. [CrossRef]
5. Anagnostis, A.; Tagarakis, A.C.; Kateris, D.; Moysiadis, V.; Sørensen, C.G.; Pearson, S.; Bochtis, D. Orchard mapping with deep learning semantic segmentation. *Sensors* **2021**, *21*, 3813. [CrossRef] [PubMed]
6. Lampridi, M.G.; Sørensen, C.G.; Bochtis, D. Agricultural Sustainability: A Review of Concepts and Methods. *Sustainability* **2019**, *11*, 5120. [CrossRef]
7. Dordas, C.A.; Lithourgidis, A.S.; Matsi, T.; Barbayiannis, N. Application of liquid cattle manure and inorganic fertilizers affect dry matter, nitrogen accumulation, and partitioning in maize. *Nutr. Cycl. Agroecosyst.* **2007**, *80*, 283–296. [CrossRef]

Proceeding Paper

Modelling Digital Circular Economy framework in the Agricultural Sector. An Application in Southern Italy [†]

Tiziana Crovella [1,*], Annarita Paiano [1], Giovanni Lagioia [1], Anna Maria Cilardi [2] and Luigi Trotta [2]

[1] Department of Economics, Management and Business Law (DEMDI), University of Bari Aldo Moro, 70121 Bari, Italy; annarita.paiano@uniba.it (A.P.); giovanni.lagioia@uniba.it (G.L.)

[2] Department of Agriculture, Rural and Environmental Development—Section Competitiveness of Agri-Food Chains, Regione Puglia, 70121 Bari, Italy; a.cilardi@regione.puglia.it (A.M.C.); l.trotta@regione.puglia.it (L.T.)

* Correspondence: tiziana.crovella@uniba.it; Tel.: +39-320-01-88148

† Presented at the 13th EFITA International Conference, online, 25–26 May 2021.

Abstract: The transition towards circular economy (CE) in agriculture requires a large amount of data in order to map the consumption of natural resources and negative externalities. This paper aims to identify a digital framework for collecting and sharing data fundamental for stakeholders with the purpose of implementing the best CE model. The methodology used is based on the guidelines of the stakeholder engagement and through a survey, and the authors have mapped the lack of data and built a set by replicable sustainability indicators. The results obtained can be used for the definition of regional policy strategies and interventions for CE model implementation.

Keywords: agriculture; circular economy; survey; data; indicators; digital framework

Citation: Crovella, T.; Paiano, A.; Lagioia, G.; Cilardi, A.M.; Trotta, L. Modelling Digital Circular Economy framework in the Agricultural Sector. An Application in Southern Italy. *Eng. Proc.* **2021**, *9*, 15. https://doi.org/10.3390/engproc2021009015

Academic Editors: Dimitrios Aidonis and Aristotelis Christos Tagarakis

Published: 24 November 2021

Publisher's Note: MDPI stays neutral with regard to jurisdictional claims in published maps and institutional affiliations.

1. Introduction

Digital transformation plays a fundamental role in all economic sectors, especially in agriculture [1]. For this reason, currently, the paradigm of data collection needs to be improved, as suggested by [2], in order to present a systematic model to enable stakeholders to transition towards CE.

2. Literature Review

The scientific interest in the fundamental digital transformation to implement the CE in agriculture is fairly recent, as the Web of Science (WoS) platform highlighted only two a few results. According to [2], the correct coordination between computer devices and agriculture can contribute to improvement in the management of farms and their practices. In 2019, other scholars [3] stated that a long-term agricultural data collection model ensured comprehensive monitoring of sustainable agriculture and the design of development plans. Given the shortage of this significant practice [4], we proposed this study in order to increase the scientific production and suggest a new framework for implementing CE in agriculture.

3. Materials and Methods

3.1. Methods

The methodology used is defined by the guidelines on stakeholder engagement (AA1000 SES) issued by [5] and based on three key items: 1. involving stakeholders to understand their expectations on governance, policies, strategies, and practices; 2. reporting of data and questions in accordance with transparency and clarity; 3. developing an innovative framework to address the sustainability issue.

3.2. Materials

The stakeholder engagement methodology is based on questionnaire administration to understand the needs identified by stakeholders and farmers during planning and managing project activities concerning the sustainability and circular economy issue. Data gathered through the questionnaire were processed and transformed through a statistical tool into a shareable framework.

3.2.1. Questionnaire

We developed a questionnaire jointly for with public administration for mapping the needs of stakeholders in terms of information and knowledge to implement CE. Table 1 displays the 21 questions, which are categorized into four areas of investigation.

Table 1. Structure of questionnaire.

Area	Questions	*n.*
General	Professional area General difficulties for farmers to implement CE Information for implementing SA	1–3
Circular Economy	CE practices in Apulia Farmers' interest in CE CE model in Apulian agriculture Obstacles in CE implementation Range of problems Fundamental keys for CE implementation Effective processes Sustainability indicators	4–11
Water Resource	Wastewater reuse in Apulian agriculture Practices of wastewater reuse Spreading olive oil mill wastewater (OOMW) Information of Regional Water Protection Plan Instruments and tools for improving CE of water resource	12–16
Planning	Relevant topic for planning and financing calls Digital data and information Collaboration for designing digital framework Advice Sharing results	17–21

3.2.2. Sample

Jointly with a regional body, we made a list of stakeholders involved in the survey. The sample was composed of 32 stakeholders from different areas that operate in private associations, public authorities, research centers, and regional agencies connected with the Apulian agricultural sector. The questionnaire was shared through a digital platform.

4. Results

Concerning question 2, half of the answers showed an absence of direct channels of information for CE in the Apulian agricultural sector and 40% complained of the lack of technical training. Two thirds of the answers for question 3 stated that Apulian farmers require information about sustainable use of natural resources. Additionally, according to 75% of the sample investigated, in the Apulian agriculture sector, some models of CE have been activated anyway. Thirty-three percent of these stakeholders were contacted about wastewater reuse in the past. All stakeholders agreed that regional farmers face difficulties in implementing an appropriate and useful model of CE, mostly (67%) due to a lack of information and specific training. According to our stakeholder sample, sustainable water use will be an important key to implement CE in Apulia, and wastewater reuse and food waste reuse followed. Notwithstanding the successful framework to implement CE will be the adoption of participatory processes between stakeholders and farmers. In terms of sustainable indicators, 67% of the sample were aware of them, whereas, according to 16% of the stakeholders, in the Apulian agriculture sector, wastewater reuse based on rainwater storage is widespread. Thirty-three percent of the sample has already had to deal with farmers who practised an agronomic use of vegetation waters. Unfortunately, no farmers know the Regional Water Protection Plan. Fifty-eight percent of the sample focused on the need for coordination between the different implementing bodies. Additionally, for planning and financing activities, the management of digital information and the use of a connected network will be fundamental. Currently, according to 92% of the sample, Apulian farmers have difficulties in retrieving collected and digital information in the short term. Nevertheless, 67% of the sample is available to collaborate for planning a participatory model for collected information useful for implementing CE in Apulia. It has to be underlined that all stakeholders agreed to share the results of this survey.

Finally, in the last step of this study, we developed some indicators for making a framework useful for all stakeholders involved in the agri-food chain (from farmers to public and private bodies) for implementing a successful CE model in agriculture. Figure 1 displays a replicable framework organized into the eight most important topics that are interconnected and fundamental to meet the needs highlighted by our sample.

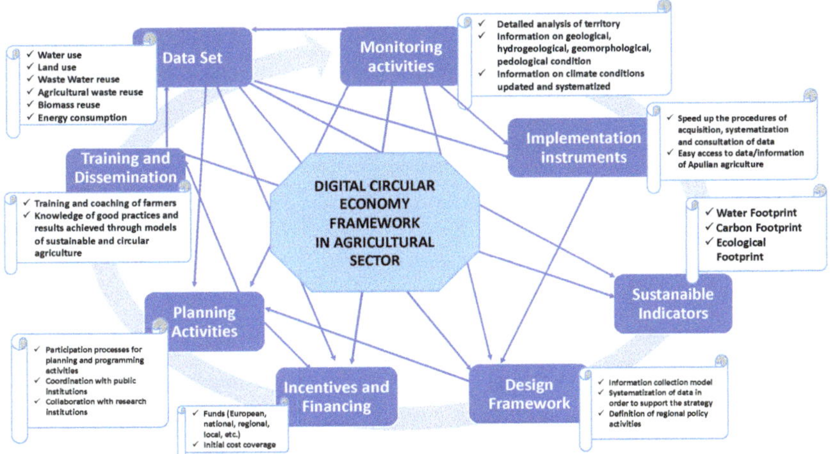

Figure 1. Framework for implementing a CE model in agriculture. Source: Authors' own work.

5. Discussion

This paper deals with some useful insights from the scientific literature in order to create a digital framework for data collection and sharing in agriculture. Furthermore, this analysis feeds responsible innovation processes for a correct transition from the linear economy to circular economy [1]. The investigation, undertaken through the use of digital tools, encompassed the consumption of natural resources, such as water resources, and the knowledge of some sustainability indicators and limitations to implementing CE in agriculture. This information is significant for planning activities, creating financing policies, and enabling the implementation of tools for Apulian agriculture. Therefore, based on the first questionnaire administered to the sample (public and private bodies and associations of agriculture), the need to conduct training and dissemination activities based on bottom-up and top-down approaches emerged. The key element for the transition to the CE is represented by the collaboration between farmers and stakeholders to design agricultural management based on a digital model of data, information, and knowledge.

6. Conclusions

We proposed this framework based on a dynamic and interdependent relationship between several fundamental steps to design a transition towards a CE approach. This digital framework improves stakeholders' decision making [6] and leads to greater cooperation in the agricultural supply chain in Southern Italy and in the Mediterranean area. It is a knowledge model that allows for overcoming the obstacles that the stakeholders of the agricultural supply chain face in the data procurement phase [7]. Furthermore, since this is the building of a replicable framework, it allows the spread of soft culture and digital learning, creating a virtuous network to be implemented in the Mediterranean area. We have thus strengthened the capacity of public administrations by combining scientific research and public/private interests for the implementation of the circular economy approach. Therefore, this new framework can spin public administration to design plans, strategies, and programs and enable stakeholders to know the most suitable sustainability indicators to implement the best CE model. In conclusion, digital transformation in agriculture supports a framework of digital-enabled circular strategies for farms and manufacturing.

Author Contributions: Conceptualization, T.C. and A.P.; methodology, T.C. and A.P.; software, T.C.; validation, A.M.C. and L.T.; formal analysis, T.C. and A.P.; investigation, T.C.; resources, A.M.C.; data curation, T.C.; writing—original draft preparation, T.C. and A.P.; writing—review, G.L., writing—editing, T.C.; visualization, T.C.; supervision, G.L. and L.T.; project administration, G.L. All authors have read and agreed to the published version of the manuscript.

Acknowledgments: This research was undertaken thanks to a partnership between the University of Bari Aldo Moro (Department of Economics, Management and Business Law), Regione Puglia (Department of Agriculture, Rural Development and Environmental), and ADISU Puglia for the MoDEC Apulia Programme 13/16.

Conflicts of Interest: The authors declare no conflict of interest.

References

1. Lajoie-O'Malley, A.; Bronson, K.; van der Burg, S.; Klerkx, L. The future(s) of digital agriculture and sustainable food system: An analysis of high-level policy documents. *Ecosyst. Serv.* **2020**, *45*, 101183. [CrossRef]
2. Chehri, A.; Chaibi, H.; Saadane, R.; Hakem, N.; Wahbi, M. A framework of optimizing the development of IoT for precision agriculture industry. 24th International Conference of Knowledge-Based and Intelligent Information & Engineering Systems. *Proc. Comp. Sci.* **2020**, *176*, 2414–2422. [CrossRef]
3. Della Chiesa, S.; la Cecilia, D.; Genova, G.; Balotti, A.; Thalheimer, M.; Tappeiner, U.; Niedrist, G. Farmers as data sources: Cooperative framework for mapping soil properties for permanent crops in South Tyrol (Northern Italy). *Geoderma* **2019**, *342*, 93–105. [CrossRef]
4. Klerkx, L.; Jakku, E.; Labarthe, P. A review of social science on digital agriculture, smart farming and agriculture 4.0: New contributions and a future research agenda. *Wagening. J. Life Sci.* **2019**, *90*, 100315. [CrossRef]
5. AccountAbility. Stakeholder Engagement Standard (AA1000SES). Available online: https://www.accountability.org/standards/ (accessed on 10 January 2021).
6. Newton, J.E.; Nettle, R.; Pryce, J.E. Farming smarter with big data: Insights from the case of Australia's national dairy herd milk recording scheme. *Agric. Syst.* **2020**, *181*, 102811. [CrossRef]
7. Wolfert, S.; Ge, L.; Verdouwa, C.; Bogaardt, M.J. Big Data in Smart Farming—A review. *Agric. Syst.* **2017**, *153*, 69–80. [CrossRef]

Proceeding Paper

A Theoretical Framework for Multi-Hazard Risk Mapping on Agricultural Areas Considering Artificial Intelligence, IoT, and Climate Change Scenarios [†]

Roberto F. Silva [1,*], Maria C. Fava [2], Antonio M. Saraiva [3], Eduardo M. Mendiondo [4], Carlos E. Cugnasca [3] and Alexandre C. B. Delbem [5]

1   Institute of Advanced Studies (IEA), University of São Paulo (USP), Sao Paulo 05508-060, Brazil
2   Institute of Exact and Technological Sciences (IEP), Federal University of Viçosa (UFV), Rio Paranaiba 38810-000, Brazil; maria.fava@ufv.br
3   Polytechnic School, University of São Paulo (USP), Sao Paulo 05508-010, Brazil; saraiva@usp.br (A.M.S.); carlos.cugnasca@usp.br (C.E.C.)
4   São Carlos School of Engineering (EESC), University of São Paulo (USP), Sao Carlos 13566-590, Brazil; emm@sc.usp.br
5   Institute of Mathematics and Computer Sciences (ICMC), University of São Paulo (USP), Sao Carlos 13566-590; Brazil; acbd@icmc.usp.br
*   Correspondence: roberto.fray.silva@gmail.com
†   Presented at the 13th EFITA International Conference, online, 25–26 May 2021.

Abstract: This work proposes a data-driven theoretical framework for addressing: (i) extreme climate events prediction through multi-hazard risk mapping using remote sensing, artificial intelligence, and hydrological models, considering multiple hazards; and (ii) environmental monitoring using on-site data collection and IoT technologies. The framework considers the possibility of evaluating multiple climate change scenarios for improving decision-making in terms of Government policies and farm planning. Its main requirements are gathered based on a literature review. Several essential metrics that can be evaluated, considering both supervised and unsupervised metrics and key performance indicators considering the triple bottom line aspects, are also proposed. The framework also adopts multi-hazard (considering several hazards) and multi-risk (considering several relevant stakeholders) aspects and can be used to simulate different scenarios, an essential task for improving decision-making.

Keywords: artific ial intelligence; climate change; environmental monitoring; IoT technologies; multi-hazard risk mapping

Citation: Silva, R.F.; Fava, M.C.; Saraiva, A.M.; Mendiondo, E.M.; Cugnasca, C.E.; Delbem, A.C.B. A Theoretical Framework for Multi-Hazard Risk Mapping on Agricultural Areas Considering Artificial Intelligence, IoT, and Climate Change Scenarios. *Eng. Proc.* **2021**, *9*, 39. https://doi.org/10.3390/engproc2021009039

Academic Editors: Dimitrios Kateris and Maria Lampridi

Published: 31 December 2021

Publisher's Note: MDPI stays neutral with regard to jurisdictional claims in published maps and institutional affil-iations.

1. Introduction

A critical impact of global warming was the increase in the occurrence and impact of extreme climate events, such as floods, droughts, heatwaves, cyclones, among others [1–5]. In this work, these are referred to as hazards. The major impacts of those hazards in agricultural areas are: crop losses, crop quality losses, environmental damage, soil nutrients loss, economic impacts, impacts on machinery and buildings, and social impacts [1–5]. Therefore, it is possible to observe that the three pillars of sustainability (economic, environmental, and social) are impacted by those hazards [4,5].

As observed in the surveys by [6–8], most of the works on the impacts of natural hazards present the following aspects: (i) evaluation of the potential impact of only one hazard; (ii) focus on urban environments; and (iii) focus on specific periods, not considering real-time monitoring. Additionally, it can be observed that only a few models in the litera-ture consider the use of Internet of Things (IoT) technologies for real-time monitoring of the occurrence of those hazards in agricultural environments. Lastly, it is crucial to observe that most works in the literature focus only on the economic impacts of climate change. These

Eng. Proc. **2021**, *9*, 39. https://doi.org/10.3390/engproc2021009039

models and frameworks do not explicitly consider the potential environmental damages and social effects of those hazards. The works by [6–8] are very relevant for multi-hazard risk analysis, a task that can be characterized as predicting and evaluating the potential impact of different types of hazards at a specific location [6–8]. This task is also referred to as event forecasting. However, those works do not consider the use of IoT technologies for real-time environmental monitoring, also referred to as ecosystem monitoring.

This work proposes a data-driven framework for multi-hazard risk mapping with a modular approach to address the following aspects: (i) domain-specific requirements of the agricultural domain; (ii) the three pillars of sustainability; (iii) both event prediction and ecosystem monitoring tasks; (iv) use of hazard-specific and general climate heterogeneous data; (v) use of hydrological and artificial intelligence (AI) models; (vi) considering different climate change scenarios; and (vii) generation of risk maps that can be used by stakeholders for decision-making. It is important to note that the results of the framework implementation can be helpful for several supply chain actors, such as the farmers, banks, traders, industries, logistics operations providers, and Government agencies. In addition, this work presents the main requirements gathered considering the two most critical hazards for sustainable agriculture (droughts and floods). Finally, it is vital to observe that the framework proposed considers the possibility of stakeholder interaction on several components to provide essential feedback loops and improve the models' predictions and the risk maps generated.

2. Methodology

The methodology used in this paper was composed of 4 main steps:

1. Requirements gathering, considering an in-depth review of the literature to identify the main functional, non-functional, and domain-specific requirements for developing a multi-hazard risk framework for agricultural areas; 2. Identification of state-of-the-art models in the literature, focusing on multi-hazard risks; 3. Evaluation of the selected models, considering fulfillment of the identified requirements; 4. Framework proposal, considering the following aspects: models used, main inputs, data lifecycle stages, different climate change scenarios, a focus on agricultural areas and problems, and considering heterogeneous data, event prediction and ecosystem monitoring tasks.

3. Identified Requirements for Developing a Multi-Hazard Risk Framework for Agricultural Areas

It is important to note that most of the requirements identified in the literature are specific to the agricultural domain. This justifies the development of a framework for multi-hazard risk mapping specific to agricultural areas and their potential problems related to extreme climate events. The nine main requirements identified, based on the works by [4–6,8–11], were:

1. Considering multiple hazards and stakeholders; 2. Incorporating data from multiple sources (news, images, sensors, and stakeholders participation); 3. Predicting events and monitoring the environment in real-time; 4. Considering different climate change scenarios; 5. Allowing for probabilistic predictions due to the stochastic nature of the extreme events; 6. Allowing for multiple temporal and spatial scales, what is essential for the different stakeholders in the agro-industrial supply chains; 7. Identifying different crop areas and soil exposure, what impacts directly on the resilience of those areas in relation to the extreme climate events; 8. Evaluating relevant socio-economic aspects, which are essential for evaluating the sustainability of the production system; 9. Being easily accessible for decision-makers by providing maps, visualizations, and dashboards and allowing for scenario evaluation.

4. Proposed Framework and Data Sources

Figure 1 illustrates the proposed framework. It is comprised of four main components:

1. Event prediction, which aims to predict the probability and potential impact of specific extreme climate events on specific areas. It considers both multiple temporal and spatial scales due to the use of spatial-temporal clustering methods. It encompasses the following tasks: (i) heterogeneous data collection, considering specific hazard-related news, weather data, satellite images, official socioeconomic data, and climate change simulations; (ii) data preprocessing and storage, considering specific tasks related to each data type; (iii) feature engineering, generating features such as drought and flood indexes that may help on extracting relevant information from spatial-temporal historical data; and (iv) implementation of hydrological, meteorological, and clustering models. These data will then allow for the identification of crop and exposed soil areas, socioeconomic aspects, extreme event probability and potential impact, multi-hazard trends, and the prediction of the impacts of climate change scenarios.

2. Use of IoT technologies for real-time environmental monitoring, using both data from computer simulations based on data from weather stations and wireless sensor networks installed on the agricultural areas.

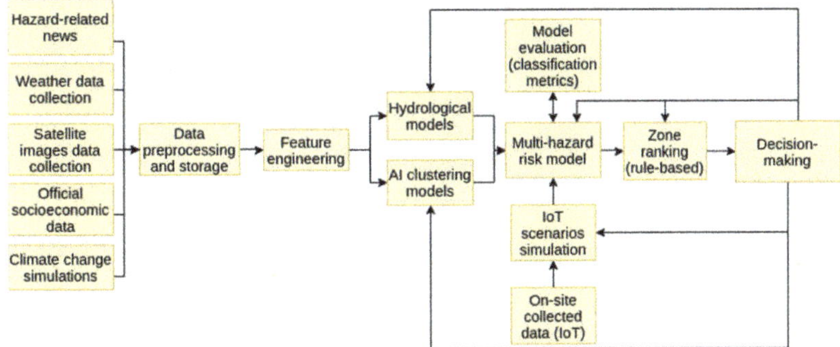

Figure 1. Proposed framework for multi-hazard risk mapping of agricultural areas.

3. Multi-hazard risk mapping, considering as inputs the results of the models implemented on component one and of the simulations on component two. This component calculates a risk index for each agricultural area using pixel-based data. Additionally, several AI models, such as deep learning models with different hyperparameters, can simulate different scenarios. These are then evaluated using: (i) traditional classification metrics, such as precision, recall, and F1-score; and (ii) key performance indicators related to economics, environmental, and social relevant aspects. Lastly, a rule-based model is used to aggregate the pixels into zones and to rank the different zones according to vulnerability criteria.

4. Decision-making, in which the final results of the simulations of component three are presented, along with their associated quality metrics, in different formats (dashboards, visualizations, maps, tables, among others that are relevant for the specific stakeholders). This component also considers several essential feedback loops, in which the stakeholders can provide additional input or change the hyperparameters used by the models in all previous components. This stakeholder feedback loop is essential for improving the quality of the information provided for decision-making and incorporating knowledge derived from external sources that the models do not consider.

5. Conclusions and Future Works

The expected increase in extreme climate events will significantly impact agricultural areas and agro-industrial supply chains. In this work, we proposed a data-driven theoretical framework for multi-hazard risk mapping, considering extreme climate events prediction and environmental monitoring using multiple models and heterogeneous data sources. The nine main requirements were presented. The framework proposed fulfilled all requirements identified and considered several aspects that improve the current models in the literature by considering the three pillars of sustainability, both event prediction and ecosystem monitoring tasks, using hazard-specific and general climate heterogeneous data, and considering different climate change scenarios. Its use could provide relevant information for different decision-makers on the supply chains, from farmers to distributors and Government agencies. Future works are related to: framework implementation and its evaluation on several case studies on different areas, crops, and climate change scenarios, and a stakeholder evaluation of its applicability.

Author Contributions: Conceptualization, R.F.S.; methodology, R.F.S. and M.C.F.; investigation, R.F.S. and M.C.F.; writing—original draft preparation, R.F.S.; writing—review and editing, M.C.F., A.M.S., E.M.M., C.E.C. and A.C.B.D.; supervision, A.M.S., E.M.M. and A.C.B.D. All authors have read and agreed to the published version of the manuscript.

Funding: This work was carried out at the Center for Artificial Intelligence (C4AI-USP), with support by: the São Paulo Research Foundation (FAPESP grant #2019/07665-4), IBM Corporation, Itaú Unibanco S.A., through the Centro de Ciência de Dados (C2D-USP), and the National Council for Scientific and Technological Development (CNPq).

Institutional Review Board Statement: Not applicable.

Informed Consent Statement: Not applicable.

Conflicts of Interest: The authors declare no conflict of interest.

References

1. Liu, X.; Zhu, X.; Pan, Y.; Li, S.; Liu, Y.; Ma, Y. Agricultural drought monitoring: Progress, challenges, and prospects. *J. Geogr. Sci.* **2016**, *26*, 750–767. [CrossRef]
2. Rasmussen, L.V.; Bierbaum, R.; Oldekop, J.A.; Agrawal, A. Bridging the practitioner-researcher divide: Indicators to track environmental, economic, and sociocultural sustainability of agricultural commodity production. *Glob. Environ. Chang.* **2017**, *42*, 33–46. [CrossRef]
3. Jung, J.; Maeda, M.; Chang, A.; Bhandari, M.; Ashapure, A.; Landivar-Bowles, J. The potential of remote sensing and artificial intelligence as tools to improve the resilience of agriculture production systems. *Curr. Opin. Biotechnol.* **2021**, *70*, 15–22. [CrossRef] [PubMed]
4. Cogato, A.; Meggio, F.; Migliorati, M.A.; Marinello, F. Extreme weather events in agriculture: A systematic review. *Sustainability* **2019**, *11*, 2547. [CrossRef]
5. Mirza, M.M.Q. Climate change and extreme weather events: Can developing countries adapt? *Clim. Policy* **2003**, *3*, 233–248. [CrossRef]
6. Gallina, V.; Torresan, S.; Critto, A.; Sperotto, A.; Glade, T.; Marcomini, A. A review of multi-risk methodologies for natural hazards: Consequences and challenges for a climate change impact assessment. *J. Environ. Manag.* **2016**, *168*, 123–132. [CrossRef] [PubMed]
7. Kappes, M.S.; Keiler, M.; Von Elverfeldt, K.; Glade, T. Challenges of analyzing multi-hazard risk: A review. *Nat. Hazards* **2012**, *64*, 1925–1958. [CrossRef]
8. Brito, M.M.; Evers, M. Multi-criteria decision-making for flood risk management: A survey of the current state of the art. *Nat. Hazards Earth Syst. Sci.* **2016**, *16*, 1019–1033.
9. Stagge, J.H.; Kohn, I.; Tallaksen, L.M.; Stahl, K. Modeling drought impact occurrence based on meteorological drought indices in Europe. *J. Hydrol.* **2015**, *530*, 37–50. [CrossRef]
10. Kim, T.W.; Jehanzaib, M. Drought risk analysis, forecasting and assessment under climate change. *Water* **2020**, *12*, 1862. [CrossRef]
11. Brémond, P.; Grelot, F.; Agenais, A.L. Flood Damage Assessment on Agricultural Areas: Review and Analysis of Existing Methods. hal-00783552. 2013. Available online: https://hal.archives-ouvertes.fr/hal-00783552 (accessed on 26 July 2021).

engineering proceedings

Proceeding Paper

Awareness Raising and Capacity Building through a Scalable Automatic Water Harvest Monitoring System to Improve Water Resource Management in Monteverde Community, Costa Rica [†]

Roberto Brenes [1], Aníbal Torres [2], José Aguilar [1] and Ronald Aguilar [1,*]

1   Escuela de Ingeniería de Biosistemas, Universidad de Costa Rica, San José 11501-2060, Costa Rica; roberto.brenesloria@ucr.ac.cr (R.B.); jose.aguilar@ucr.ac.cr (J.A.)
2   Monteverde Institute, Monteverde 69-5655, Costa Rica; antorres@mvinstitute.org
*   Correspondence: ronaldesteban.aguilar@ucr.ac.cr
†   Presented at the 13th EFITA International Conference, online, 25–26 May 2021.

Abstract: An outreach project is conducted for the protection of water resources in Monteverde, Costa Rica. An automatic monitoring system for rainwater harvesting (AMSWH) was implemented into a water-harvesting system in the Monteverde Institute (MVI). Through information and communications technology, the project aims to promote water-harvesting systems in Monteverde community by demonstrating the quantifiable benefits in water use (e.g., capture rainwater for use in toilets, vegetable gardens and washing machines), thereby saving drinkable water. During 2021, implemented flow meters and distance sensors in the AMSWH have indicated that 5000 L of potable water has been saved by the MVI.

Keywords: ICT adoption; rain-harvesting monitoring; water resources protection; water balance prototyping; biosystems engineering

Citation: Brenes, R.; Torres, A.; Aguilar, J.; Aguilar, R. Awareness Raising and Capacity Building through a Scalable Automatic Water Harvest Monitoring System to Improve Water Resource Management in Monteverde Community, Costa Rica. *Eng. Proc.* **2022**, *9*, 45. https://doi.org/10.3390/engproc2021009045

Academic Editors: Dimitrios Aidonis and Aristotelis Christos Tagarakis

Published: 24 January 2022

Publisher's Note: MDPI stays neutral with regard to jurisdictional claims in published maps and institutional affiliations.

1. Introduction

This project is conducted in the communities of Santa Elena and Monteverde, Monte Verde, Costa Rica. Straddling the continental divide, the Monteverde area contains varied abiotic conditions and rich biodiversity, including the unique cloud forest. The mountain system where Monte Verde is located is considered an aquifer recharge area, which is strategic for the country due to its geographical location, supplying water for energy production on the Caribbean slope and for human consumption as well as agricultural activities on the Pacific slope.

This being said, the Monte Verde area faces problems related to water resources. First, population growth and increased tourism through the years have created pressure on water resources. For example, during the high-tourism season (January–May), the demand is $76 \text{ m}^3/\text{day}$ of water, while during the low-tourism season (August–November) demand decreases by 20% [1]. The high season coincides with the months of lowest precipitation, which stresses water resources even more due to agriculture activities in the watershed's lowlands. Secondly, the produced graywater is not treated. A common practice is to discharge graywater to storm water systems, creeks or rivers, negatively impacting water bodies. It was estimated that 1400 houses discharge about $685 \text{ m}^3/\text{day}$ of wastewater. Finally, there is a lack of proper storm water infrastructure to manage runoff [1]. Therefore, during precipitation events, instead of infiltration and groundwater recharge occurring, water runoff provokes landslides, floods and pollution.

To promote social awareness and implementation of water resource management on house and building scales, the University of Costa Rica, Department of Biosystems Engineering (UCR-BE), and the Monteverde Institute (MVI) conducted the social project entitled Unitary Biosystems for the Protection of Water Resources in Monteverde, Costa

Rica. Integration biosystems, such as water harvesting, water sowing and wastewater treatment, are offered to the community to generate awareness for saving drinkable water and aquifer water conservation. In particular, by using a scalable prototype with information and communications technology (ICT), the work reported in this paper aims to achieve a viable product: an automatic monitoring system, the AMSWH, to quantify the amount of rainwater that it is harvested, used and sown. The AMSWH demonstrates the rainwater mass balance through ICT using low-cost solutions. In this case, the MVI's AMSWH captures rainwater for use in toilets, vegetable gardens and washing machines, thereby reducing the consumption of potable water on campus; finally, the surplus water is carried into a water sowing system. The quantifiable benefits of these harvesting/sowing systems for measuring the conservation of water resources can thus be demonstrated to the community.

2. Materials and Methods

The rainwater-harvesting system at the MVI (Figure 1) has four collection tanks, with a total capacity of over 2000 L. Rainwater is collected from a 108 m^2 roof. A screened gutter and first flush diverter help to keep leaves and other solids out. Tanks are equipped with auxiliary valves for cleaning and discharging the water. The collected water can be discharged for use in three toilets (0.007 m^3 capacity each), a vegetable garden (6 m^2 area) and two washing machines (0.070 m^3 capacity each). For each use, water flow sensors measure water consumption. Additionally, ultrasonic sensors monitor the water level in the collection tanks. Once the full capacity of the tanks is reached, the surplus water goes to the sowing system for groundwater recharge.

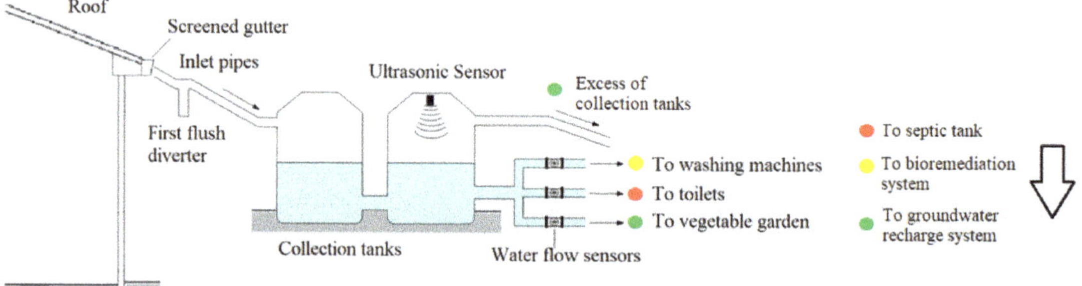

Figure 1. General diagram of the water-harvesting system at MVI.

The AMSWH has two main components: the electronic system based on the ESP32 Development Board and the Blynk platform [2]. The ESP32 gathers data from the sensors and sends it to the Blynk mobile app. The monitoring data is displayed in a dashboard created with Blynk.

3. Results

3.1. Electronic System

The electronic system is integrated by the Sparkfun ESP32 Development Board "ESP32 Thing", the mechanical flow meter "YF-S201", the "AJ-SR04M" waterproof ultrasonic distance sensor (JSN-SR04T equivalent) and bi-directional logic level converters. The "AJ-SR04M" monitors the water level in the tanks. The "YF-S201" senses the water flowing from the tanks to the different uses (e.g., toilets, vegetable gardens and washing machines). The bi-directional logic level converter is needed as the ESP32 uses 3.3 V logic. The firmware was written in PlatformIO, using the Arduino framework, following the Blynk documentation guidelines [3].

3.2. Blynk Platform

The Blynk mobile app is used for data visualization. Via Wi-Fi, the dashboard created with Blynk shows the data sent by the ESP32. Sensed data can be visualized with graphs. For example, users can visualize the volume and level of the water collected in the tanks. Additionally, the app shows the water consumed in the toilets, vegetable garden and washing machines (Figure 2). Finally, users can send data reports in CSV files to a predefined e-mail address for additional data management.

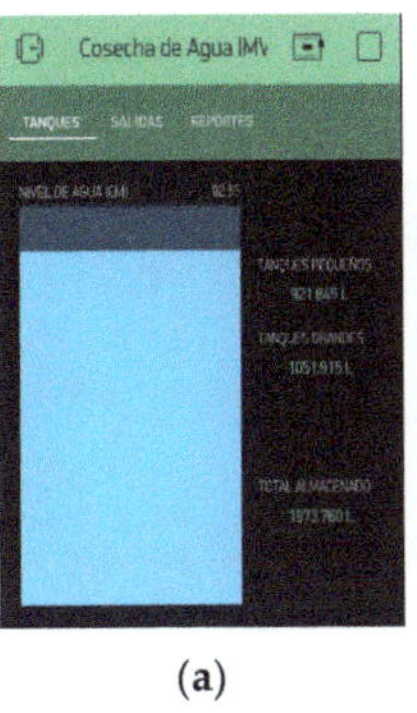

(a)

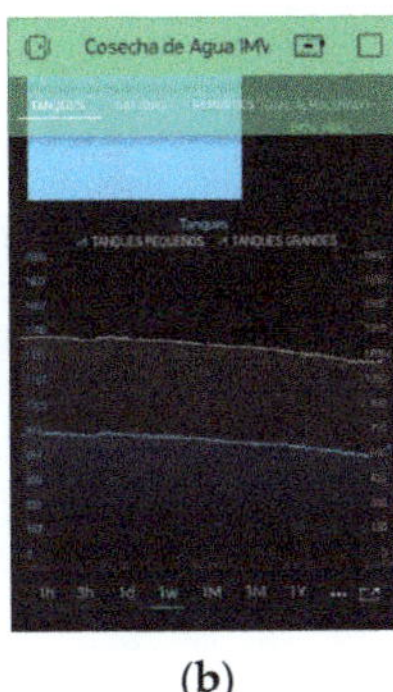

(b)

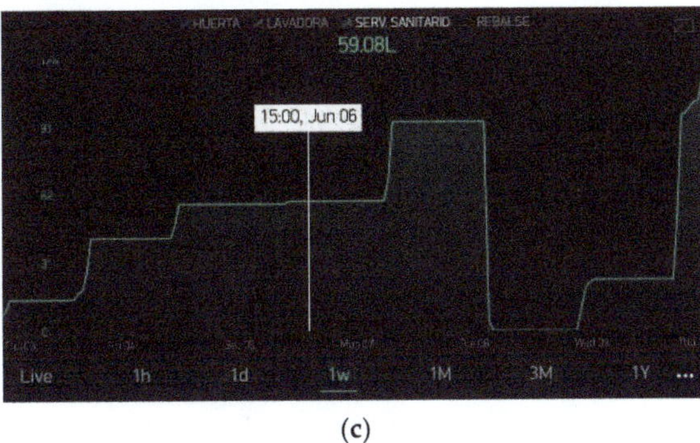

(c)

Figure 2. Blynk mobile app screen captures (**a**) real-time display of water level in the tanks; (**b**) a historical graph of water volume in the tanks; and (**c**) a historical graph of water consumption.

3.3. Water Harvest Monitoring

The AMSWH monitors and generates data on water collection and consumption. Additionally, the AMSWH provides system status. Sensors used here represent a low-cost component, available locally. Since December 2020, the MVI has used an estimated 5000 L of rainwater in toilets, vegetable gardens and washing machines. The monitoring system shows live data feeds (Figure 2a) and historical behavior of the harvesting system (Figure 2b,c).

4. Discussion

In terms of the AMSWH construction, the firmware for this prototype is adaptable to other ESP32 Development Boards and platforms, such as Adafruit IO or ESP-Dash. Many ESP32 Development Boards expose strapping pins that have a specific behavior at booting, as shown in the ESP32 datasheet [4]. Using these pins can cause trouble when resetting the board or flashing new firmware.

In terms of the outreach project conducted by the UCR-BE and the MVI, a viable product of an automatic monitoring system, the AMSWH, to quantify the amount of rainwater being harvested, used and sown, was achieved. The use of ICT is beneficial for adopting this technology to the community because the AMSWH quantifies and showcases the benefits of implementing a water-harvesting system. The quantifiable benefits of these harvesting/sowing systems for measuring water resource conservation can thus be demonstrated to the community. For example, from January to June 2021, 5000 L of potable water has been saved by the MVI, as collected rainwater has been used in toilets, vegetable gardens and washing machines.

For adoption in remote locations, the ESP32 ULP co-processor allows the implementation of the AMSWH as long as Wi-Fi usage scheduling can be employed to improve battery life. For example, the signals from the mechanical water flow meters could be handled before they reach the ESP32 using an interface circuit. Although the signal's frequency is below 1KHz, it is better to avoid using many interrupt service routines in the firmware.

Author Contributions: Conceptualization, R.B., J.A., A.T. and R.A.; methodology, R.B. and R.A.; writing—original draft preparation, R.B. and R.A.; funding acquisition, R.A. All authors have read and agreed to the published version of the manuscript.

Funding: This project was funded by the Outreach Vice-Rectory of the University of Costa Rica, project number ED-3498.

Institutional Review Board Statement: Not applicable.

Informed Consent Statement: Not applicable.

Data Availability Statement: Not applicable.

Acknowledgments: The authors acknowledge the support given by the UCR-BE and the Outreach Vice-Rectory of the University of Costa Rica, in addition to the Monteverde Institute, to promote the continuity of this outreach project during the 2021 pandemic. In particular, the authors acknowledge the technical support given by Gabriela Blanco, Ensio Pérez, Jorge Mora and Henry Bolaños. Finally, the authors acknowledge Fern Perking for her contribution to reviewing the language.

Conflicts of Interest: The authors declare no conflict of interest. The funders had no role in the design of the study; in the collection, analyses or interpretation of data; in the writing of the manuscript; or in the decision to publish the results.

References

1. Sandí, A. (ASADA Santa Elena, Monteverde, Puntarenas, Costa Rica). Personal communication, 2021.
2. Blynk Inc. Available online: https://blynk.io/ (accessed on 6 June 2021).
3. Blynk Docs. Available online: http://docs.blynk.cc/#blynk-firmware-configuration (accessed on 6 June 2021).
4. Espressif Systems. Available online: https://www.espressif.com/en/support/documents/technical-documents? (accessed on 6 June 2021).

engineering proceedings

MDPI

Proceeding Paper

Developing a Method to Simulate and Evaluate Effects of Adaptation Strategies to Climate Change on Wheat Crop Production: A Challenging Multi-Criteria Analysis [†]

Christophe Gigot [1,*], Doriane Hamernig [1,*], Violaine Deytieux [2,*], Ibrahima Diallo [2], Olivier Deudon [1], Emmanuelle Gourdain [1], Jean-Noël Aubertot [3], Marie-Hélène Robin [3], Marie-Odile Bancal [4], Laurent Huber [4] and Marie Launay [5]

[1] ARVALIS–Institut du Végétal, Station Expérimentale, 91720 Boigneville, France; o.deudon@arvalis.fr (O.D.); e.gourdain@arvalis.fr (E.G.)
[2] INRAE, UE115 Domaine Expérimental d'Epoisses, 21000 Dijon, France; ibrahima-koula.diallo@inrae.fr
[3] INRAE-INPT-ENSAT-EI-Purpan, Université de Toulouse, UMR 1248 AGIR, 31326 Castanet-Tolosan, France; jean-noel.aubertot@inrae.fr (J.-N.A.); mh.robin@purpan.fr (M.-H.R.)
[4] INRAE, UMR1402 ÉcoSys, AgroParisTech, Université Paris-Saclay, 78850 Thiverval-Grignon, France; marie-odile.bancal@inrae.fr (M.-O.B.); laurent.huber@inrae.fr (L.H.)
[5] INRAE, US1116 AgroClim, 84914 Avignon, France; marie.launay@inrae.fr
[*] Correspondence: c.gigot@arvalis.fr (C.G.); d.hamernig@arvalis.fr (D.H.); violaine.deytieux@inrae.fr (V.D.)
[†] Presented at the 13th EFITA International Conference, online, 25–26 May 2021.

Citation: Gigot, C.; Hamernig, D.; Deytieux, V.; Diallo, I.; Deudon, O.; Gourdain, E.; Aubertot, J.-N.; Robin, M.-H.; Bancal, M.-O.; Huber, L.; et al. Developing a Method to Simulate and Evaluate Effects of Adaptation Strategies to Climate Change on Wheat Crop Production: A Challenging Multi-Criteria Analysis. *Eng. Proc.* **2021**, *9*, 21. https://doi.org/10.3390/engproc2021009021

Academic Editors: Dimitrios Kateris and Maria Lampridi

Published: 25 November 2021

Publisher's Note: MDPI stays neutral with regard to jurisdictional claims in published maps and institutional affiliations.

Abstract: Adaptation of cropping management strategies is necessary to ensure the sustainability of our agriculture, which is facing threats arising from climate change. A methodology is proposed to find out and compare the most promising adaptation strategies in this context considering both biotic and abiotic stresses. A set of pre-selected strategies were evaluated based on economic, plant health and environmental criteria. A dedicated workflow combining the STICS crop model, epidemiological models and multi-criteria analysis was designed, implemented and tested for a wheat production situation. Flexible by design, this methodology can consider different criteria weights to be used as an exchange support with stakeholders.

Keywords: climate change; wheat production; multi-criteria analysis; crop model; epidemiological model; adaptation strategy

1. Introduction

To combat climate change (CC), stakeholders have requested the development of cropping and breeding strategies that limit the impacts of increasing abiotic stress on plant production. This often overlooks plant health risk that is either directly induced by CC or indirectly through adaptations of agricultural practices. In this context, the OPERATE project ("crop disease response to climate change adaptation") aims to address this issue for arable crops by evaluating the impact of possible adaptation strategies to CC on three sets of criteria (economics, plant health and environment). A methodology was developed, implemented and tested on a wheat crop situation in a climatic context located in the French region of Haut de France.

2. Approach

The first step was to precisely define a set of potential adaptation strategies to abiotic stress that may be experienced by wheat crop in the next decades. Two groups of agronomical levers were proposed based on preliminary surveys conducted with stakeholders [1]: (i) compensation for limited water resources using different irrigation approaches or mulches to maintain soil moisture; and (ii) avoidance of unfavorable conditions shifting sowing dates and/or mobilizing earliness of wheat varieties. Other agronomical levers were considered due to their potential impact on disease development (biotic stress): (i) tillage or no tillage; (ii) disease resistance of wheat varieties; and (iii) nature of the previous crop. The prospective impacts of the adaptation strategies on the economic, plant health and environmental criteria were simulated, mobilizing climate change, crop and disease models. The required steps to identify and characterize the most promising strategies were implemented within the operational workflow described in Figure 1.

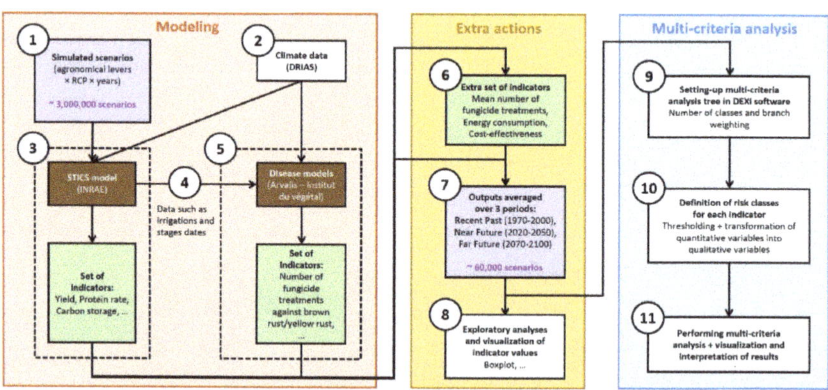

Figure 1. Workflow implemented to simulate impacts of candidate adaptation strategies to climate change and identify the most promising ones for wheat crop situation.

In this workflow, all possible combinations of agronomical levers, prospective CC scenarios and each year between 1970 and 2100 were used as inputs of the crop and disease models, which resulted in a total of about 3,000,000 scenarios (step 1 in Figure 1). Two CC scenarios were considered: either the moderate Representative Concentration Pathway (RCP [2]) 4.5 or the alarmist RCP 8.5, using prospective weather data (Météo-France, CERFACS, IPSL–2014) from the DRIAS platform [3] (step 2). The models were combined in a seamless way to simulate the various effects of adaptation strategies. Simulations with STICS crop model [4] were conducted to estimate production (e.g., yield, nitrogen grain contents) and environmental (e.g., nitrate loss, N_2O emissions) indicators (step 3). In addition to inputs in common with STICS, disease models developed by Arvalis-Institut du végétal [5–9] used output data from STICS (irrigation and phenology, step 4) (i) to assess risks relative to *Septoria tritici* blotch, yellow and brown rusts, eyespot and *Fusarium* head blight (step 5) and (ii) to compute an indicator of fungicide use needed to maintain damage-free crops (step 6).

Supplementary indicators (energy consumption and semi-net margin) were estimated through the CRITER tool [10] (step 6). To focus on tendencies, all the indicators' values were averaged on subsets of years corresponding to three distinct climatic periods: reference past (1970–2000); near future (2020–2050); and far future (2070–2100) (step 7)–for one location (Estrées-Mons, France). Exploratory analyses were performed at this stage to (i) visualize and check the outputs; and (ii) identify expected and possible unexpected effects of levers (step 8). This allows us to highlight necessary compromises to reach a "good" adaptation relevancy to CC (e.g., antagonist effect of sowing date on potential yield and sparing use of fungicides). To identify promising strategies, multi-criteria analysis was performed using the software DEXi [11]. A qualitative decisional tree was designed to aggregate all the relevant indicators of the previously mentioned models into only one general criterion (adaptation relevancy to CC) with five levels, ranging from "very poorly adapted" to "very well adapted". A set of criteria weights was chosen for the aggregation, but such parameters could be modified depending on the aims of the stakeholders (step 9). To be able to use DEXi with the indicators, thresholding was performed on the quantitative indicators to obtain qualitative criteria with balanced classes that the qualitative decisional tree can handle (step 10). Outputs of multi-criteria analysis were plotted to facilitate their interpretation. In addition, clustering methods on the principal components from multiple correspondence analysis (MCA) were applied on the best scenarios (with high adaptation relevancy to CC) to characterize the diversity of promising strategies (step 11).

3. Example of Application

Some outputs of a multi-criteria analysis performed on bread wheat crop situation using a specific decision tree (not detailed) are shown hereinafter as a proof of concept of the method, without any actual field-scale application at this time.

3.1. Ranking of Adaptation Scenario Relevancy

This multi-criteria analysis allowed for ranking the scenarios according to their adaptation relevancy to CC. The Figure 2 shows how two kinds of irrigation strategies, automatic irrigation when water comfort status (water comfort status corresponds to actual evapotranspiration divided by maximum evapotranspiration) is below 0.7 (IRA) and four fixed irrigation dates (IRF), may impact the final aggregated criterion "adaptation relevancy to CC" in reference past (RP) and far future (FF) for RCP 8.5. In this example, the IRA strategy seems to be promising because it makes it possible to drastically reduce the number of very poorly (- -) and poorly (-) adapted scenarios in the future, contrary to the IRF strategy.

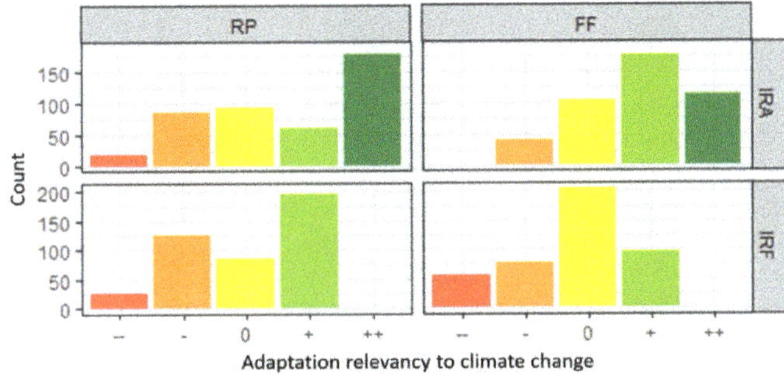

Figure 2. Impact on the adaptation relevancy to climate change (CC) of the irrigation strategies IRA (automatic irrigation when under significant water stress) and IRF (four fixed irrigation dates) for reference past (RP) and far future (FF) under the most alarmist CC scenario (Representative Concentration Pathway 8.5). Five categories of adaptation relevancy are defined: very poorly adapted (- -), poorly adapted (-), moderately adapted (0), well adapted (+), very well adapted (++). The y-axis corresponds to the number of tested scenarios that fall into each of these categories.

3.2. Clustering to Identify Similar Scenarios in Terms of Adaptation Relevancy

To characterize the diversity of promising strategies, clustering methods on principal components from MCA can be applied on the scenarios with the best levels of adaptation relevancy to CC (well adapted and very well adapted, i.e., + and ++ in Figure 2). Using five clusters of similar promising strategies suggested by this proof-of-concept multi-criteria analysis (data not shown), it was possible to identify a cluster characterized by (i) relatively high economic and plant health scores, but medium environmental scores and (ii) a prevalence of corn as previous crop while other levers are not very discriminant. Another cluster was characterized by (i) high economic and environmental scores, but poor plant health scores and (ii) no tillage, high prevalence of canola as previous crop, and low global disease resistance which, as expected, are combined with a dominant profile of late sowing dates to escape diseases.

4. Conclusions and Perspectives

This flexible method potentially allows us to deal with a large range of locations, soil types, crop production, adaptation strategies, and decision trees with contrasted criteria weights. Using an indicator considering interactions between plant diseases to be more consistent with field situations would be interesting. The outputs can serve as support in discussions with stakeholders about most promising strategies to be considered to limit the impacts of CC on crop production, or conversely, the most damaging strategies to avoid in the context of CC. Getting feedback from stakeholders is required to improve the relevance and legitimacy of this method.

Author Contributions: Conceptualization and validation, J.-N.A., M.-O.B., O.D., V.D., C.G., E.G., D.H., L.H., M.L. and M.-H.R.; methodology, M.-O.B., V.D., C.G., E.G., D.H. and M.L.; software O.D., V.D., C.G. and D.H.; investigation, I.D., V.D., C.G. and D.H.; data curation, M.-O.B., O.D., V.D., C.G., D.H. and M.L.; writing—original draft preparation, C.G. and D.H.; writing—review and editing, M.-O.B., O.D., V.D., C.G., E.G., D.H., M.L. and M.-H.R.; supervision, E.G. and V.D.; funding acquisition and project administration, M.-O.B., L.H. and M.L. All authors have read and agreed to the published version of the manuscript.

Funding: This research is part of the OPERATE project ("crop disease response to climate change adaptation") which received financial support from the ACCAF metaprogram of INRAE.

Data Availability Statement: All relevant data have been included in this manuscript.

Acknowledgments: The authors are grateful to Dominique Ripoche for its expertise and support in crop modeling with STICS. They also thank the anonymous reviewers for their comments.

Conflicts of Interest: The authors declare no conflict of interest.

References

1. Calluaud, M.; de Goulaine, B.; Ferrand, A. *Construction d'une Méthode d'évaluation Multicritère de Stratégies d'adaptation Au Changement Climatique pour Les Filières Blé, Pomme de Terre et Tournesol*; AgroSup: Dijon, France, 2019.
2. Moss, R.H.; Edmonds, J.A.; Hibbard, K.A.; Manning, M.R.; Rose, S.K.; Van Vuuren, D.P.; Meehl, G.A. The next Generation of Scenarios for Climate Change Research and Assessment. *Nature* **2010**, *463*, 747. [CrossRef] [PubMed]
3. Lémond, J.; Dandin, P.; Planton, S.; Vautard, R.; Pagé, C.; Déqué, M.; Franchistéguy, L.; Geindre, S.; Kerdoncuff, M.; Li, L.; et al. DRIAS: A Step toward Climate Services in France. *Adv. Sci. Res.* **2011**, *6*, 179–186. [CrossRef]
4. Brisson, N.; Mary, B.; Ripoche, D.; Jeuffroy, M.H.; Ruget, F.; Nicoullaud, B.; Gate, P.; Devienne-Barret, F.; Antonioletti, R.; Durr, C.; et al. STICS: A Generic Model for the Simulation of Crops and Their Water and Nitrogen Balances. *I. Theory and Parameterization Applied to Wheat and Corn. Agronomie* **1998**, *18*, 311–346. [CrossRef]
5. Pierre, A.; Gourdain, E.; Couleaud, G.; Le Brix, X.; Moreau, F.; Deudon, O.; Heritier, E.; Simonneau, D.; Piraux, F. Elaboration d'un Modèle d'alerte de La Présence d'une Maladie Fongique: Application Au Pathosystème Puccinia Striiformis sur Blé Tendre d'hiver. In Proceedings of the AFPP—11ème Conférence Internationale sur les Maladies des Plantes, Tours, France, 7–9 September 2015.
6. Gouache, D.; Léon, M.S.; Duyme, F.; Braun, P. A Novel Solution to the Variable Selection Problem in Window Pane Approaches of Plant Pathogen–Climate Models: Development, Evaluation and Application of a Climatological Model for Brown Rust of Wheat. *Agric. For. Meteorol.* **2015**, *205*, 51–59. [CrossRef]
7. Gouache, D.; Bensadoun, A.; Brun, F.; Pagé, C.; Makowski, D.; Wallach, D. Modelling Climate Change Impact on Septoria Tritici Blotch (STB) in France: Accounting for Climate Model and Disease Model Uncertainty. *Agric. For. Meteorol.* **2013**, *170*, 242–252. [CrossRef]
8. Georgel, F.; Huguet, B.; Couleaud, G.; Bouttet, D.; Gauchet, D.; Moreau, F.; Piraux, F.; Gourdain, E. Evaluer Le Risque Piétin Verse En Ile-de-France. In Proceedings of the AFPP—6ème Conférence sur les Moyens Alternatifs de Protection pour une Production Intégrée, Lille, France, 21 March 2017.
9. Gourdain, E.; Grignon, G. Piloter L'intervention Fusariose des Épis sur Blé. In Proceedings of the Végéphyl—12ème Conférence Internationale sur les Maladies des Plantes, Tours, France, 11–12 December 2018.
10. Hirschy, M.; Ravier, C.; Lorin, M. *CRITER 5.4, un Outil de Caractérisation des Performances de Systèmes de Culture, Manuel D'utilisateur*; INRAE: Paris, France, 2015.
11. Bohanec, M. *DEXi: Program for Multi-Attribute Decision Making, User's Manual, Version 5.04*; IJS Report DP-13100; Jožef Stefan Institute: Ljubljana, Slovenia, 2020.

engineering
proceedings

Proceeding Paper

Identification and Modeling Carbon and Energy Fluxes from Eddy Covariance Time Series Measurements in Rice and Rainfed Crops [†]

Víctor Cicuéndez [1,*], Javier Litago [1], Víctor Sánchez-Girón [2], Laura Recuero [3], César Sáenz [1] and Alicia Palacios-Orueta [2,3]

1. Departamento de Economía Agraria, Estadística y Gestión de Empresas, ETSIAAB, Universidad Politécnica de Madrid (UPM), Avda. Complutense 3, 28040 Madrid, Spain; javier.litago@upm.es (J.L.); cesar.saenzf@alumnos.upm.es (C.S.)
2. Departamento de Ingeniería Agroforestal, ETSIAAB, Universidad Politécnica de Madrid (UPM), Avda. Complutense 3, 28040 Madrid, Spain; victor.sanchezgiron@upm.es (V.S.-G.); alicia.palacios@upm.es (A.P.-O.)
3. Centro de Estudios e Investigación Para la Gestión de Riesgos Agrarios y Medioambientales (CEIGRAM), Universidad Politécnica de Madrid (UPM), C/Senda del Rey 13 Campus Sur de prácticas de la ETSIAAB, 28040 Madrid, Spain; laura.recuero@upm.es
* Correspondence: victor.cicuendez.lopezocana@alumnos.upm.es
† Presented at the 13th EFITA International Conference, online, 25–26 May 2021.

Abstract: Gross primary production (GPP) represents the carbon (C) uptake of ecosystems through photosynthesis and it is the largest flux of the global carbon balance. Our overall objective in this research is to identify and model GPP dynamics and its relationship with meteorological variables and energy fluxes based on time series analysis of eddy covariance (EC) data in two different agroecosystems, a Mediterranean rice crop in Spain and a rainfed cropland in Germany. Crops exerted an important influence on the energy and water fluxes dynamics existing a clear feedback between GPP, meteorological variables and energy fluxes in both type of crops.

Keywords: gross primary production (GPP); energy fluxes; time series analysis; feedback; Granger-causality tests

Citation: Cicuéndez, V.; Litago, J.; Sánchez-Girón, V.; Recuero, L.; Sáenz, C.; Palacios-Orueta, A. Identification and Modeling Carbon and Energy Fluxes from Eddy Covariance Time Series Measurements in Rice and Rainfed Crops. *Eng. Proc.* **2021**, *9*, 9. https://doi.org/10.3390/engproc2021009009

Academic Editors: Dimitrios Aidonis and Aristotelis Christos Tagarakis

Published: 19 November 2021

Publisher's Note: MDPI stays neutral with regard to jurisdictional claims in published maps and institutional affiliations.

1. Introduction

The understanding of the CO_2 flux between the biosphere and atmosphere is essential for assessing the carbon cycle and its influence on global climate change. Gross primary production (GPP) represents the C uptake of ecosystems through photosynthesis and it is the largest flux of the global carbon balance. Crop GPP contributes approximately 15% of global carbon dioxide fixation [1].

Climate determines carbon, water and energy fluxes on a seasonal to interannual time scale creating a complex pattern of CO_2 exchange between the atmosphere and the ecosystem. Meanwhile, the ecosystem itself is responsible for strong feedback processes to climate [2]. Jung et al. [3] showed the strong connections between GPP and energy fluxes at a global scale; however, the relationships between fluxes should be studied at a local scale in different ecosystems.

The eddy covariance (EC) technique has been established as a valid method to measure meteorological variables as well as carbon, water and energy fluxes between ecosystems and the atmosphere [4]. The high temporal resolution of EC measurements makes the statistical time series analysis (TSA) an excellent method to analyze and study these data. TSA in the time and frequency domains provides tools and methodologies to model and forecast these variables and their relationship based on their dynamics [5]. Carbon, energy and water fluxes could be analyzed by TSA and be used as an excellent indicator of the phenology of ecosystems monitoring several characteristics of the ecosystems such as, cycles, evolution and trends.

Our overall objective in this research was to identify and model the GPP dynamics and its relationship with meteorological variables and energy fluxes based on time series analysis of EC data in two different agroecosystems, a Mediterranean rice crop in Spain and a rainfed cropland in Germany.

2. Material and Methods

2.1. Study Site

Two EC flux towers integrated into the FLUXNET micrometeorological tower sites [6] have been selected for this research.

The first one is situated in Gebesee (Germany) in an experimental agricultural area which has been cultivated since 1970 mainly with cereals, potato and sugar beet. More details on the site can be found in Anthoni et al. [7]. The study period analyzed of this flux tower is from 1 January 2001 to 31 December 2014 in which has been cultivated rainfed crops such as wheat, barley and potato. The area is characterized by a temperate climate (Cfb) according to Köppen-Geiger classification.

The second one is placed in El Saler-Sueca in Valencia (Spain) in a rice crop [8]. The area is characterized by a Mediterranean climate (Csa) and the available data of this flux tower site is from 1 January 2004 to 31 December 2010.

The following pre-processed daily variables were used: (1) air temperature (Ta, °C); (2) vapor pressure deficit (VPD, hPa); (3) precipitation (PPT, mm day^{-1}); (4) soil water content (SWC, % vol); (5) latent heat (LE, W m^{-2}) (6) sensible heat (H, W m^{-2}); and (7) gross primary productivity (GPP_T, g C m^{-2} day^{-1}).

2.2. Statistical Methods

Buys–Ballot tables [9] were used to study the intra-annual variability of meteorological data, energy fluxes and GPP. Secondly, Granger causality tests [10] were used to study dynamic relationships between variables. The statistical analysis was computed through Statgraphics 18 (StatPoint Technologies Inc., Warrenton, VA, USA), Eviews 10 University edition (IHS Global Inc., Irvine, CA, USA) and SAS software (SAS 9.4 Software, SAS Institute Inc., Cary, NC, USA).

3. Results and Discussion

In Germany, H and LE increases with temperature during spring. However, while LE increases together with GPP during the growing period due to the link between growth and ETP, H reaches its maximum during May when the crop begins to cover the ground and the energy inverted to change temperature stabilizes. The majority of Net radiation is converted into latent heat. After harvest, sensible heat reaches its maximum in August because the crop has been disappeared and does not influence on temperature. Latent heat decreases because there is no crop and no ETP. Gao et al. [11] also showed that during growing season latent heat was the main consumer of net radiation, while sensible heat was dominant during the non-growing season in a rainfed maize crop.

In Spain, sensible heat reaches its maximum in mid-April when there is no crop and between flooded periods. Latent heat has two cycles that coincides with flooded periods. The first one is related to evaporation and the second one to ETP. Other studies highlighted that latent heat is the main component of available energy in rice crops [12].

Figure 1 shows the Granger causality tests between H and LE and GPP in both directions (i.e., variable cause GPP and GPP cause variable). In Germany, sensible heat causes GPP at very short term (first three lags). Then, both directions show similar F-test values. In Spain, from lag 2 to 10 GPP causes H more than in the other direction. In both sites, GPP causes Latent heat more than backwards. However, in Germany F-tests are much higher than in Spain especially in the sense GPP causes LE.

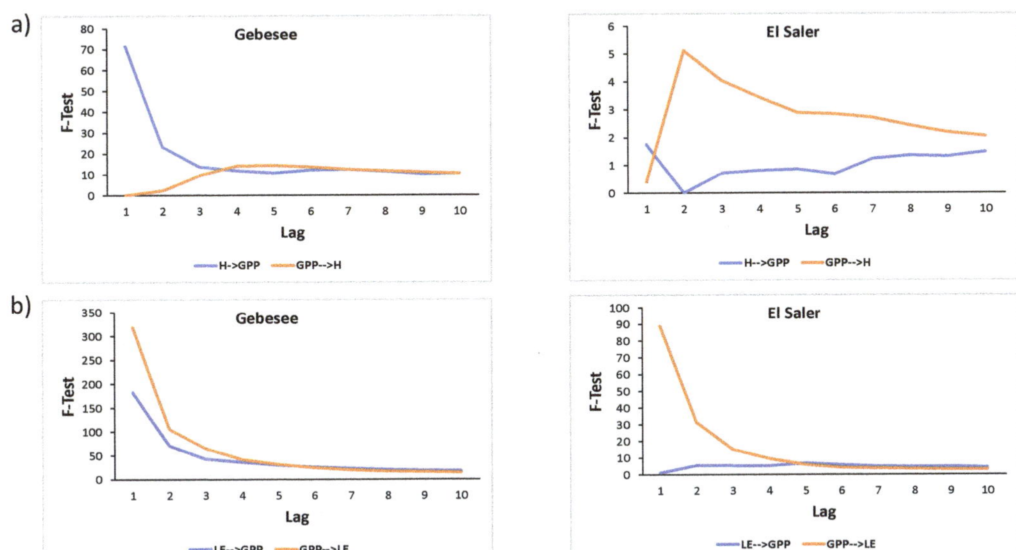

Figure 1. Granger causality test between gross primary production GPP_T (g C m^{-2} day^{-1}) and (**a**) sensible heat H (W m^{-2}) and (**b**) latent heat LE (W m^{-2}) in the two croplands (Left: Germany (DE), Right: Spain (ES)).

These feedback processes that vegetation exerts on the energy partitioning have been assessed by several authors [13,14].

4. Conclusions

Differences in GPP values and dynamics between sites are due to different crops and their interaction with the meteorological variables in each climate. GPP depends mainly on temperature especially in irrigated crops. In terms of energy fluxes, latent heat is more coupled to GPP in rainfed crops and the sensible heat depends especially on ground cover in both crops. There is a clear feedback between GPP, meteorological variables and energy fluxes in both type of crops that we have quantify by Granger causality tests. In this sense, GPP causes temperature and latent heat, especially in rainfed crops.

Author Contributions: Conceptualization, V.C., J.L., V.S.-G., A.P.-O.; methodology, V.C., J.L., V.S.-G., L.R., A.P.-O.; software, V.C., J.L.; formal analysis, V.C., J.L., A.P.-O.; investigation, V.C., J.L., V.S.-G., L.R., C.S., A.P.-O.; resources, J.L., A.P.-O.; writing—original draft preparation, V.C., J.L., V.S.-G., L.R., C.S., A.P.-O.; writing—review and editing, V.C., J.L., V.S.-G., L.R., C.S., A.P.-O.; supervision, A.P.-O.; funding acquisition, A.P.-O. All authors have read and agreed to the published version of the manuscript.

Funding: This research was funded by the Universidad Politécnica de Madrid within the project RP2013210028: "Generación de un observatorio de dinámica de la vegetación a distintas escalas a partir de series de tiempo de imágenes de teledetección".

Acknowledgments: We would like to thank the FLUXNET2015 data set and the European Fluxes Database Cluster for the assignment and permission to download the data of Gebesee and El Saler-Sueca, respectively.

References

1. Malmström, C.M.; Thompson, M.V.; Juday, G.P.; Los, S.O.; Randerson, J.T.; Field, C.B. Interannual variation in global-scale net primary production: Testing model estimates. *Glob. Biogeochem. Cycles* **1997**, *11*, 367–392. [CrossRef]
2. Arora, V. Modeling vegetation as a dynamic component in soil-vegetation-atmosphere transfer schemes and hydrological models. *Rev. Geophys.* **2002**, *40*, 1006. [CrossRef]
3. Jung, M.; Reichstein, M.; Margolis, H.A.; Cescatti, A.; Richardson, A.D.; Arain, M.A.; Arneth, A.; Bernhofer, C.; Bonal, D.; Chen, J.; et al. Global patterns of land-atmosphere fluxes of carbon dioxide, latent heat, and sensible heat derived from eddy covariance, satellite, and meteorological observations. *J. Geophys. Res.* **2011**, *116*, G00J07. [CrossRef]
4. Aubinet, M.; Vesala, T.; Papale, D. (Eds.) *Eddy Covariance: A Practical Guide to Measurement and Data Analysis*; Springer: Dordrecht, The Netherlands, 2012; ISBN 978-94-007-2350-4.
5. Huesca, M.; Litago, J.; Palacios-Orueta, A.; Montes, F.; Sebastián-López, A.; Escribano, P. Assessment of forest fire seasonality using MODIS fire potential: A time series approach. *Agric. For. Meteorol.* **2009**, *149*, 1946–1955. [CrossRef]
6. Pastorello, G.; Trotta, C.; Canfora, E.; Chu, H.; Christianson, D.; Cheah, Y.-W.; Poindexter, C.; Chen, J.; Elbashandy, A.; Humphrey, M.; et al. The FLUXNET2015 dataset and the ONEFlux processing pipeline for eddy covariance data. *Sci. Data* **2020**, *7*, 225. [CrossRef] [PubMed]
7. Anthoni, P.M.; Knohl, A.; Rebmann, C.; Freibauer, A.; Mund, M.; Ziegler, W.; Kolle, O.; Schulze, E.-D. Forest and agricultural land-use-dependent CO_2 exchange in Thuringia, Germany. *Glob. Chang. Biol.* **2004**, *10*, 2005–2019. [CrossRef]
8. SUECA. Available online: http://ceamflux.com:9090/sueca/index.html (accessed on 20 May 2021).
9. Buys-Ballot, C.H.D. *Les Changements périodiques de température, dépendants de la nature du soleil et de la lune, mis en rapport avec le pronostic du temps, déduits d'observations néerlandaises de 1729 à 1846*; Kemink: Utrecht, The Netherlands, 1847.
10. Granger, C.W.J. Investigating Causal Relations by Econometric Models and Cross-spectral Methods. *Econometrica* **1969**, *37*, 424–438. [CrossRef]
11. Gao, X.; Mei, X.; Gu, F.; Hao, W.; Gong, D.; Li, H. Evapotranspiration partitioning and energy budget in a rainfed spring maize field on the Loess Plateau, China. *CATENA* **2018**, *166*, 249–259. [CrossRef]
12. Liu, B.; Cui, Y.; Luo, Y.; Shi, Y.; Liu, M.; Liu, F. Energy partitioning and evapotranspiration over a rotated paddy field in Southern China. *Agric. For. Meteorol.* **2019**, *276–277*, 107626. [CrossRef]
13. Wang, P.; Li, X.; Tong, Y.; Huang, Y.; Yang, X.; Wu, X. Vegetation dynamics dominate the energy flux partitioning across typical ecosystem in the Heihe River Basin: Observation with numerical modeling. *J. Geogr. Sci.* **2019**, *29*, 1565–1577. [CrossRef]
14. Forzieri, G.; Miralles, D.G.; Ciais, P.; Alkama, R.; Ryu, Y.; Duveiller, G.; Zhang, K.; Robertson, E.; Kautz, M.; Martens, B.; et al. Increased control of vegetation on global terrestrial energy fluxes. *Nat. Clim. Chang.* **2020**, *10*, 356–362. [CrossRef]

Proceeding Paper

Rainfed and Supplemental Irrigation Modelling 2D GIS Moisture Rootzone Mapping on Yield and Seed Oil of Cotton (*Gossypium hirsutum*) Using Precision Agriculture and Remote Sensing [†]

Agathos Filintas *, Aikaterini Nteskou, Persefoni Katsoulidi, Asimina Paraskebioti and Marina Parasidou

Department of Agricultural Technologists, Campus Gaiopolis, University of Thessaly, 41500 Larisa, Greece; aikantes@uth.gr (A.N.); perskats@uth.gr (P.K.); asimpara@uth.gr (A.P.); maripara5@uth.gr (M.P.)
* Correspondence: filintas@uth.gr
† Presented at the 13th EFITA International Conference, Online, 25–26 May 2021.

Abstract: The effects of two irrigation (IR1: rainfed; IR2: rainfed + supplemental drip irrigation), and two fertilization (Ft1, Ft2) treatments were studied on cotton yield and seed oil by applying a number of new agro-technologies such as: TDR sensors; soil moisture (SM); precision agriculture; remote-sensing NDVI (Sentinel-2 satellite sensor); soil-hydraulic analyses; geostatistical models; SM-rootzone, and modelling 2D GIS mapping. A daily soil-water-crop-atmosphere (SWCA) balance model was developed. The two-way ANOVA statistical analysis results revealed that irrigation (IR2 = best) and fertilization treatments (Ft1 = best) significantly affected yield and oil content. Supplemental irrigation, if applied during critical growth stages, could result in substantial improvement on yield (+234.12%) and oil content (+126.44%).

Keywords: geostatistical modelling; cotton yield; 2D TDR-GIS soil moisture mapping; precision agriculture; GIS and NDVI; rainfed and supplemental irrigation; soil-hydraulic analyses

Citation: Filintas, A.; Nteskou, A.; Katsoulidi, P.; Paraskebioti, A.; Parasidou, M. Rainfed and Supplemental Irrigation Modelling 2D GIS Moisture Rootzone Mapping on Yield and Seed Oil of Cotton (*Gossypium hirsutum*) Using Precision Agriculture and Remote Sensing. *Eng. Proc.* **2021**, *9*, 37. https://doi.org/10.3390/engproc2021009037

Academic Editors: Dimitrios Kateris and Maria Lampridi

Published: 28 December 2021

Publisher's Note: MDPI stays neutral with regard to jurisdictional claims in published maps and institutional affiliations.

1. Introduction

Cotton plant grows as a perennial crop but is cultivated annually and considered the world's largest non-food crop [1]. Global cottonseed production has ranged between 35 and 59 million tonnes in the last three decades [2]. Regarding water, on a global scale, the agricultural sector accounts for 70% of global freshwater withdrawals [1,3,4]. In Europe, it accounts for around 59% of total water use, and approximately 284,000 million m^3 of water is abstracted annually to meet the demands of the European economy [5]. At present, many countries worldwide are experiencing a scarcity of fresh water [1,3,4] for potable and irrigation use. Global water demand is projected to increase by 55% between 2000 and 2050, from 3500 to 5425 km^3. Evidence has shown that climate change will have an adverse impact on world water resources and food production, with a high degree of regional variability and scarcity [6]. The irrigation water amount has always been the main factor limiting crop production in much of the world, where rainfall is insufficient to meet crop water requirements [1,4,7]. With the ever-increasing competition for finite water resources worldwide, and the steadily rising demand for agricultural commodities, the call to improve the efficiency and productivity of water use for crop production, to ensure future food and crop products security and address the uncertainties associated with climate change, has never been more urgent [8]. The global challenge for the coming decades will be increasing crop production with less water. Precision agriculture (PA) is a management strategy that helps farmers to improve crop production, efficiency and productivity of water use. The aim of this study was to determine the effects of rainfed (IR1) and rainfed plus supplemental irrigation (IR2) and fertilization (Ft1, Ft2) treatments on cotton yield and seed oil content, by applying new agro-technologies.

2. Materials and Methods

2.1. Experimental Plot Design, Soil Sampling and Laboratory Soil and Hydraulic Analysis

The 2.15 ha field had a factorial split plot design with the main factor being the 2 irrigation treatments: (a) IR1: rainfed; (b) IR2: rainfed + supplemental drip irrigation. The sub factor was 2 fertilization treatments: Ft1: N-P-K = 91.00-19.62-37.35 Kg·ha^{-1}, and Ft2: N-P-K = 71.40-13.08-24.90 Kg·ha^{-1}. A GPS receiver was used to identify the locations of soil samples that were collected at a depth of 0–30 cm, and then analyzed at the laboratory. The soil's pH was measured in a 1:2 soil/water extract, with a glass electrode and a pH meter. Soil organic matter was analyzed by chemical oxidation with 1 mol·L^{-1} $K_2Cr_2O_7$ and titration of the remaining reagent with 0.5 mol·L^{-1} $FeSO_4$ [9]. The cation-exchange capacity (CEC) was analyzed by (i) saturation of cation exchange sites with Na by "equilibration" of the soil with pH 8.2, 60% ethanol solution of 0.4N NaOAc-0.1N NaCI; and (ii) extraction with 0.5N $MgNO_3$. The total Na and CI were determined in the extracted solution [9]. The soil's nitrate and ammonium nitrogen were extracted with 0.5 mol L^{-1} $CaCl_2$, and estimated by distillation in the presence of MgO and Devarda's alloy, respectively. Available phosphorus P (Olsen method) was extracted with 0.5 mol L^{-1} $NaHCO_3$, and measured by spectroscopy [9]. The potassium exchangeable K forms were extracted with 1 mol L^{-1} CH_3COONH_4, and measured with a flame photometer. Field capacity (FC) and wilting point (WP) were measured with the porous ceramic plate method, with 1/3 Atm for FC and 15 Atm for WP [1]. Cotton (*Gossypium hirsutum*, var. *Armonia*) was seeded at the end of April, and harvested on 10th (1st harvest) and 25th of October (2nd harvest).

2.2. Soil Moisture Measurements, Digital 2D GIS Moisture Maps Utilizing GIS, Precision Agriculture and Geostatistics

The TDR (time domain reflectometry) method was used to measure the soil moisture, because it gives accurate results within an error limit of ±1% [1,7,10,11]. A TDR instrument and probes with 5 sensors each were used [1,7,12], placed at 0–15, 15–30, 30–45, 45–60 and 60–75 cm depths for measuring volumetric water content ($\theta vi, \ldots , \theta vn$) ($i = 1, 2, \ldots , n$ and $n = 5$) of the root zone. Data were imported daily in a GIS geodatabase, utilizing precision agriculture and geostatistics [1,7,12] in order to model and produce soil moisture 2D maps of the cotton's root zone profile.

2.3. Remote Sensing Crop's NDVI, Evapotranspiration and Net Irrigation Requirements

Climatic data were obtained from a nearby meteorological station. The effective rainfall was calculated according to USDA-SCS (1970) [13]. The NDVI (Normalized Difference Vegetation Index) [1] was calculated every week using remote-sensing (RS) data (Sentinel-2 satellite sensor) for studying spatial crop development and coefficients. The reference evapotranspiration was computed based on the F.A.O. Penman–Monteith method [1,7,8]. The crop evapotranspiration (ETc) and actual evapotranspiration (ETa) were computed using crop coefficients obtained from remote-sensing NDVI vegetation index [1,7,8]. The net irrigation requirement (NIR) was calculated using a soil-water-crop-atmosphere balance model (Equation (1)) [1,7]:

$$NIR = ETc - Pe - GW - \Delta\theta v_{(TDR)} \tag{1}$$

where: NIR = net irrigation requirement (mm); ETc = evapotranspiration (mm); Pe = effective rainfall (mm); GW = groundwater contribution from water table (mm); $\Delta\theta v_{(TDR)}$ = change in TDR sensors measuring soil–water content $\theta v_{(TDR)}$ (mm).

3. Results and Discussion

3.1. Study Area, Soil-Hydraulic Analysis and 2D Moisture Maps Utilizing GIS, Precision Agriculture and Geostatistics

The experiment was conducted in a farm field located at Livadia, in Central Greece. The study area was characterized by a typical Mediterranean climate [1,4,7] with a cold winter, hot summer, and low precipitation in spring and summer (Figure 1a). The results found from TDR sensors and analyses were used as input variables to delineate the digital soil moisture profile on a 2D GIS map of the cotton's root zone (Figure 1b). SM is the major factor for crops' enhanced growth and production [1,7]. The spatial analysis revealed an excellent moisture distribution.

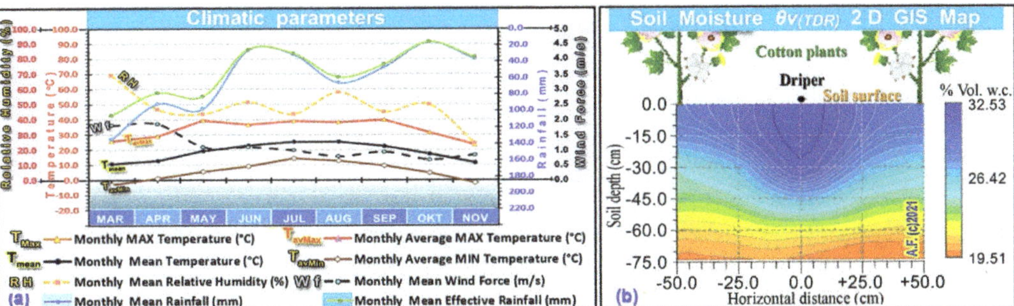

Figure 1. (**a**) Climatic parameters' time variation; (**b**) a soil moisture 2D GIS map of cotton's root zone soil profile.

The results of the laboratory soil and hydraulic analysis revealed that the field's soil was suitable for cotton growth [1,8,9,13], and was characterized as sandy clay loam (SCL) [1,9,13]. Soil organic matter was 1.62% ($\pm$0.15), the bulk specific gravity of the soil was 1.42g cm^{-3} ($\pm$0.02), plant available water was 0.130 cm cm^{-1} ($\pm$0.02), the pH at (1:2) soil/water extract was 8.06 ($\pm$0.21), and the cation-exchange capacity of the soil was 19.72 cmol kg^{-1} ($\pm$1.05) [a sufficient level]. The N-NO$_3$ was found at 9.05 mg kg^{-1} ($\pm$2.23) [a marginal level] and the N-NH$_4$ was found at 3.07 mg kg^{-1} ($\pm$1.02) [a low level]. The phosphorus P-Olsen was found at 17.60 mg kg^{-1} ($\pm$2.15) [a sufficient level], and the potassium K-exchangeable was found at high concentration levels (520 mg kg^{-1} ($\pm$21.15)).

3.2. Daily Soil-Water-Crop-Atmosphere (SWCA) Balance Model and NDVI Vegetation Index

The NDVI vegetation index [1] was calculated using RS data (the Sentinel-2 satellite sensor) for monitoring spatial crop development and coefficients for the SWCA model (Figure 2a,b). The daily TDR moisture measurements, 2D GIS SM mapping, the accurate estimations of ETc, the monitoring of the system inflows and estimated surface outflows, and NDVI vegetation index mapping, resulted in a reliable daily soil-water-crop-atmosphere model [1,7] for the four crop-growth stages of cotton.

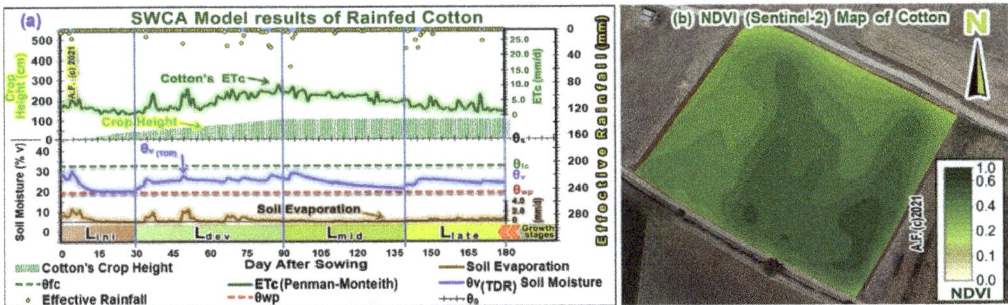

Figure 2. (**a**) Treatment IR1 results of the daily soil-water-crop-atmosphere model for the four cotton growth stages; (**b**) cotton's field NDVI vegetation map (using Sentinel-2 satellite sensor data) in early July, after the blossom of the first flowers of the crop.

3.3. Statistical Analysis, Cotton's Yield and Seed Oil Content Results

The two-way ANOVA statistical analysis ($p = 0.05$) using IBM-SPSS (v.26) [1,7,14,15] revealed that the irrigation treatments (IR2: rainfed + supplemental irrigation [best]) and the fertilization treatments (Ft1: [best]) significantly affected the cotton yield and seed oil content. Whenever a prolonged shortage of soil moisture occurred during the most sensitive growth stages (flowering (L_{mid} stage) and grain filling (L_{late} stage)) (see Figure 2a), the rainfed crop growth was poor, and yield was consequently low [1,7,8]. Supplemental irrigation, if applied during the cotton's critical growth stages, could result in substantial improvements in yield (+234.12%) and seed oil content (+126.44%).

4. Conclusions

The prolonged shortage of soil moisture in dry rainfed areas often occurs during the most sensitive growth stages (flowering (L_{mid} stage) and grain filling (L_{late} stage)) of many crops. As a result, rainfed crop growth is poor and yield is consequently low. Supplemental irrigation, if applied properly by the use of new agro-technologies during the critical crop growth stages (L_{mid} and L_{late}), may constitute a method that can result in substantial improvement in yield (+234.12%) and seed oil content (+126.44%), in addition to water use efficiency, and the sustainable management of environment and water resources.

Author Contributions: Conceptualization, A.F.; methodology, A.F.; software, A.F.; validation, A.F., A.N., P.K., A.P. and M.P.; formal analysis, A.F.; investigation, A.F. and A.N.; resources, A.F., A.N., P.K., A.P. and M.P.; data curation, A.F., A.N., P.K., A.P. and M.P.; writing—original draft preparation, A.F.; writing—review and editing, A.F.; visualization, A.F.; supervision, A.F.; project administration, A.F.; funding acquisition, A.F. and A.N. All authors have read and agreed to the published version of the manuscript.

Funding: This research received no external funding.

Institutional Review Board Statement: Not applicable.

Informed Consent Statement: Not applicable.

Data Availability Statement: Data are available on reasonable request to the corresponding author.

Acknowledgments: The technical and human support provided by the local farmers of Livadia region in Central Greece is gratefully acknowledged.

Conflicts of Interest: The authors declare no conflict of interest.

References

1. Filintas, A. Land Use Evaluation and Environmental Management of Biowastes, for Irrigation with Processed Wastewaters and Application of Bio-Sludge with Agricultural Machinery, for Improvement-Fertilization of Soils and Crops, with the Use of GIS-Remote Sensing, Precision Agriculture and Multicriteria Analysis. Ph.D. Thesis, University of the Aegean, Mitilini, Greece, 2011.
2. USDA. *Cotton: World Markets and Trade*, 2021st ed.; United States Department of Agriculture: New York, NY, USA, 2021; p. 30.
3. FAO. *Coping with Water Scarcity: An Action Framework for Agriculture and Food Security*; FAO: Rome, Italy, 2012; p. 100.
4. Stamatis, G.; Parpodis, K.; Filintas, A.; Zagana, E. Groundwater quality, nitrate pollution and irrigation environmental management in the Neogene sediments of an agricultural region in central Thessaly (Greece). *Environ. Earth Sci.* **2011**, *64*, 1081–1105. [CrossRef]
5. EEA. *Use of Freshwater Resources in Europe, CSI 018*; European Environment Agency (EEA): Copenhagen, Denmark, 2019.
6. Islam, S.M.F.; Karim, Z. World's Demand for Food and Water: The Consequences of Climate Change. In *Desalination-Challenges and Opportunities*; Farahani, M.H.D.A., Vatanpour, V., Taheri, A.H., Eds.; IntechOpen: London, UK, 2019; Chapter 4; pp. 1–27. [CrossRef]
7. Filintas, A.; Wogiatzi, E.; Gougoulias, N. Rainfed cultivation with supplemental irrigation modelling on seed yield and oil of Coriandrum sativum L. using Precision Agriculture and GIS moisture mapping. *Water Supply* **2021**, *21*, 2569–2582. [CrossRef]
8. Allen, R.; Pereira, L.; Raes, D.; Smith, M. *Crop Evapotranspiration*; Drainage & Irrigation paper N°56; FAO: Rome, Italy, 1998.
9. Page, A.L.; Miller, R.H.; Keeney, D.R. *Methods of Soil Analysis Part 2: Chemical and Microbiological Properties*; Agronomy, ASA and SSSA: Madison, WI, USA, 1982; p. 1159.
10. Dioudis, P.; Filintas, A.; Koutseris, E. GPS and GIS based N-mapping of agricultural fields' spatial variability as a tool for non-polluting fertilization by drip irrigation. *Int. J. Sustain. Dev. Plan.* **2009**, *4*, 210–225. [CrossRef]

11. Dioudis, P.; Filintas, A.; Papadopoulos, A. Corn yield response to irrigation interval and the resultant savings in water and other overheads. *Irrig. Drain.* **2009**, *58*, 96–104. [CrossRef]
12. Filintas, A.; Dioudis, P.; Prochaska, C. GIS modeling of the impact of drip irrigation, of water quality and of soil's available water capacity on Zea mays L, biomass yield and its biofuel potential. *Desalination Water Treat.* **2010**, *13*, 303–319. [CrossRef]
13. USDA-SCS. *Irrigation Water Requirements*; Technical R. No. 21; USDA Soil Conservation Service: Washington, DC, USA, 1970.
14. Norusis, M.J. *IBM SPSS Statistics 19 Advanced Statistical Procedures Companion*; Pearson: London, UK, 2011.
15. Hatzigiannakis, E.; Filintas, A.; Ilias, A.; Panagopoulos, A.; Arampatzis, G.; Hatzispiroglou, I. Hydrological and rating curve modelling of Pinios River water flows in Central Greece, for environmental and agricultural water resources management. *Desalination Water Treat.* **2016**, *57*, 11639–11659. [CrossRef]

Proceeding Paper

Production of Fertilizer from Seawater with a Remote Control System [†]

Federico Hahn [1,*], Carlos Juárez González [2] and Canek Mota Delfín [2]

[1] Irrigation Department, Universidad Autonoma Chapingo, Texcoco 56230, Mexico
[2] IAUIA, Universidad Autonoma Chapingo, Texcoco 56230, Mexico; carlos.juarez.gonzalez904@gmail.com (C.J.G.); kincanekmd@gmail.com (C.M.D.)
[*] Correspondence: fhahn@correo.chapingo.mx; Tel.: +52-553-242-3033
[†] Presented at the 13th EFITA International Conference, online, 25–26 May 2021.

Abstract: Seawater is abundant and full of nutrients known as ORMUS. Inorganic fertilizers have become scarce and expensive, so alternatives to feed plants are being studied. An automatic tank on a fishing boat was designed to extract salts from seawater, as follows: Sodium hydroxide is applied to seawater and agitated within a tank until its pH reaches 10.78. Salts begin to deposit, and the sodium mixed with the water stays at the surface. Water with sodium is removed after 3 h with a low-pressure pump. Clean water is added to the salty solution at the bottom of the tank to remove more sodium. Water at the top is sucked by the pump again, and the process is repeated once more. After the white salt (ORMUS) lying at the bottom of the tank is removed, the fertilization extraction process can start again. The automatic system regulates the agitator speed, pump filling and suction timing, and bottom valve opening.

Keywords: fertilizer extraction; sodium hydroxide; ORMUS; remote control

Citation: Hahn, F.; González, C.J.; Delfín, C.M. Production of Fertilizer from Seawater with a Remote Control System. *Eng. Proc.* **2021**, *9*, 29. https://doi.org/10.3390/engproc2021009029

Academic Editors: Dimitrios Kateris and Maria Lampridi

Published: 3 December 2021

Publisher's Note: MDPI stays neutral with regard to jurisdictional claims in published maps and institutional affiliations.

1. Introduction

Water scarcity is expected to increase as water needs intensify, due to population growth, climate change, and agricultural demands [1]. Spain has invested large amounts of capital to increase the availability of water over the course of this century. Desalination consumes high amounts of energy, and may cause deleterious effects on marine ecosystems [1]. The amount of water available for irrigation is limited, and treated seawater can be used for tree plantations close to the sea. Water for irrigation can mix seawater with fresh water at a 1:30 dilution [2], which is known as diluted seawater (DSW). Soil ammonium concentration in wetlands increases with salinization caused by saltwater incursion. Long-term saltwater exposure of intact soil shows an exchange of salt cations that increases the reactive N being released.

The existence of inorganic fertilizers is decreasing. Cover cropping provides plant-available nitrogen in organic systems economically, as well as manure-based compost [3]. Dry organic fertilizers such as fishery waste and guano are widely used, but their quantity is limited, and mineralization of their N can take between four and eight weeks [3]. Sea salt has been applied as a source of sea mineral solids for foliar and soil treatments [4,5]. In a total replacement scenario, the increase in water cost EUR ha^{-1}) with respect to production costs is far more relevant than fertilizer cost for all crops; water cost ranges from 2.6% for lettuce to 35.1% for lemons [5]. Fertilizer increase represents less than 1% of the production costs for all crops. Irrigation of faba bean plants with diluted seawater (3.13 to 6.25 dS m^{-1}) led to significant reductions in shoot length, shoot dry weight, number of leaves per plant, and photosynthetic pigments [6]. A bittern solution, which is a byproduct of a solar salt production process, can be used as a reliable source of Mg and other nutrients required for plants; 390 ppm is the ideal bittern dosage [7].

A fertilizer extraction system on a canoe, with process control, was implemented. The control system is managed remotely through a LoRa system, and the main actions are controlled within a loop at the boat. Monitored data are transmitted to the shore to keep track of the process and the amount of fertilizer obtained.

2. Equipment Design

The fertilizer extraction system uses a stainless steel tank fixed over the boat base, with an output cone at the bottom (Figure 1a). The closed tank presents an injection/extraction pump, a powder dosifier, and an agitator fixed at its surface. The injection pump sucks water from the sea, filling the tank with seawater, and when it turns in the opposite direction it extracts the salt water produced by the chemical reaction and returns it to the sea. The 0.5 kg h^{-1} volumetric dosifier (model Kö 2-1, Werner Koch Maschinentechnik GMBH, Germany) feeds the powder to the tank during each event. The chamber of the dosing roll is filled equally before it is emptied at a constant speed. The agitator is composed of a two-pitched-blade turbine that is fixed to the motor blade. This turbine produces axial and radial flows with an intermediate shearing effect, and its flow can be controlled with a speed controller.

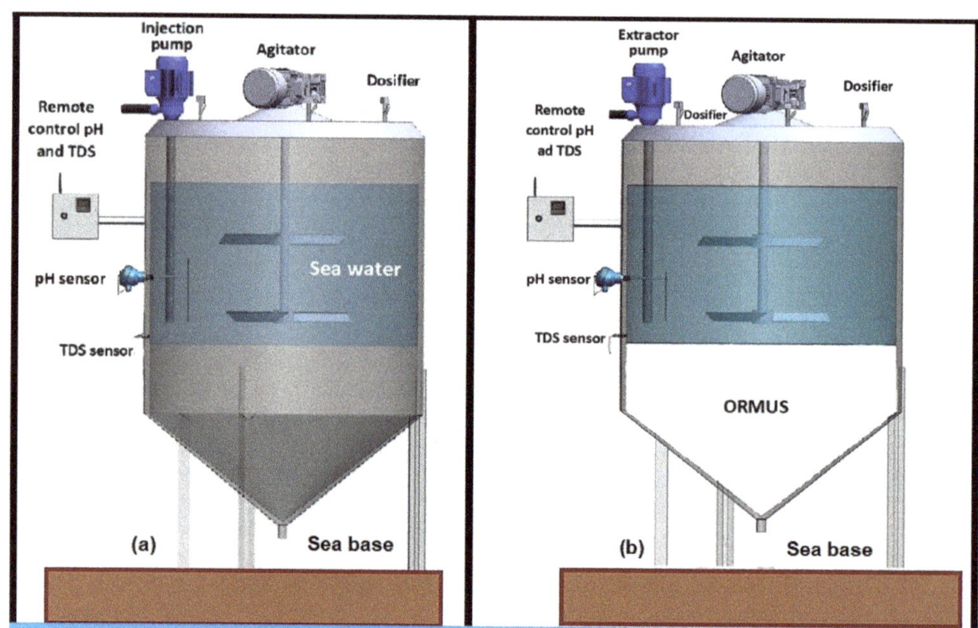

Figure 1. Process equipment (**a**) when seawater is injected, and (**b**) after ORMUS is produced.

3. Instrumental Equipment

The monitoring equipment for the processing tank uses pH (HI-11310, Hanna, Woonsocket, RI, USA) and TDS (HI-7634-00, Hanna Instruments, Woonsocket, RI, USA) sensors (Figure 1b). A pH controller (BL 931700, Hanna, Woonsocket, RI, USA) with a set point fixed at 10.78 controls the sodium hydroxide dosage. Measurements are directly acquired from a pH electrode in the range from 2 to 16, with a resolution of 0.01; the cable of the glass probe is 1 m long. A two-point calibration is performed daily and calibrated through trimmers on the front panel. The TDS sensor is calibrated by the HI-70030 solution at $12880 \text{ } \mu\text{S cm}^{-1}$, and by the HI-70038 reference at $6.44 \text{ } \mu\text{S cm}^{-1}$.

Total dissolved solids (TDS) combine the sum of all ion particles that are smaller than 2 microns. Freshwater can have a maximum conductivity of 2000 mg L^{-1}, while saline water conductivity is always over 3700 mg L^{-1}. A TDS controller (model BL-983319-0, Hanna Instruments, USA) is employed with a 0.5 conversion factor and temperature compensation. Water TDS increase when seawater is injected. Once the sodium hydroxide is applied, pH increases up to 10.78. Salts separate from the sodium water, and the salts gravimetrically deposit due to their higher density. Once the higher TDS value is reached, the pump at the top extracts the salty water with sodium and returns it to the sea. As the electrical conductivity at the top of the tank becomes 0 as it is full of air, it is refilled with purified water and agitated again. This process is repeated 3 times.

An ESP32-WROOM-32 microcontroller board (Espressif, China) controls the data acquisition and transmits the values through the LoRa transmitter. Controllers for pH and TDS are used in real time to turn off the dosifier and start removing the sodium-concentrated water.

4. Results and Discussion

Dosing of powder implies that it is homogeneous and does not carry small stones. The sodium hydroxide powder was dry, and its diameter was smaller than 1 mm. When the dosifier applied the powder over the tank water surface, it was too light to mix. Therefore, powder was introduced using a metallic screw and mixed with the water 20 cm above the bottom of the tank. Three different pH probes were installed at that height within the periphery of the tank, and their values varied considerably. The pH at the place where sodium hydroxide left the screw increased instantly, and was higher than on the opposite side. For example, seawater pH increased to 8.5 at the output of the screw, but on the opposite side pH was only 7. The closed loop was difficult to implement until another two dosifiers were added.

In the first experiment, a remote controller turned on the pump and filled the tank with 100 L of seawater. The dosifier added sodium hydroxide until the seawater pH reached 10.78 near the dosifier screw. The agitator turned slowly, and heating was noted within the tank, with an endothermic reaction taking place. The pH controller turned off the dosifier and, after three minutes, the temperature became constant within the tank. Under constant temperature, the agitator was stopped for a period of three hours.

With continuous agitation at 150 rpm, pH quickly reached 10.78 (Figure 2). If timers were used with the agitators at 150 rpm (purple line Figure 2), pH remained constant until the dosifier worked again. It is better to provide 2 kg of NaOH and wait until pH reaches 10, and then slowly add the powder until it reaches 10.78, reducing overheating of the sensors. A fuzzy algorithm was used to implement the dosage.

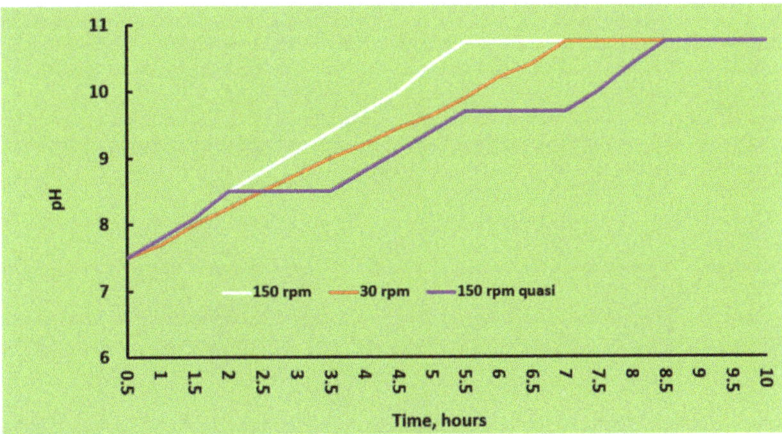

Figure 2. pH monitoring at different agitator speeds.

An event-driven controller was used instead a time-driven one, as fewer events took place at times when the controlled process did not require high accuracy [8]. Event-driven controllers can achieve a compromise between processor load and control performance. Three classes of control algorithms can occur according to the way in which actions take place in time. These classes can be classified as synchronous, semi-synchronous, and asynchronous. In this process, TDS and pH are measured synchronously within the tank, but the activation of different actuators is asynchronous [8]. At a distance of 5 km from where the boat was located, the wireless system with LoRa worked properly. A spreading factor of 12, a frequency of 915 MHz, and a coding rate of 4/6 were used, obtaining an RSSI of -75 at 5 km. This result is consistent with LoRa experiments using normal antennas, where 22 km receptions over seawater were reported [9].

Large desalination plants using reverse osmosis in Spain, processing more than 100,000 m^3 day^{-1}, can obtain freshwater from seawater at a cost of EUR 0.36–0.53 m^{-3} [1]. Nutritional deficiencies (mainly Ca_2^+ and Mg_2^+) when using diluted seawater (DSW) can be fixed by blending DSW with other hard water sources [7]. The combined use of desalinated seawater and brackish water reduces the salinization of underground water, enabling the reclamation contaminated aquifers [10]. The water obtained from our prototype can also yield other interesting byproducts. Although fertilizer costs with respect to production only account for 1%, fertilizers are scarce in Mexico, and must be imported.

Author Contributions: Conceptualization, F.H., C.M.D. and C.J.G.; writing—original draft preparation, C.J.G. and C.M.D.; writing—review and editing, F.H. and C.J.G.; supervision, F.H. All authors have read and agreed to the published version of the manuscript.

Funding: This research was funded by UNIVERSITY AUTONOMA CHAPINGO, grant number 20014-DTT-65.

Conflicts of Interest: The authors declare no conflict of interest.

References

1. March, H.; Saurí, D.; Rico-Amorós, A.M. The End of Scarcity? Water Desalination as The New Cornucopia for Mediterranean Spain. *J. Hydrol.* **2014**, *519*, 2367–2660. [CrossRef]
2. Sgherri, C.; Kadlecova, Z.; Pardossi, A.; Navari-Izzo, F.; Izzo, R. Irrigation with diluted sea water improves the nutritional value of cherry tomatoes. *J. Agric. Food Chem.* **2008**, *56*, 3391–3397. [CrossRef] [PubMed]
3. Hartz, T.K.; Smith, R.; Gaskell, M. Nitrogen Availability from Liquid Organic Fertilizers. *HortTechnology* **2010**, *20*, 169–172. [CrossRef]
4. Mekki, B.B.; Orabi, S.A. Response of prickly lettuce to uniconazole and irrigation with diluted seawater. *Am. Eurasian J. Agric. Environ. Sci.* **2007**, *2*, 611–618.
5. Álvarez, V.; Gallego-Elvira, B.; Maestre-Valero, J.F.; Martin-Gorriz, B.; Soto-Garcia, M. Assessing concerns about fertigation costs with desalinated seawater in south-eastern Spain. *Agric. Water Manag.* **2020**, *239*, 106257. [CrossRef]
6. Sadak, S.H.; Abdelhamid, M.T.; Schmidhalter, U. Effect of foliar application of aminoacids on plant yield and physiological parameters in bean plants irrigated with seawater. *Acta Biol. Colomb.* **2015**, *20*, 141–152.
7. Wijayarathne, U.; Wasalathilake, K.; Vidanage, P. Development of a Multi-Nutrient Fertilizer from Liquid Waste of Solar Salt Manufacturing Process. In Proceedings of the 2015 Moratuwa Engineering Research Conference (MERCon), Moratuwa, Sri Lanka, 7–8 April 2015; pp. 24–28.
8. Sandee, J.H.; Heemels, W.P.; van den Bosch, P.P. Event driven control as an opportunity in the multidisciplinary development of embedded controllers. In Proceedings of the American Control Conference, Portland, OR, USA, 8–10 June 2005; Volume 3, pp. 1776–1781.
9. Jovalekic, N.; Drndarevic, V.; Pietrosemoli, E.; Zennaro, I. Experimental Study of LoRa Transmission over Seawater. *Sensors* **2018**, *18*, 2853. [CrossRef] [PubMed]
10. Reca, J.; Trillo, C.; Sánchez, J.A.; Martínez, J.; Valera, D. Optimization model for on-farm irrigation management of Mediterranean greenhouse crops using desalinated and saline water from different sources. *Agric. Syst.* **2018**, *166*, 173–183. [CrossRef]

engineering proceedings

Proceeding Paper

Farm-Scale Greenhouse Gas Emissions' Decision Support Systems †

Evangelos Alexandropoulos, Vasileios Anestis * and Thomas Bartzanas

Department of Natural Resources Management and Agricultural Engineering, Agricultural University of Athens, Iera Odos 75, 11855 Athens, Greece; vagalexandr1991@aua.gr (E.A.); t.bartzanas@aua.gr (T.B.)
* Correspondence: vanestis@aua.gr
† Presented at the 13th EFITA International Conference, online, 25–26 May 2021.

Abstract: In this paper, 15 farm-scale Green House Gas-based (GHG-based) decision support (DS) tools were evaluated based on a number of criteria (descriptive evaluation), as well as the parameters requested as inputs and the outputs, all of which are considered important for the estimation procedure and the decision support approach. The tools were grouped as emission calculators and tools providing indicators in terms of more than one pillar of sustainability. The results suggest an absence of automatic consultation in decision support in most of the tools. Furthermore, dairy and beef cattle production systems are the most represented in the tools examined. This research confirms a number of important functionalities of modern GHG-based DS tools.

Keywords: GHG-based decision support; GHG emissions' estimation; multi-pillar assessment; calculators

Citation: Alexandropoulos, E.; Anestis, V.; Bartzanas, T. Farm-Scale Greenhouse Gas Emissions' Decision Support Systems. *Eng. Proc.* **2021**, *9*, 22. https://doi.org/10.3390/engproc2021009022

Academic Editors: Dimitrios Kateris and Maria Lampridi

Published: 25 November 2021

Publisher's Note: MDPI stays neutral with regard to jurisdictional claims in published maps and institutional affiliations.

1. Introduction

Nowadays, the need to communicate high-quality estimates of greenhouse gases' (GHG) emissions as well as the effect of mitigation strategies at the livestock farm level to various stakeholders has become more and more intense [1]. Furthermore, high-quality measurements of GHG emissions at the livestock farm level are, in practical terms, almost impossible. In this respect, the role of farm-scale GHG-based Decision Support Systems (DSSs) is expected to increase in importance [2].

Until today, most of the research that has been conducted provides information about how GHG emissions are estimated and how these are involved in the sustainability assessment of an agricultural system [3]. In this review, various aspects of DSSs related to GHG emissions' estimation and modeling, their use, as well as the information requested and provided were analyzed. As a result, basic characteristics of a modern GHG-based DSS are suggested.

2. Materials and Methods

A literature search was conducted and resulted in a selection of relevant, previously published review papers [4], papers focusing on specific tools, project reports, and case studies. Fifteen tools with potential for GHG emissions mitigation strategy selection were finally identified and used based on this literature search and proposals from the partners of a European research project. Two evaluation sections can be distinguished: (a) based on descriptive criteria; (b) based on checklists.

3. Results and Discussion

3.1. Evaluation Based on Descriptive Criteria

Table 1 shows the criteria and sub-criteria based on which the studied DS tools were descriptively evaluated.

163

Table 1. Decriptive evaluation criteria and their evaluation parameters.

Criterion	Sub-Criteria
Degree of accessibility	Availability. Registration required. Complexity in access.
User Friendliness	Level of expertise. Degree of information provision—website and user interface. Provision of manuals/guidance. Degree of simplicity. Presentation of results. Availability of results for downloading or saving. Error management. Design of user interface.
Stakeholders	Agriculture sub-sector represented. Target group. Stakeholders' involvement.
Methodology of sustainability	Level of sustainability assessment. Types of gas emissions (farm level). Estimation methodology. Decision support approach.
Modeling aspects	Software. Type of modeling. Modeling method transparency.

3.2. Checklists

3.2.1. Inputs of DS Tools

Table 2 presents the input categories and the types of inputs whose existence or non-existence was checked for the various DS tools studied.

Table 2. Input categories and types of inputs checked in the tools evaluated.

Input Categories	Types of Inputs
Soil-related	Soil type. Year of soil analysis. Soil organic matter content. Soil pH. Soil drainage. Soil component percentages (clay, sand, silt).
Crop-related	Crop type. Crop diversity. Crop rotation. Residue management. Use of fertilizers. Composition of fertilizers. Use of pesticides. Cutting frequency. Irrigation type. Irrigation frequency.
Climate-related	Country of tool's validity. Agro-ecological zone. Climate type.
Livestock-related	Livestock species. Livestock breed. Livestock Age. Stage in livestock development.
Manure management-related	On-farm manure management. Manure field application.
Livestock feed management-related	Livestock ration. Cost of feed.

3.2.2. Outputs of DS Tools

The categories of outputs that were checked in the DS tools evaluated were the following: (a) Environmental Impact Category Indicators (EICIs); (b) Emission per source; (c) CO_2 (emission to air); (d) CH_4 (emission to air); (e) N_2O (emission to air); (f) Total GHG emissions (in CO_2 eq); (g) NH_3 (emission to air); and (h) Feed consulting.

3.3. DS Tools' Evaluation

A short description of the DS tools in the context of the evaluation criteria (Section 3.1) and the inputs and outputs (Section 3.2) is given in Table 3.

The use of scores for sustainability indicators is the major difference in the results presentation between the emission calculators and the all-pillar DS tools. Dairy and beef cattle production seems to be the livestock sub-sector that is most represented in the DS tools examined. Nevertheless, in the majority of these DS tools, stakeholders are not involved in their development. Scenario analysis, contribution analysis and progress monitoring seem to be the decision support approaches that are used by the majority of the DS tools examined. With respect to the provided outputs, a minority of emission calculators further provide livestock feed consultation as a type of output. Furthermore, the evolution to consulting decision support would be innovative and of importance for the wider use of such tools [5].

Table 3. Evaluation of the DS tools examined.

a/a	DS Tool	Evaluation
1	Cool Farm Tool v2.0	Online. Quantitative modeling approach. Includes the majority of inputs. Various Livestock systems. GHG emissions and farm costs. It does not provide NH_3 emissions and consulting.
2	FarmAC	Emission calculator. Quantitative modeling approach. Focus on cattle GHG emissions. Majority of soil inputs not included. Provides NH_3 and herd energy requirements estimations.
3	Overseer	Country-specific license purchased online tool. Quantitative modeling approach. Includes almost all inputs. Provides feed consulting and management functions, N-balance and GHG emissions.
4, 5	Carbon Navigator—Beef and Dairy	Developed for the Irish beef and dairy production systems. Quantitative modeling approach. Focus on beef and Dairy GHG emissions. It includes economic analysis.
6	KSNL	A questionnaire-form online multi-pillar tool made for German agricultural production systems. Semi-quantitative approach. Provides a multi pillar sustainability assessment with various multi pillar indicators.
7	SMART	A sustainability assessment tool based on SAFA indicators. Semi-quantitative approach. It is not freely accessible. Provides a sustainability assessment with various multi pillar indicator.
8	SAFA	A free educational, sustainability assessment tool (FAO). Semi-Quantitative approach. Three indicators for GHG emissions
9	RISE 3.0	A free informative multi-pillar sustainability assessment tool which estimates GHG and NH_3 emissions. Both online and offline functionality. Semi-quantitative modeling approach.
10	BEK	MS Excel spreadsheet tool. For German agricultural production systems. Semi-quantitative modeling approach. Provides a template for GHG emissions' calculations for various processes.
11	AgBalance®	An LCA-based tool. Quantitative modeling approach. Contribution analysis. Refers to crop production processes (no livestock-related inputs). Provides GHG and NH_3 emissions' estimation.
12	DLG	A sustainability assessment tool developed for German agricultural production systems. Semi-quantitative modeling approach. Includes the GHG emissions in the sustainability assessment.
13	HOLOS	A tool for Canadian livestock systems. Quantitative modeling approach. Does not include many of the soil-related inputs. Provides GHG and NH_3 emissions' estimates and a feed estimation report.
14	EX-ACT (version 8.0)	A freely accessible MS Excel spreadsheet tool (FAO). Quantitative modeling approach. Does not include the majority of soil related inputs and the majority of livestock inputs. Estimates GHG emissions.
15	GLEAM 2.0 (GLEAM-I):	A freely accessible MS Excel spreadsheet tool (FAO). LCA-based. Quantitative modeling approach. Provides GHG estimates for various livestock systems based on three input categories herd, feed and manure.

4. Conclusions

A modern GHG-based DSS for livestock systems would need to include clearly defined system boundaries and recently published emission estimation algorithms (e.g., the 2019 refinement of the IPCC 2006 Guidelines and Tier 2 approaches). It would need to consider GHG and ammonia emissions from all sources at the farm level (including feed crop production in case this in under the control of the livestock farmer) as well as soil carbon sequestration, by respecting the N and C cycles. Inputs from all the categories described for the emission calculator tools would be required in this respect. It would finally need to: (a) have an online user interface; (b) be easily accessible; (c) target inexperienced users and provide detailed guidelines regarding its use (but also be transparent with respect to the methodology followed); (d) provide easily comprehensible errors and easy handling of them; (e) involve stakeholders' opinions before its release; and (f) have multi-national validity.

Author Contributions: Conceptualization, E.A. and V.A.; methodology, E.A.; investigation, E.A.; writing—original draft preparation, E.A. and V.A.; writing—review and editing, E.A., V.A. and T.B.; supervision, T.B.; project administration, V.A. and T.B.; funding acquisition, T.B. All authors have read and agreed to the published version of the manuscript.

Funding: This research was co-funded by the European Regional Development Fund of the European Union and Greek national funds through the Operational Program Competitiveness, Entrepreneurship and Innovation 2014–2020, under the call "European R&T Cooperation–Action of Granting Greek Bodies that Successfully Participated in Joint Calls for Proposals of the European ERA-NET Networks 2019b" (MELS project-funded under the Joint Call 2018 ERA-GAS, SusAn and ICT-AGRI on 'Novel technologies, solutions and systems to reduce the greenhouse gas emissions in animal production systems'-project code: T11EPA4-00076).

Data Availability Statement: The study did not report any additional data.

Acknowledgments: The authors would like to acknowledge the contribution of the partners of the MELS project (www.mels-project.eu) in collecting and proposing country-specific GHG-based DS tools to be evaluated.

Conflicts of Interest: The authors declare no conflict of interest. The funders had no role in the design of the study; in the collection, analyses, or interpretation of data; in the writing of the manuscript, or in the decision to publish the results.

References

1. Schils, R.L.M.; Ellis, J.L.; de klein, C.A.M.; Lesschen, J.P.; Petersen, S.O.; Sommer, S.G. Mitigation of greenhouse gases from agriculture: Role of models. *Acta Agric. Scand. A Anim. Sci.* **2012**, *62*, 212–224. [CrossRef]
2. Vibart, R.; de Klein, C.; Jonker, A.; van der Weerden, T.; Bannink, A.; Bayat, A.R.; Crompton, L.; Durand, A.; Eugène, M.; Klumpp, K.; et al. Challenges and opportunities to capture dietary effects in on-farm greenhouse gas emissions models of ruminant systems. *Sci. Total Environ.* **2021**, *769*, 144989. [CrossRef] [PubMed]
3. Ahmed, M.; Ahmad, S.; Waldrip, H.M.; Ramin, M.; Raza, M.A. Whole Farm Modeling: A Systems Approach to Understanding and Managing Livestock for Greenhouse Gas Mitigation, Economic Viability and Environmental Quality. In *Animal Manure: Production, Characteristics, Environmental Concerns, and Management, Volume 67*; American Society of Agronomy: Madison, WI, ASA, 2020; pp. 345–371. [CrossRef]
4. Arulnathan, V.; Heidari, M.D.; Doyon, M.; Li, E.; Pelletier, N. Farm-level decision support tools: A review of methodological choices and their consistency with principles of sustainability assessment. *J. Clean. Prod.* **2020**, *256*, 120410. [CrossRef]
5. Rose, D.C.; Sutherland, W.J.; Parker, C.; Lobley, M.; Winter, M.; Morris, C.; Twining, S.; Ffoulkes, C.; Amano, T.; Dicks, L.V. Decision support tools for agriculture: Towards effective design and delivery. *Agric. Syst.* **2016**, *149*, 165–174. [CrossRef]

Proceeding Paper

Society's View on Autonomous Agriculture: Does Digitalization Lead to Alienation? †

Olivia Spykman [1,*], Agnes Emberger-Klein [2], Andreas Gabriel [1] and Markus Gandorfer [1]

[1] Institute for Agricultural Engineering and Animal Husbandry, Bavarian State Research Center for Agriculture, Kleeberg 14, 94099 Ruhstorf a.d. Rott, Germany; andreas.gabriel@lfl.bayern.de (A.G.); markus.gandorfer@lfl.bayern.de (M.G.)

[2] Department of Marketing and Management of Biogenic Resources, Weihenstephan-Triesdorf University of Applied Sciences, TUM Campus Straubing for Biotechnology and Sustainability, Am Essigberg 3, 94315 Straubing, Germany; agnes.emberger-klein@hswt.de

* Correspondence: olivia.spykman@lfl.bayern.de

† Presented at the 13th EFITA International Conference, online, 25–26 May 2021.

Abstract: Digital and autonomous technologies enter the agricultural market at an increasing rate, yet little is known about society's view on this development, although the public is an important stakeholder. By means of a discrete choice experiment (n = 675), societal preferences for different weed control technologies and tractor types of different degrees of autonomy are investigated. The model applied focuses on emotion-related covariates. The results indicate preferences for conventional or autonomous tractors and for methods of weed control that reduce the need for herbicides. Additionally, positive associations with images of robots correlate with the rejection of conventional tractors in the discrete choice experiment.

Keywords: autonomous weed management; digital farming technologies; discrete choice experiment; field crop robot; public acceptance

Citation: Spykman, O.; Emberger-Klein, A.; Gabriel, A.; Gandorfer, M. Society's View on Autonomous Agriculture: Does Digitalization Lead to Alienation? . *Eng. Proc.* **2021**, *9*, 12. https://doi.org/10.3390/engproc2021009012

Academic Editors: Dimitrios Aidonis and Aristotelis Christos Tagarakis

Published: 24 November 2021

1. Introduction

Increasingly autonomous machines represent a groundbreaking development in agricultural technology, yet their success depends as much on economics and farmer acceptance as on the approval of society. To prevent negative societal reactions to the on-going fourth agricultural revolution, some researchers [1] advocate for transferring the concept of Responsible Research and Innovation to agricultural technology development. Additionally, expert interviews point to the importance of public acceptance, or rather the lack thereof, as a barrier to sustainable innovations in the German agricultural sector [2]. More recent research shows that the German population is still largely undecided on the use of digital technologies in agriculture, including autonomous equipment for weed management [3]. A farmer survey in Bavaria, Germany, however, has indicated that fear of creating an image of alienated agriculture significantly reduces the intent of farmers to invest in field crop robots [4], pointing to the importance of societal approval for the adoption of autonomous farming equipment. Therefore, knowledge is needed on society's perspective on autonomous cropping equipment under specific consideration of underlying emotional factors to better understand the perception of autonomous weed management.

2. Materials and Methods

A discrete choice experiment (n = 675) with three attributes, each subdivided into three levels, covering methods of weed management (conventional spraying, spot-spraying, and hoeing), tractor type (conventional tractor, autonomous tractor, and small swarm robots), and changes in consumer price of the end product (none, moderate and strong), respectively, is analyzed using a Hierarchical-Bayes estimation in Lighthouse Studio 9.5.3

(Sawtooth Software, Provo (UT), USA). This experiment was part of a larger online survey (n = 2012) conducted among the German adult population in June 2018 (see [3] for more details). In nine repetitions of the experiment, participants were presented three choice cards, i.e., combinations of one level per attribute, and asked to select the choice card they considered most attractive or alternatively the "none"-option. Four mean-centered covariates were included in the model: Likert-type items "Digital farming technologies alienate the farmer from his/her soil and animals" and "Family farming structures seem valuable and should be preserved" as well as scores of spontaneous associations with images of a large autonomous tractor and a swarm of small robots, respectively, during field work. These covariates are expected to provide an emotional dimension to the experiment, adding to its socioeconomic and technology-focused evaluation [5]. The Likert-type items are each measured on 5-point Likert-type scales (from 1 = "fully disagree" to 5 = "fully agree") and correspond to the concern identified in the farmer survey [4] and the disagreement between society's romanticized view of agriculture and the actual level of technology use on farms in Germany [6], respectively. The image association scores are measured on a scale from −3 to +3, with scores given depending on the entered associations being rated negative (−1), neutral or empty (0), or positive (+1) (cf. [3]). Three associations per image were possible. These two covariates permit a comparison of spontaneously formed image associations with stated choice preferences, which allows a comparison of implicit and explicit associations with different methods of crop protection (cf. [7]). The results of the model are presented as theta-weights, which indicate by how much the part-worth utility of an attribute level changes if a participant evaluates a given covariate better or worse than the sample average.

3. Results

Participants in the experiment considered the method of weed management more important (46%) than changes in the end product price (32%). The tractor type is considered least important (22%). The evaluation of a given product combination is thus predominantly guided by the method of weed management. The sample-average part-worth utilities for each attribute level may be found in Table 1 (column "Constant"). The further columns provide the respective theta-weights for each covariate and attribute level. The overall fit of the model is good (RLH = 0.63).

Table 1. Theta-coefficients of covariates in HB-estimation.

	Attribute/Level	Constant	Family Farms Worth Preserving	Farmer Alienation from Soil/Animals	Image of Large Tractor	Image of Swarm Robots
Weed Mngmt	Conventional [H]	−1.530 *	−0.314 *	0.053	0.145 *	−0.173 *
	Spot-Spraying [H]	−0.018	−0.163 *	−0.138 *	0.125 *	0.038
	Hoeing	1.548 *	0.477 *	0.085	−0.270 *	0.136
Tractor Type	Conventional	0.138 *	−0.014	0.150 *	−0.139 *	−0.243 *
	Autonomous	0.088 *	0.002	−0.046	0.052	0.013
	Small Swarm Robots	−0.226 *	0.012	−0.103 *	0.086 *	0.230
Price Increase	None	0.734 *	−0.265 *	−0.196 *	−0.014	−0.014
	Moderate	0.325 *	0.097	−0.027	0.062	0.029
	Strong	−1.058 *	0.167 *	0.223 *	−0.048	−0.015
	NONE	−1.169 *	−0.613 *	0.081	−0.395 *	−0.242

Mean root likelihood (RLH): 0.63; [H] herbicide-based method; * significant at 90% confidence level, 2-sided test.

The importance of the weed management attribute breaks down into significant rejection of conventional spraying and preference for hoeing at the sample-level. Spot-spraying, a method using reduced amounts of herbicides, ranks between the two in terms of utility provided, presenting a possible compromise. The rejection of conventional spraying increases among those considering family farming structures important or positively evaluating the image of swarm robots. Those viewing images of large autonomous tractors positively, on the other hand, draw greater utility from conventional spraying than the sample average. For hoeing, these trends reverse for proponents of family farms and large autonomous tractors. The compromised solution of spot-spraying is rejected significantly by family-farm proponents and those agreeing that digitalization alienates farmers from their soils and animals. Conversely, supporters of large-scale tractors based on image associations also draw above-average utility from spot-spraying.

Regarding the tractor type, the sample draws most utility from conventional tractors, while in comparison finding large autonomous tractors acceptable and favoring swarm robots the least. This evaluation is even more pronounced in those participants who see digitalization as a reason for farmer alienation. Conversely, those evaluating images of large autonomous tractors and swarm robots positively reject conventional tractors at an above-average rate. Those scoring positive on image associations with the large autonomous tractor also rate swarm robots significantly more positively, but not autonomous tractors. A similar trend cannot be observed for positive associations with the image of swarm robots.

The price increase attribute indicates that the sample draws the most utility from no price change but also finds moderate price changes acceptable compared to strong price changes. Strong price changes are rejected. Those individuals agreeing that family farms are worth preserving or that digitalization leads to alienation, however, indicate clearly lower utility from no price change and higher utility from strong price change than the sample average. The image evaluations have no significant impact on the part-worth utilities for this attribute. The "none"-option's negative part-worth utility indicates that participants saw more value in choosing a choice card than the "none"-option. This effect is even more pronounced for those valuing family farms highly or having positive associations with the image of a large autonomous tractor.

4. Discussion and Conclusions

The results of this discrete choice experiment indicate that not all emotion-related items in the model equally impact the evaluation of digital farming technologies. Preferences for family farming structures are associated with the rejection of herbicide-based weed management methods and a greater willingness to accept higher prices. Agreement that digitalization leads to farmer alienation is consequently associated with the rejection of technologies, such as spot-spraying and swarm robots, as well as an above-average preference for conventional tractors. This group also indicates a greater willingness to accept higher prices. The spontaneous image associations match the stated preferences only to a certain degree: Participants with positive image associations tend to reject conventional tractors. However, only those with positive views on large autonomous tractors draw greater utility from swarm robots, which is inconsistent. Positive associations with large autonomous tractors are further linked to increased utility from herbicide-based methods of weed control and reduced utility from hoeing. Those indicating positive associations with swarm robots only indicate decreased utility from conventional spraying.

The inconsistency between image associations and part-worth utilities in the tractor type corresponds to previously found discrepancies between implicit and explicit evaluations [7] and may indicate that participants are driven more by a rejection of the conventional tractor rather than a clear preference for an alternative. Conversely, the type of alternative viewed positively does seem to correlate with participants' opinions on weed management. Those rating the image of swarm robots positively strongly reject conventional spraying, sharing the average sample's preference for precision agriculture methods, such as spot-spraying and automatic hoeing. Yet, participants rating the image of large autonomous tractors positively support herbicide use, as they draw decreased utility, relative to the whole sample, only from hoeing. More thorough investigation of underlying motives behind such relationships may provide a better understanding of these groups in society. The relationship between Likert-type items and part-worth utilities indicates that hesitancy towards certain aspects of digital farming tools may be due to misconceptions. For example, proponents of family farm structures reject herbicide-based measures. It is possible that they are also proponents of organic farming, a variable not included in the present model. Organic farming and family farming, however, are not synonymous, so this insight may be used to address clearly the differences between the two and explain the utility of different methods of weed management when informing the public. Similarly, those fearing alienation through digitalization project this specifically on spot-spraying and swarm robots. However, precision tools, such as spot-spraying, may provide the farmer with data of his/her soils. Similarly, swarm robots, while physically removing the farmer from the field, can also collect valuable information otherwise inaccessible to the farmer, thus leading to better knowledge about rather than alienation from his/her fields. These aspects should also be communicated to the public to avoid rejection of helpful technologies due to misunderstandings and to reduce the already existing gap between agricultural reality and society's image thereof (cf. [6]).

Author Contributions: Conceptualization, A.G. and M.G.; formal analysis, A.E.-K.; investigation, A.G. and M.G.; data curation, A.G. and O.S.; writing—original draft preparation, O.S.; writing—review and editing, O.S., A.G. and A.E.-K. All authors have read and agreed to the published version of the manuscript.

Funding: This research was funded by the Bavarian State Ministry for Food Agriculture and Forestry, grant number D/17/01.

Institutional Review Board Statement: The study was conducted in accordance with the General Data Protection Regulation (EU) for anonymous surveys. No personal data were processed.

Informed Consent Statement: Participation to the survey was based on explicit consent.

Data Availability Statement: The data presented in this study are available on request from the corresponding author.

Acknowledgments: We thank Sebastian Schleicher for his assistance in survey design and data collection.

Conflicts of Interest: The authors declare no conflict of interest. The funders had no role in the design of the study; in the collection, analyses, or interpretation of data; in the writing of the manuscript, or in the decision to publish the results.

References

1. Rose, D.C.; Chilvers, J. Agriculture 4.0: Broadening Responsible Innovation in an Era of Smart Farming. *Front. Sustain. Food Syst.* **2018**, *2*, 87. [CrossRef]
2. König, B.; Kuntosch, A.; Bokelmann, W.; Doernberg, A.; Schwerdtner, W.; Busse, M.; Siebert, R.; Koschatzky, K.; Stahlecker, T. Nachhaltige Innovationen in der Landwirtschaft: Komplexe Herausforderungen im Innovationssytem. *Vierteljahresh. Wirtsch.* **2012**, *81*, 71–92. [CrossRef]
3. Pfeiffer, J.; Gabriel, A.; Gandorfer, M. Understanding the public attitudinal acceptance of digital farming technologies: A nationwide survey in Germany. *Agric. Hum. Values* **2020**, *38*, 107–128. [CrossRef]
4. Spykman, O.; Gabriel, A.; Ptacek, M.; Gandorfer, M. Farmers' perspectives on field crop robots—Evidence from Bavaria, Germany. *Comput. Electron. Agric.* **2021**, *186*, 106176. [CrossRef]
5. Spykman, O.; Emberger-Klein, A.; Gabriel, A.; Gandorfer, M. Feldroboter aus Sicht der Gesellschaft—Auswertung eines Discrete Choice Experiments. In Proceedings of the 41st Annual GIL Meeting, Potsdam, Germany, 8–9 March 2021; Meyer-Aurich, A., Gandorfer, M., Hoffmann, C., Weltzien, C., Bellingrath-Kimura, S., Floto, H., Eds.; Gesellschaft für Informatik e.V.: Bonn, Germany, 2021; pp. 295–300.
6. Zander, K.; Isermeyer, F.; Bürgelt, D.; Christoph-Schulz, I.; Salamon, P.; Weible, D. *Erwartungen der Gesellschaft an die Landwirtschaft*; Thünen-Institut: Braunschweig, Germany, 2013.
7. Römer, U.; Schaak, H.; Mußhoff, O. The perception of crop protection: Explicit vs. implicit associations of the public and in agriculture. *J. Environ. Psychol.* **2019**, *66*, 101346. [CrossRef]

engineering proceedings

Proceeding Paper

Have City Dwellers Lost Touch with Modern Agriculture? In Quest of Differences between Urban and Rural Population †

Andreas Gabriel * and Markus Gandorfer

Bavarian State Research Center for Agriculture, Institute for Agricultural Engineering and Animal Husbandry, Kleeberg 14, 94099 Ruhstorf a.d. Rott, Germany; markus.gandorfer@lfl.bayern.de

* Correspondence: andreas.gabriel@lfl.bayern.de

† Presented at the 13th EFITA International Conference, online, 25–26 May 2021.

Abstract: Frequently, urbanization and loss of an urban population's connection to agriculture are given as the main reasons for decreasing societal acceptance of modern-day agriculture. An online survey of the German population in 2018 provided two selective subsamples of rural (n = 337) and urban residents (n = 560). Comparing group differences with regard to (a) the general societal perspective on agriculture and (b) positions on the use of digital farming technologies shows only little evidence of significant contributions of the used predictor items. Thus, no generalized tendency can be found that city dwellers are more opposed to agricultural developments based on different attitudes and perceptions.

Keywords: attitudes; digital farming technologies; modern agriculture; rural; urban

Citation: Gabriel, A.; Gandorfer, M. Have City Dwellers Lost Touch with Modern Agriculture? In Quest of Differences between Urban and Rural Population. *Eng. Proc.* **2021**, *9*, 25. https://doi.org/10.3390/engproc2021009025

Academic Editors: Dimitrios Kateris and Maria Lampridi

Published: 29 November 2021

Publisher's Note: MDPI stays neutral with regard to jurisdictional claims in published maps and institutional affiliations.

1. Introduction

Urban society's knowledge of, understanding of, and relationship to modern agriculture is highly questioned by various agricultural stakeholders in Germany. In agricultural media, these issues are increasingly being taken up and discussed with politicians and the agricultural and food industry with regard to how to deal with the urban–rural discrepancy. Frequently, urbanization and loss of an urban population's connection to agricultural production are given as the main reasons for the perceived differences. Due to assumed alienation from the agricultural sector of people living in urban areas, it seems difficult to reconcile the benefits and legitimization of modern agricultural production with related social and environmental issues (e.g., animal welfare, insect mortality).

Nowadays, more than three quarters of the German population live in urban settlement areas [1]. Whereas at the beginning of the last century almost 40% of the population practiced agriculture themselves, less than 2% are currently employed in agriculture and related sectors [2]. Nevertheless, people are also consumers who are becoming increasingly responsible and sophisticated when it comes to transparent and sustainable food production [3]. Easier and faster access to information and higher media attention accelerate this process. The current literature does not clearly indicate whether attitudes toward and perceptions of contemporary agriculture differ between urban and rural populations [4,5]. However, there is evidence that the physical distance of urban dwellers from agriculture affects their knowledge on agriculture [6]. Many stakeholders and decision makers in the agricultural sector attribute agriculture's image problems to the disconnection of society from modern agriculture. However, is urbanization really the main reason for the lack of societal acceptance?

2. Materials and Methods

To explore this question, we draw on data from a 2018 online consumer survey querying societal acceptance of digital farming technologies. In addition to partially pre-quoted information on the sociodemographics of 2012 respondents, several items were asked about general attitudes toward agriculture, self-assessed knowledge about agricultural production processes and equipment, and trust in farmers' responsible acting. Furthermore, we investigated respondents' attitudes toward digital farming technologies (DFT) using Likert scales. Additionally, the survey participants should give spontaneous short associations with four shown pictures of digital technologies in practice (milking robot, cow-feeding robot, autonomous tractor sowing, swarm robots on the field). The web sources for the images used can be found in the supplementary material of this article. More details on the surveyed items and pictures can be found in a recent publication by Pfeiffer et al. [7].

For the purpose of this analysis, the population densities of the individual postal code areas were classed on the basis of the official divisions between urban and rural (150,000 inhabitants per km^2). We can thus compare a subsample of 560 respondents who live in larger cities and highly populated areas with a subsample of 337 respondents who live in towns with fewer than 100,000 inhabitants and a low regional population density. Both subsamples were tested for differences in sociodemographic characteristics and in deviation of median responses concerning the queried Likert scales and picture association ratings. Two binominal logistic regression models were designed to indicate the probability of group classification by predictors indicating (a) the general societal perspective on agriculture and (b) positions on the use of digital farming technologies (DFT).

3. Results

Members of the urban subsample have a significantly weaker relation to agriculture and show a slightly higher level of education. Differences can be measured in specific statements regarding self-assessed knowledge of current agricultural methods and trust in responsible practice of farmers.

Figure 1 represents the mean values of the two subsamples' responses to each surveyed rating scale. The 5-point scales range from "I fully agree" or "very high" to "I fully disagree" or "very low". Unsurprisingly, regarding self-assessed knowledge, both subsamples are very restrained when it comes to current practice and technology use. In this respect, urban dwellers are even more cautious than the rural population, as shown by the statistical comparison. There are almost no differences between urban and rural residents on statements about general attitudes toward agriculture. Both groups agree with statements on animal welfare, environmental protection, and the life quality of farmers. For the two items concerning trust in farmers' practice, the mean values are more in the middle of the scale range, with the rural population extending more trust to farmers.

A further scale measured different attitudes toward the benefits of digital farming technologies. Differences can be observed between the items but not between the subsamples. Both subsamples agree that the technologies primarily support the farming family and are undecided as to whether new technologies lead to the alienation of farmers from their land or animals. Finally, the mean ratings of the maximum three associations per shown picture were compared on a scale of triple positive to triple negative. Robotics in the context of dairy production was viewed more negatively than robotics in the context of arable farming. In particular, the image of a cow surrounded by a milking robot led to frequent negative associations. The city dwellers are more positive concerning field robotics, while rural residents are less critical when they are confronted with pictures of autonomous animal husbandry technologies.

Two logistic regression models were designed to indicate the probability of classification to the two groups and the explained variance of the indicating predictors. Regarding the first model (social perspective on agriculture), it is found that differences through items of self-assessed knowledge on modern agriculture, attitudes toward agriculture, and trust in farmers offer no explanatory contribution to the classification of the respondents to one of the two groups. For the second model (use of DFT), the variance explained by covariates indicating 'knowledge of agricultural machines', several 'attitudes towards DFT' and 'spontaneous associations' of shown pictures of selected technologies are also low (Table 1).

Only better knowledge of machines and equipment by the rural population and the respective direction of associations (positive, negative) with pictures of a large autonomous tractor and a feeding robot, respectively, provide statistically significant contribution to the classification. The odds indicate the corresponding probability of classification when changing these parameters by one scale step.

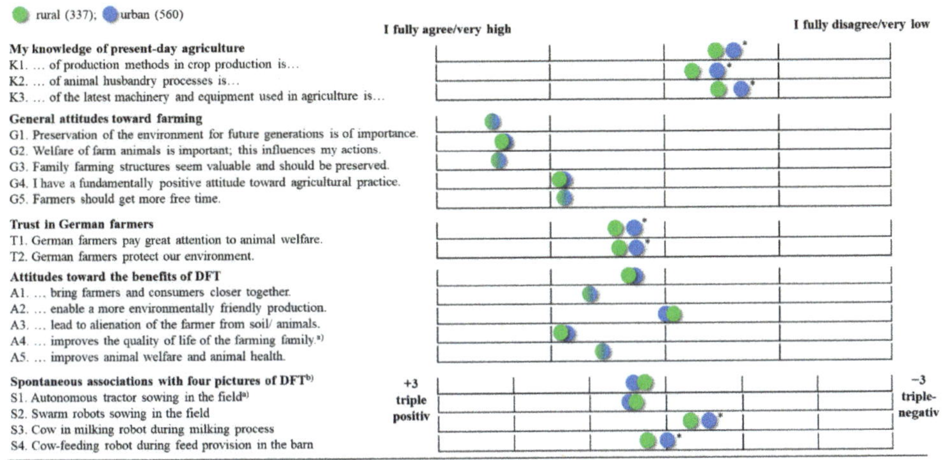

Figure 1. Mean values of subsamples; * = rejection of H0; nonparametric Mann–Whitney U test; [a] unequal variances; [b] aggregated value of associations from −3 'triple negative' to +3 'triple positive'.

Table 1. Binominal logistic regression model 2: positions on the use of DFT.

Predictors	B	SE	Wald	*p*	Odds
K3. Knowledge of machinery and equipment used	0.214	0.076	8.008	0.005 *	1.239
A1. DFT brings farmers and consumers closer together	0.095	0.094	1.018	0.313	1.100
A2. DFT enables a more environmentally friendly product	−0.177	0.118	2.254	0.133	0.838
A3. DFT leads to alienation of farmers from soil/animals	−0.125	0.073	2.958	0.085	0.883
A4. DFT improves the quality of life of a farm family	0.133	0.111	1.450	0.229	1.143
A5. DFT improves animal welfare and animal health [T]	0.437	0.427	1.046	0.306	1.548
S1. Autonomous tractor sowing in the field	0.246	0.061	5.778	0.016 *	1.158
S2. Swarm robots sowing in the field	0.080	0.064	1.580	0.209	1.084
S3. Cow in milking robot during milking process	−0.071	0.052	1.875	0.171	0.932
S4. Cow-feeding robot during feed provision	−0.180	0.059	9.351	0.002 **	0.835
Constant	−0.311	0.584	0.285	0.594	0.732

$\chi^2(10) = 33.111$, $p < 0.000$; Nagelkerke's $R^2 = 0.049$ (Cox and Snell $R^2 = 0.036$) I overall percentage of accuracy classifications (contribution of predictor variables) = 63.1% (+0.7%) I [T] variable was 1/x-transformed due to lacking linearity (tested and assessed using the Box-Tidwell procedure [8]). Bonferroni correction was applied to all 10 model terms [9] I * ($p < 0.05$); ** ($p < 0.01$).

4. Discussion and Conclusions

The two clearly distinguished subsamples of rural and urban residents reveal differences in the level of education and physical contact with agriculture. In addition, lower rates of self-assessed knowledge and trust in modern agriculture can be shown among city dwellers. City dwellers were also significantly more critical of the images concerning animal husbandry than the comparison group. The two logistic models regarding general societal perspective on agriculture and positions on the use of digital farming technologies, however, show that most predictors provide no explanatory contribution to the classification into the two groups. In both models, the classification performance of the factors used is very low. On the basis of the two subsamples, it can be concluded that there is no generalized tendency that the urban population is more opposed to the development of modern agriculture based on different attitudes and perceptions in contrast to their rural counterparts. Blanket pointing at the urban population to explain the agricultural sector's current societal problems is thus misleading, particularly since the ongoing urbanization in Germany creates a growing fraction of society that, on average, holds higher income, spends more on food, and demands high-quality food products. Therefore, agricultural communication strategies should address the entire population to explain state-of-the-art agriculture and to communicate modern food production and the various services (e.g., ecosystem services) provided by the agricultural sector.

Supplementary Materials: Used pictures of DFT in an online consumer survey are available online at: https://www.caseih.com/emea/de-at/News/Pages/2016-08-30-Case-IH-stellt-auf-der-Farm-Progress-Show-neues-Traktorkonzept-vor.aspx (S1), https://www.fendt.com/int/fendt-mars (S2), https://www.schweizerbauer.ch/landtechnik/firmen--personen/20000-melkroboter-von-lely-in-betrieb-19341.html (S3), and https://melktechnik-center.com/Fuetterungstechnik/FMR-Roboter/ (S4).

Author Contributions: Conceptualization, A.G. and M.G.; formal analysis, A.G.; investigation, A.G.; data curation and validation, A.G.; writing—original draft preparation, A.G.; writing—review and editing, A.G. and M.G.; visualization, A.G.; supervision, M.G.; funding acquisition, M.G. All authors have read and agreed to the published version of the manuscript.

Funding: Research was funded by the Bavarian State Ministry for Food Agriculture and Forestry (grant number D/17/01).

Institutional Review Board Statement: The study was conducted in accordance with the General Data Protection Regulation (EU) for anonymous surveys. No personal data were processed.

Informed Consent Statement: All survey participants declared their explicit consent.

Data Availability Statement: The data presented in this study are available on request from the corresponding author.

Acknowledgments: The authors acknowledge Sebastian Schleicher for his student assistance in survey design and data collection. Thanks to Florian Klein for the assistance with data processing.

Conflicts of Interest: The authors declare no conflict of interest. The funders had no role in the design of the study; in the collection, analyses, or interpretation of data; in the writing of the manuscript; or in the decision to publish the results.

References

1. Bangel, C.; Faigle, P.; Gortana, F.; Loos, A.; Mohr, F.; Speckmeier, J.; Stahnke, J.; Venohr, S.; Blickle, P. Stadt, Land, Vorurteil. Available online: https://www.zeit.de/feature/deutsche-bevoelkerung-stadt-land-unterschiede-vorurteile (accessed on 17 April 2020).
2. Deutscher Bauernverband. *Situationsbericht 2018/19: Trends und Fakten zur Landwirtschaft*; Deutscher Bauernverband: Berlin, Germany, 2018; pp. 1–282.
3. Grunert, K.G.; Bredahl, L.; Brunsø, K. Consumer perception of meat quality and implications for product development in the meat sector. *Meat Sci.* **2004**, *66*, 259–272. [CrossRef]
4. Sharp, J.S.; Tucker, M. Awareness and concern about largescale livestock and poultry: Results from a statewide survey of Ohioans. *Rural Sociol.* **2005**, *70*, 208–228. [CrossRef]
5. Freudenburg, W.R. Rural-Urban differences in environmental concern: A closer look. *Sociol. Inq.* **1991**, *61*, 167–198. [CrossRef]
6. Arcury, T.A.; Christianson, E.H. Rural-urban differences in environmental knowledge and actions. *J. Environ. Educ.* **1993**, *25*, 19–25. [CrossRef]
7. Pfeiffer, J.; Gabriel, A.; Gandorfer, M. Understanding the public attitudinal acceptance of digital farming technologies: A nationwide survey in Germany. *Agr. Hum. Values* **2021**, *38*, 107–128. [CrossRef]
8. Box, G.E.; Tidwell, P.W. Transformation of the independent variables. *Technometrics* **1962**, *4*, 531–550. [CrossRef]
9. Tabachnick, B.G.; Fidell, L.S. *Using Multivariate Statistics*, 7th ed.; Pearson Education: London, UK, 2018.

Proceeding Paper

Requirements Identification for a Blockchain-Based Traceability Model for Animal-Based Medicines †

Rodrigo S. Aranda [1,*], Roberto F. Silva [2] and Carlos E. Cugnasca [1]

[1] Polytechnic School, University of São Paulo (USP), São Paulo 05508-010, Brazil; carlos.cugnasca@usp.br
[2] Institute of Mathematics and Computer Sciences (ICMC), University of São Paulo (USP), São Carlos 13566-590, Brazil; roberto.fray.silva@gmail.com
*   Correspondence: rodrigo.aranda@usp.br
†   Presented at the 13th EFITA International Conference, online, 25–26 May 2021.

Abstract: In this paper, the traceability of heparin medicine will be studied. Currently, the registration of traceability data is conducted in a decentralized manner. With blockchain implementation, the traceability systems that use this data could become semi-automated, increasing the quality, security, and confidence of the information generated in the supply chain. This paper presents the essential requirements and activities wherein information must be collected within the heparin drug supply chain, focusing on the animal raw materials production link and its requirements. Blockchain technology is proposed to increase traceability and reliability in relation to the current situation. It also fulfills all the requirements identified if used as part of a traceability system. These requirements are: the existence of a consensus mechanism; anonymity; protocol, efficiency, and consumption; immutability; ownership and management; and approval time. We conclude the paper by presenting the mapping of requirements and entities and critical activities for adopting blockchain technology to support the traceability of raw materials from animals used in heparin production.

Keywords: blockchain; livestock; supply chain; medicine; traceability

Citation: Aranda, R.S.; Silva, R.F.; Cugnasca, C.E. Requirements Identification for a Blockchain-Based Traceability Model for Animal-Based Medicines. *Eng. Proc.* **2021**, *9*, 11. https://doi.org/10.3390/engproc 2021009011

Academic Editors: Charisios Achillas and Lefteris Benos

Published: 24 November 2021

Publisher's Note: MDPI stays neutral with regard to jurisdictional claims in published maps and institutional affiliations.

1. Introduction

Livestock systems contribute significantly to national economies worldwide, and the value of livestock production accounts for 20 to 40 percent of the total agricultural output in developing and developed countries, respectively [1]. The food chains worldwide contain a considerable number of actors in their links. Some of their main links or entities are: farmers, processing industries, shipping companies, wholesalers and retailers, distributors, and retailers [2]. The drug supply chain is also global and complex, presenting problems related to counterfeit products and raw materials [3]. Blockchain is a technology that could improve decision-making on these supply chains and guarantee product quality. Drug supply chains can use this technology to establish product and raw materials' provenance in the context of fraud and counterfeit products [3]. Blockchain has shown its potential for transforming traditional industry with its key characteristics: decentralization, persistence, anonymity, and auditability [4]. It can also monitor social and environmental responsibilities and facilitate real-time management of the supply chain [2]. However, few works in the literature evaluate the use of this technology on animal-based drug supply chains. In this paper, heparin, an important anticoagulant that uses components extracted from cattle lungs and intestines [5], was used as a case study for evaluating the use of blockchain technology. The main objective of this paper was to: (i) identify the main requirements for implementing blockchain on this supply chain, (ii) identify the main entities or links and services involved, and (iii) map the requirements and the entities identified, allowing for better decision-making.

Eng. Proc. **2021**, *9*, 11. https://doi.org/10.3390/engproc2021009011   https://www.mdpi.com/journal/engproc

2. Use of Blockchain Technology for Supply Chain Traceability

Bitcoin is considered the first real-life application of blockchain technology [6]. Blockchain leads to technology development that allows for ubiquitous financial transactions among distributed untrusted parties [2]. In supply chains, its use could allow for trustworthy data gathering, storage, and visualization, even if some agents are untrustworthy. The primary use of this information is for traceability purposes [7]. Blockchain can be described as a digital transaction ledger maintained by a network of multiple computing machines that are not relying on a trusted third party [2]. It is essential to observe that it allows transactions to occur in a decentralized manner [4]. Blockchain traceability solutions are widely recognized in food and agricultural supply chains [8]. Several serious problems such as tampering of products, delay, and fraud exist in the traditional supply chain management system, and blockchain technology can provide essential features such as decentralization, transparency, product traceability, and trust-less environment anonymity and immutability [7]. Agriculture and food supply chains are well interlinked since agriculture products are almost always used as inputs in some multi-actor distributed supply chains [2], and this concept was applied in this article considering animal-based medicines. Animal-based medicines are described as medicines with one or more critical components extracted from animal products. The primary animal-based medicine for most countries is heparin, widely used to inhibit blood coagulation [5].

3. Methodology

The methodology used in this paper was composed of three main steps: (i) requirements gathering, considering an in-depth literature survey on the following domains: blockchain models used for product traceability, animal products supply chains, medicines supply chains, and animal-based medicines supply chains. The main entities of the heparin supply chain were also identified in this step; (ii) identification of main supply chain blockchain models, considering both theoretical models and models currently being implemented in real-life scenarios; (iii) model assessment, considering a mapping of the requirements identified in Step 1 with each supply chain entity. Three forms of traceability were explored: pre-production, production, and distribution.

4. Supply Chain Entities of the Heparin Supply Chain

Supply chain management is related to managing the production, flow of goods, data, and finances and oversees the processes from origin to consumption [7]. A supply chain often intersects business functions and national boundaries with an extensive network of trading partners, and these interactions increase the vulnerability of the supply chain and can lead to its disruption [9]. It is essential to observe that the heparin supply chain operates differently from other animal products supply chains because product delivery is subjected to various kinds of comprehensive regulations and rules to guarantee product quality [7]. Blockchain technology can track the physical movements of medicines and monitor their quality and authenticity through all supply chains [8].

Figure 1 represents the main entities and processes of the animal-based medicines' supply chain, focusing on the heparin supply chain. Its main entities are: (i) farmers and consultants involved in planning, breeding, feeding, and health maintenance of animals; (ii) the slaughterhouse, which is responsible for slaughtering and packaging, and storing the animal organs such as intestines and lungs; (iii) the pharmaceutical industry, which is responsible for the production of the medicine; (iv) distributors, retail, and hospitals, which are responsible for acquiring, storing, and supplying the medicines; and (v) patients, which will consume the medicine.

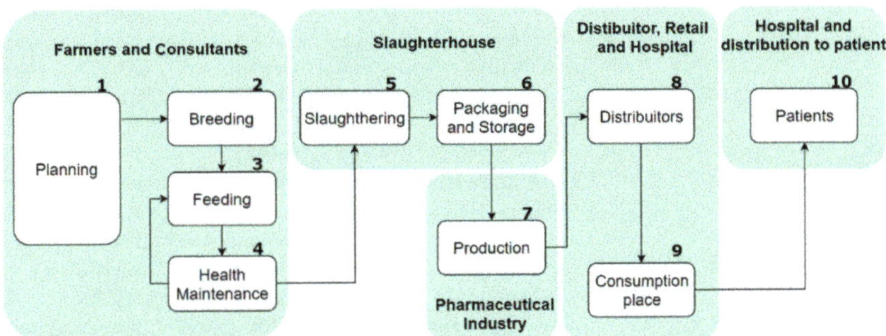

Figure 1. Supply chain of animal-based medicines.

5. Mapping of Requirements and Entities for Using Blockchain Technologies on the Heparin Supply Chain

Table 1 illustrates the main requirements for implementing blockchain technology for heparin traceability. The first is a consensus mechanism, which decides the block-validation process, and provides consensus among all the nodes [7]. This guarantees that the information is trustworthy. There are several common approaches to reach a consensus in a blockchain, such as: proof of work, proof of stake, practical byzantine fault tolerance, delegated proof of stake, Ripple, Tendermint, and others [4]. The second is anonymity, which guarantees that it is impossible to establish, a priori, to which users are involved in a specific transaction [6]. The third is protocol, efficiency, and consumption, and it is linked to convenience, energy use, computational resources consumption, and the costs of operating the system [4]. The fourth is immutability, which is related to the capacity of guaranteeing that the information cannot be changed once it is added to an information block in the blockchain [7]. The fifth is ownership and management, and it is linked to aspects of governance, maintenance, and policies to operate the blockchain [10]. Lastly, the approval time is linked to how long it takes to create a block or register a transaction in the blockchain [9].

Table 1. Mapping of the main requirements and entities for blockchain-based traceability for the heparin supply chain.

Requirements	Farmers and Consultants	Slaughtering	Pharmaceutical Industry	Distributor, Retail and Hospital	End Users
Consensus mechanism			X	X	
Anonymity	X	X			X
Protocol, efficiency, and consumption			X	X	
Immutability	X	X	X	X	X
Ownership and management	X	X	X	X	X
Approval time			X	X	

Source: based on the works by [2,4,6–10].

It is important to observe that these requirements are addressed differently by the different blockchain models and that their importance is relative to each supply chain entity. Therefore, we also included in Table 1 which requirements are considered the most important for each supply chain entity. It is critical to observe: (i) immutability and ownership are essential for all entities; (ii) consensus mechanism, protocol, efficiency, and consumption, and approval time are more relevant for the pharmaceutical industry and the distribution, retail, and hospital entities; (iii) anonymity is essential for farmers and consultants, slaughtering, and end-users; and (iv) end-users consider critical the requirements of anonymity, immutability, and ownership and management.

Lastly, it was possible to identify three main types of traceability: (i) pre-production, related to all aspects that guarantee raw materials quality, from farm to the pharmaceutical industry; (ii) production, related to all the production aspects and the internal traceability in the pharmaceutical industry; and (iii) distribution, related to all the aspects related to transporting, storing, and distributing the products to the end consumers. The blockchain technology used must gather information from all those traceability systems.

6. Conclusions and Future Works

Animal-based medicine supply chains are essential for healthcare. However, several problems related to counterfeiting exist in those supply chains. In this work, we evaluated the use of blockchain for improving product traceability, information visibility, and end users' trust in those products, focusing on the heparin supply chain. The main entities, processes, main requirements for those entities and traceability types necessary for guaranteeing product quality were identified. A mapping between requirements and entities was also conducted based on an in-depth literature review to identify the most critical requirements for each entity and traceability type. It is essential to observe that the methodology used can be applied in other supply chain studies, such as plant-based medicines or high-cost medicines. Future works are related to: (i) using computer simulation to evaluate the potential impacts of adopting blockchain on the heparin supply chain; and (ii) implementing a pilot project to evaluate this technology in real-life scenarios.

Author Contributions: Conceptualization, R.S.A. and R.F.S.; methodology, R.S.A. and R.F.S.; investigation, R.S.A.; writing—original draft preparation, R.S.A.; writing—review and editing, R.F.S. and C.E.C.; supervision, C.E.C. All authors have read and agreed to the published version of the manuscript.

Funding: This work was carried out with support from Itaú Unibanco S.A., linked to the Centro de Ciência de Dados (C2D) of the Polytechnic School of the University of São Paulo (USP).

Institutional Review Board Statement: Not applicable.

Informed Consent Statement: Not applicable.

Data Availability Statement: Not applicable.

Conflicts of Interest: The authors declare no conflict of interest.

References

1. FAO. *Transforming the Livestock Sector through the Sustainable Development Goals*; World Livestock: Rome, Italy, 2018; p. 222.
2. Kamilaris, A.; Fonts, A.; Prenafeta-Boldú, F. The rise of blockchain technology in agriculture and food supply chains. *Trends Food Sci. Technol.* **2019**, *91*, 640–652. [CrossRef]
3. Schmidt, C.G.; Wagner, S.M. Blockchain and supply chain relations: A transaction cost theory perspective. *J. Purch. Supply Manag.* **2019**, *25*, 100552. [CrossRef]
4. Zheng, Z.; Xie, S.; Dai, H.; Chen, X.; Wang, H. An overview of blockchain technology: Architecture, consensus, and future trends. In *2017 IEEE International Congress on Big Data*; IEEE: Piscataway, NJ, USA, 2017; pp. 557–564.
5. Bolten, S.N.; Rinas, U.; Scheper, T. Heparin: Role in protein purification and substitution with animal-component free material. *Appl. Microbiol. Biotechnol.* **2018**, *102*, 8647–8660. [CrossRef] [PubMed]
6. Parino, F.; Beiró, M.G.; Gauvin, L. Analysis of the Bitcoin blockchain: Socio-economic factors behind the adoption. *EPJ Data Sci.* **2018**, *7*, 38. [CrossRef]
7. Dwivedi, S.K.; Amin, R.; Vollala, S. Blockchain based secured information sharing protocol in supply chain management system with key distribution mechanism. *J. Inf. Secur. Appl.* **2020**, *54*, 102554. [CrossRef]
8. Sunny, J.; Undralla, N.; Pillai, V.M. Supply chain transparency through blockchain-based traceability: An overview with demonstration. *Comput. Ind. Eng.* **2020**, *150*, 106895. [CrossRef]
9. Min, H. Blockchain technology for enhancing supply chain resilience. *Bus. Horiz.* **2019**, *62*, 35–45. [CrossRef]
10. Mettler, M. Blockchain technology in healthcare: The revolution starts here. In Proceedings of the IEEE 18th International Conference on E-Health Networking, Applications and Services (Healthcom), Ostrava, Czech Republic, 17–20 September 2018; pp. 1–3.

Proceeding Paper

Tailored Digitization for Rural Development [†]

Valeria Sodano

Department of Political Science, University of Naples Federico II, Via Rodinò 22/A, 80138 Napoli, Italy;
vsodano@unina.it
† Presented at the 13th EFITA International Conference, online, 25–26 May 2021.

Abstract: The widespread use of many digital technologies along the food supply chain might have negative effects on rural development and on small and medium farms. One conclusion of this paper is that in order for rural areas to exploit all the benefits from digitization, avoiding the associated risks, there should be more agricultural extension services to farmers and more open data portals and platforms. This is in order to develop technologies specifically tailored for the economic, natural and social environment of rural areas, and therefore to be able to promote their modernization without giving up their cultural heritages.

Keywords: rural policy; innovation; food chain; digitization

Citation: Sodano, V. Tailored Digitization for Rural Development. *Eng. Proc.* **2021**, *9*, 14. https://doi.org/10.3390/engproc2021009014

Academic Editors: Dimitrios Aidonis and Aristotelis Christos Tagarakis

Published: 24 November 2021

1. Introduction

Goals of rural development (RD) policies include: reducing economic and social disparities with respect to urban areas (as regards income, education, modernization, and access to new technologies); fighting depopulation; fighting unemployment; and maintaining all the public goods and positive externalities provided by rural areas such as cultural heritage, landscape, environmental services.

Over the last years, increasing emphasis has been put on agriculture digitization as the main instrument through which to promote rural development. In recent years, the appeal to digitization has become one of the cornerstones of the rural development programs proposed by the main international institutions working in the field of development and agri-food policy [1–3].

The aim of this work is to shed light on some possible risks and shortcomings of an excessive emphasis on digitization benefits. The paper focuses on the EU case. A list of digital technologies are tentatively assessed towards their possible role in achieving EU RD goals and some suggestions are provided in order to make agriculture digitization more suitable for RD.

2. Methodology

The qualitative assessment of the role of digitization for RD was performed by cross-referencing the information and opinions gathered from the European Union documents on RD strategies, and some recent articles and reports drawing attention to the possible risks of agriculture digitization.

In the EU, RD policies are set within the Common Agricultural Policy (CAP), which is renewed about every seven years. The new CAP programming period is scheduled from 2021 to 2027 (with a current one-year delay due to the COVID pandemic). The new CAP proposal indicates nine key objectives [4], which are built around the three key words found in the presentation of the new CAP: 1—modernization through digitization; 2—simplification through digitization, 3—compatibility with the 10 priorities of the Commission, among which "a connected digital single market" and "A Europe fit for the digital age" [1].

These three key words clearly indicate the importance given to digital innovations for the pursuing of CAP goals, as has been furthermore stressed through the signature of

Eng. Proc. **2021**, *9*, 14. https://doi.org/10.3390/engproc2021009014

member states to the Declaration "A smart and sustainable digital future for European agriculture and rural areas", launched in Brussels on 9 April 2019 [5]. According to the Declaration "digital technologies such as artificial intelligence (AI), robotics, blockchain, high performance computing (HPC), Internet of Things (IoT) and 5G have the potential to increase farm efficiency and improve production, and also to contribute to making farming systems more sustainable from an economic, social and environmental point of view, as it is the case in other sectors".

Besides the new CAP goals, agriculture digitization has been already promoted by the EU research policy. Over the last years, through Horizon 2020, more than €200 million for research and innovation were allocated to the deployment of digital technologies for the agricultural sector.

Risks and shortcomings of agriculture digitization [6,7] have been associated with at least three characteristics of the digitalization processes observed so far: (1) the disproportionate innovation benefiting large farms and intensive conventional agriculture with respect to small farms and alternative agricultural practices; (2) the processes of consolidation along the food supply chain induced by digitalization, with unprecedent concentration of economic power, monopolization and increased power imbalance throughout the chain; and (3) security and political issues associated with data access and property rights.

(1) Most digital innovations available so far (such as guidance systems, semi- autonomous tractors and harvest robot) have been developed for large-scale industrial farming. Agricultural technologies are affected by economies of scale, creating disparity between large and small-scale farmers, with a corresponding inequality between developed and developing countries [3]. Transformative digital innovations and technologies are often not designed for the scale at which smallholder farmers operate. Moreover, the currently proposed digital innovations might negatively affect alternative agricultural systems, such as agroecology and organic farming. Digital innovations might marginalize the cultural and ecological knowledge of small-scale farmers, with their knowledge being replaced by data analytics and/or AI. The main risks from digitization for alternative agriculture practices reside in the possibilities for small farmers of losing their knowledge and skills, as well their right to repair their equipment or access sensitive data.

(2) Since most applications suit large farmers and intensive agriculture, digitization may push toward further consolidation in the food system. Agriculture digitization favors not only concentration at horizontal level, but also vertical coordination architectures along the food supply chain characterized by strong power imbalances, so that a few powerful actors are able to exploit the whole added value of the supply chain. In other terms, it fuels chains of oligopolies (and double marginalization inefficiencies) possibly associated with monopsonies and imbalanced marketing channel structures. Agrochemical industry has been the first to ramp on digital power followed recently by the largest internet companies which have started to integrate agribusiness apps in their platforms [8,9].

(3) Security and political issues are associated with the absence, so far, of a clear and effective regulation on data protection and exploitation rights in presence of data harvesting and processing practices led by the biggest world's internet and agribusiness MNCs. For farmers, key issues include who controls access to, and sharing of, data that are generated on and about farms, and how the value that is created from that data is re-distributed [10]. One concern is due to the fact that farmers do not own their data, which limits the ability of farmers to transfer historical data between technology providers, or to choose who services their machinery. Table 1 resumes the risks associated with the main digital innovations.

Table 1. Digital innovations in the food supply chain and associated risks.

Digital Innovations		Risks			
Sector	**Applications**	**Not Benefiting Small Farmers**	**Hindering Alternative Agricultural Practices**	**Fostering Consolidation and Power Imbalances**	**Data Governance Concerns**
1 Agricultural inputs	Financial services	*Very Likely/but some opportunities*	*Very Likely/but some opportunities*	*Likely*	*Very likely*
	Genome-edited seeds	*Very likely*	*Very likely*	*Very likely*	*Very likely*
	Smart tractors and sensors	*Very likely*	*Very Likely/but some opportunities*	*Very likely*	*Very likely*
	farm robotics	*Very likely*	*Very likely*	*Very likely*	*Very likely*
	Farm management platforms	*Likely/but some opportunities*	*Likely/but some opportunities*	*Very likely*	*Very likely*
2 Primary commodity trade	Digital marketplaces	*Likely/but some opportunities*	*Likely/but some opportunities*	*Likely*	*Likely*
	Digital freight management	*Likely*	*Likely*	*Likely*	*Likely*
3 Food processing	robotics	*Very likely*	*Very likely*	*Very likely*	*likely*
	3D food printing	*Likely*	*Likely*	*Likely*	*Likely*
	Smart packaging	*Very likely*	*Very likely*	*Very likely*	*likely*
4 Food distribution	Quality sensors and analytics	*Likely*	*Likely*	*Likely*	*Likely*
	Automated warehouses	*Likely*	*Likely*	*Very likely*	*likely*
	Smart shopping—E-commerce platforms	*Very Likely/but some opportunities*	*Very Likely/but some opportunities*	*Likely*	*Very likely*
5 Food chain organization and regulation	Digital tools for commodity chain traceability and transparency	*Very Likely/but some opportunities*	*Very Likely/but some opportunities*	*Likely*	*Very likely*

3. Discussion and Conclusions

A sustainable development of rural areas includes the following objectives: the diffusion of agricultural practices with low environmental impact; maintaining a fabric of efficient small and medium-sized enterprises; the maintenance of traditional knowledge that links production activities to the specific environmental and socio-cultural needs of a territory. The results of the analysis summarized in the table show how the digitization of agriculture can hinder the pursuit of these objectives, as it: (1) creates competitive disadvantages for small farms, which risk exiting the market; (2) hinders the spread of alternative agricultural practices such as organic farming and agroecology, and consequently destroys the agronomic knowledge related to them, often based on adaptation to the particular pedoclimatic conditions of a territory; and (3) supports further concentration at each stage of the food chain, increasing power asymmetries.

Given the risks of digitization with respect to the sustainability of RD, there is the urge by government bodies to carry out interventions aimed at managing the risks while creating opportunities for better outcomes of digital innovation. Risks of excessive concentration and power imbalances should be tackled by fostering competition policies, either using the current legislation, either by designing new forms of interventions also in an international cooperation framework (therefore through international bodies, treaties and agreements). Data governance concerns should urgently be addressed by implementing effective interventions in the field of privacy norms, intellectual property rights, data access democratization and public data management services. Discussions about data governance should entail a whole-of-government and multi-stakeholder approach, where the perspectives of all stakeholders and sectors are properly represented. As highlighted in the table, small farmers and alternative agricultural practices might benefit from digitization with respect to two main domains: the possibility to directly and effectively approach those socially responsible customers who are interested in local traditional products, short supply chains and sustainable production; the possibility to better communicate their commitment to sustainability to such customers through transparency and traceability; the possibility of adapting farm management platforms as to benefits agronomic innovations consistent with principles of organic farming and agroecology. Obstacles to the exploitation of such opportunities reside in: the lack of digital skills; the high costs in accessing digital services due to high costs of software as well hardware and communication networks; the ill-designed public platforms for accessing governmental administrations, which do not take into account the organization diversity between small and large farms; the scant supply of apps and digital services specifically addressed to small farmers and alternative agriculture (since such products are designed in a way as to fit conventional agriculture and large farmers). State interventions useful to overcome such obstacles could be: public agricultural extension services; public research investments in organic agriculture and agroecology; public research investments in digital innovations tailored for small farms and alternative agriculture; grants to small farms to cover costs of digitization; simplifications of procedures for accessing online administrative services by small farms; and public digital skills training aimed at small and alternative farmers so they can learn to assess and implement the best practices and technologies for their farm business.

Funding: This research received no external funding.

Institutional Review Board Statement: Not applicable.

Informed Consent Statement: Not applicable.

Data Availability Statement: No new data were created or analyzed in this study. Data sharing is not applicable to this article.

Conflicts of Interest: The authors declare no conflict of interest.

References

1. EU. A Europe Fit for the Digital Age. 2019. Available online: https://ec.europa.eu/info/strategy/priorities-2019-2024 (accessed on 15 April 2021).
2. OECD. *Digital Opportunities for Better Agricultural Policies*; OECD: Paris, France, 2019.
3. FAO. *Digital Technologies in Agriculture and Rural Areas*; Briefing Paper; FAO: Rome, Italy, 2019.
4. EC. Key Policy Objectives of the Future CAP, 2018. Key Policy Objectives of the Future CAP | European Commission. 2018. Available online: https://ec.europa.eu/info/food-farming-fisheries/key-policies/common-agricultural-policy/new-cap-20 23-27/key-policy-objectives-new-cap_en (accessed on 15 April 2021).
5. Declaration. A Smart and Sustainable Digital Future for European Agriculture and Rural Areas. Available online: https://smartagrihubs.eu/latest/news/2019/August/DD3Declarationonagricultureandruralareas-signedpdf-%281%29.pdf (accessed on 15 April 2021).
6. Sodano, V. Innovation Trajectories and Sustainability in the Food System. *Sustainability* **2019**, *11*, 1271. [CrossRef]
7. Prause, L.; Hackfort, S.; Lindgren, M. Digitalization and the third food regime. *Agric. Hum. Values* **2020**, *38*, 641–655. [CrossRef] [PubMed]
8. GRAIN. Digital Fences: The Financial Enclosure of Farmlands in South America. 2020. Available online: https://www.welthungerhilfe.org/news/latest-articles/2019/digitalisation-in-agriculture/ (accessed on 15 April 2021).
9. GRAIN. *Digital Control How Big Tech Moves into Food and Farming*; GRAIN: Girona, Spain, 2021.
10. Jouanjean, M.A.; Casalini, F.; Wiseman, L.; Gray, E. *Issues around Data Governance in the Digital Transformation of Agriculture: The Farmers' Perspective*; OECD Papers, No. 146; OECD: Paris, France, 2020; pp. 1–38.

Proceeding Paper

Analysis of the Concept of Feasibility in Sustainable Agricultural Systems †

Lavinia Popescu and Adela Sorinela Safta *

Doctoral School Economics II, Bucharest University of Economic Studies, 010374 Bucharest, Romania; popesculavinia14@stud.ase.ro
* Correspondence: saftaadela19@stud.ase.ro
† Presented at the 13th EFITA International Conference, online, 25–26 May 2021.

Abstract: The paper addresses through the new type concept the analysis of the feasibility of agricultural systems adaptable to environmental requirements. The development of a conceptual system for the feasibility of soil techniques through the responsible management of agricultural technological mechanisms can be expected to mitigate the impact of climate change. This paper analyzes the role of agricultural technologies based on the implementation of sustainable agriculture; as well as their interconnectivity in the local environment. In the methodological analysis was performed by compiling statistical data on agricultural production and effects of greenhouse gases (GHG) on agriculture

Keywords: feasibility; climate change; agricultural; greenhouse gas emissions

Citation: Popescu, L.; Safta, A.S. Analysis of the Concept of Feasibility in Sustainable Agricultural Systems. *Eng. Proc.* **2021**, *9*, 13. https://doi.org/10.3390/engproc2021009013

Academic Editors: Dimitrios Aidonis and Aristotelis Christos Tagarakis

Published: 24 November 2021

Publisher's Note: MDPI stays neutral with regard to jurisdictional claims in published maps and institutional affiliations.

1. Introduction

There is a growing interest and numerous observations on land use that show different balances in feasible ecosystems by applying good management practices that have the role of streamlining the performance of soil with controlled fertilizers. Thus, the separate approach through case studies and the history of the soil, the agricultural systems are part of the entire value chain by reducing the negative impacts. Objectives such as the conservation and increase of biological diversity in competition with the projections caused by climate change can help reduce the negative impacts on rural areas in a separate approach. Reducing pollution, degrading soil and greenhouse gas emissions, maintaining biodiversity and maintaining balance by improving soil fertility call into question how much we actually rely on fertilizers and how we reduce consumption while maintaining the same yield [1] (Swinnen, 2015).

The assumption of carbon storage in soil could be an option to mitigate climate change for agriculture, by managing the land and changing its use. In this study, we examine the possibility of adapting the methods used to estimate soil carbon changes.

2. The Issue of Climate Change

In the first instance, the evolution of agricultural systems has always had an effect on other systems. The resettlement of key pawns that form the study of agricultural systems. It has become vital to investigate agricultural models with innovative alternative practices that reduce greenhouse gas emissions, biodiversity loss, and soil erosion. The interdependent relationship between carbon sequestration in soil and climate change led to the idea of the need to find new methods of conserving agriculture, as argued by Popescu, L. et al. (2021) [2]. On the other hand, the sustainability of agricultural ecosystems can be applied separately, depending on the ecosystem, to give efficiency, the process being cyclical, therefore, in conditions of climate risk and the agricultural system suffers from the natural ecosystem [3]. The planet offers countless benefits to our individual well-being and the well-being of our societies and economies. According to investigations conducted in the Intergovernmental Panel on Climate Change (IPCC), global warming is a definite fact highlighted by the increase in global average air and ocean temperatures and more. As a consequence, in addition to the elements mentioned, climate change also attracts by changing the frequency and/or intensity of extreme risks of natural phenomena. On the other hand, other studies have shown that soil respiration with CO_2 is reduced in chronic fertilization. Against this background, a number of agricultural systems have emerged as having the potential to increase soil carbon, through land use, productivity. It is important to consider the emission of these greenhouse gasses (GHGs) when evaluating various practical agricultural products for their carbon sequestration potential, as well as among the most comprehensive attempts to estimate the global potential for GHG mitigation is a paper by the IPCC Working Group on Mitigation (Smith et al., 2012) [4,5].

3. Results

With the introduction of cross-compliance in agricultural systems in response to climate change, it was initially introduced as a formality in the Common Agricultural Policy (CAP) in the previous multiannual financial framework, with an emphasis on increasing the sustainability of agricultural systems to facilitate climate change mitigation. Agro-ecosystems can be divided into cultivated land and meadows as indicated in Table 1. Spatial nutrition enhances agro-ecosystems and therefore studies the environment with the aim of synthesizing the succession of technological flows related to the main agri-food sectors as claimed. The data in Table 1 show ecosystem extent trends for all tier II ecosystem categories between 2000–2018, and suggest that the area of N entry category of <50 kg/ha has increased, while the surface share of all Nitrogen (N) entry levels decreased between 2000 and 2018. This confirms that the overall contribution of N to agricultural land decreased during the study period and that areas of especially high nutrient pressures decreased to some extent. Differentiation of grasslands into subtypes by ecological characteristics and sensitivity to N input is necessary to make the accounting of spatial conditions as strong as analytically possible.

Table 1. Ecosystem extent trends for all tier II ecosystem categories between 2000–2018, EU28.

Tier I	Title 2	Area 2000 (km^2)	Area 2018 (km^2)	Net Change (%)
	Arable land	1,105,007	1,095,846	−0.8
	Rice fields	6500	6559	0.9
Cropland	Permanent crops	117,835	119,990	1.8
	Mixed farmland	375,523	373,727	−0.5
	Total cropland	1,604,865	1,596,122	−0.5
	Modified grassland	398,203	394,608	−0.9
Grassland	Semi-natural grassland	106,646	106,029	−0.6
	Total grassland	504,849	500,637	−0.8

Source: Eurostat data 2020.

Nutrient enrichment is a key indicator of pressure for the condition of the ecosystem, negatively affecting all terrestrial and aquatic ecosystems. The entry of N into agroecosystems is a consequence mainly of agriculture but also of the atmospheric deposition of emissions from transport, energy and industry. As shown in Figure 1a greenhouse gases are assessed spatially and the level at which we should act to counteract the effects of climate pollution from agricultural sources can avoid the constraints of effective soil reduction. It becomes evident that the feasible adaptation of processes in agricultural practice that generally reduce soil degradation could keep carbon dioxide underground, the spatial exposure achieved in Figure 1a,b Ecosystem categories between 2000–2018, Net change.

It becomes clear that the biosphere helps a great deal in reducing CO_2 emissions from the atmosphere. Figure 2 shows the many different types of data that need to be brought together in a common spatial framework, ranging from biodiversity monitoring data to agricultural statistics [6,7].

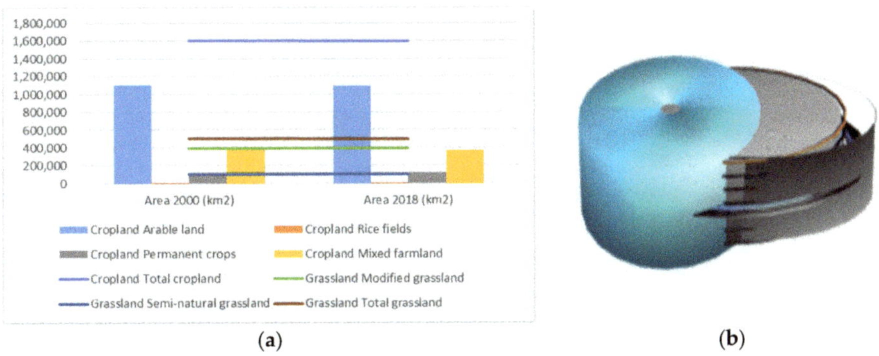

(a) (b)

Figure 1. Ecosystem categories between 2000–2018, Net change, Source: Authors' research. (**a**) graphic presentation; (**b**) spatial presentation.

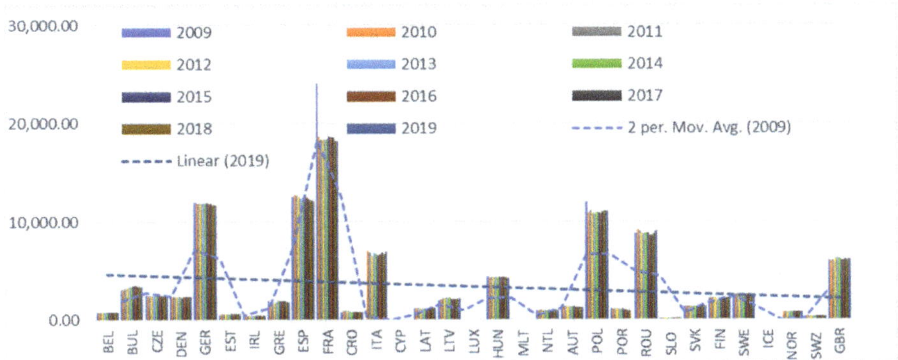

Figure 2. Utilised agricultural area by categories. Sources: European Commission, Eurostat and Directorate General for Agriculture and Rural Development.

4. Discussion and Conclusions

Investigating carbon storage and soil use feasibility strategies creates premises for reducing greenhouse gas emissions. In terms of the agricultural context, the economic system of sustainability creates the perception of a future benefit for the environment, with agricultural producers being obliged to comply with environmental requirements.

The interdependent relationship between carbon sequestration in soil and climate change has led to the need of finding new methods of conserving agriculture. The implementation of sustainable agricultural production can lead to an increase in carbon sequestration in soil through proper land management, thus making it possible to improve its properties [8]. The evolutions of carbon sequestration in soil are an important method and a priority.

On the other hand, the vision of the Common Agricultural Policy is an open door for agricultural producers to strive for sustainable management of natural resources burdened by the pressures of climate change. Thus, farmers are active participants in the sustainability of agricultural ecosystems step by step. This concept of standardization of agriculture can be adapted to the needs of the ecosystem to give efficiency. The resumption of annual agricultural practices does not necessarily lead to the standardization of agricultural processes, the innovation consisting even in changing these paradigms. The disadvantage of climate imbalances must be approached from another angle because in conditions of climate risk and the agricultural system suffers and farmers must be able to respond to these changes [9]. Thus, the monitoring of ecosystem forecasting trends as a condition of predictability gives rise to the need of investigating disruptive links, such as forcible cases due to climate change. Forecasting as a condition of feasibility of production methods based on the integration of natural resources and the biosphere is authors' vision of integration in the agricultural ecosystem, as an integral part of the rural space. Therefore, the relationships investigated in the present research should also be analyzed in dynamics, by conducting a longitudinal study, in order to increase the predictive power of the proposed conceptual model of feasibility. The empirical research undertaken in this paper has highlighted the fact that there is a need to position the effects in terms of the causes that produced them, so the relationship between greenhouse gases is influenced by the environment in which they are generated. Hence the need to revise production models predictable to the risk of climatic conditions in agricultural systems. This vision arises from the obligations imposed by the new CAP conditionality on meeting the target of reducing greenhouse gas emissions from agriculture [10]. Therefore, future research could aim at transforming multidimensional agricultural processes to ensure adaptation efforts. This vision is also based on the practice of intensifying feasibility models in agriculture to increase yields, while respecting environmental conditions. This has led to far-reaching implications for shaping the predictable behavior of producers either among local entrepreneurs or farmers. Taking into account the properties of the soil, we investigated through the S_{phor} coefficient the relationship of conditionality between climate and production monitoring as vital characteristics for the emissions perspective.

Author Contributions: Methodology by L.P. and developed by A.S.S. All authors have read and agreed to the published version of the manuscript.

Funding: This research received no external funding.

Conflicts of Interest: The authors declare no conflict of interest.

References

1. Colen, L.; Swinnen, J. Economic Growth, Globalisation and Beer Consumption. *J. Agric. Econ.* **2015**, *67*, 186–207. [CrossRef]
2. Popescu, L.; Safta, A.S. The Causal Relationship of Agricultural Standards, Climate Change and Greenhouse Gas Recovery. *Environ. Sci. Proc.* **2021**, *4*, 21. [CrossRef]
3. Aznar-Sánchez, J.A.; Piquer-Rodríguez, M.; Velasco-Muñoz, J.F.; Manzano-Agugliaro, F. Worldwide research trends on sustainable land use in agriculture. *Land Use Policy* **2019**, *87*, 104069. [CrossRef]
4. IPCC. *IPCC Guidelines for National Greenhouse Gas Inventories*; Workbook; Intergovernmental Panel on Climate Change: Paris, France, 1997. Available online: https://www.ipcc.ch/ (accessed on 12 April 2021).
5. Smith, P.; Davies, C.A.; Ogle, S.; Zanchi, G.; Bellarby, J.; Bird, N.; Boddey, R.M.; McNamara, N.P.; Powlson, D.; Cowie, A.; et al. Towards an integrated global framework to assess the impacts of land use and management change on soil carbon: Current capability and future vision. *Glob. Chang. Biol.* **2012**, *18*, 2089–2101. [CrossRef]
6. European Commission. Environment Action Programme to 2020. Available online: https://ec.europa.eu/environment/action-programme/ (accessed on 12 April 2021).
7. Eurostat Data. 2020. Available online: https://ec.europa.eu/eurostat/documents/7870049/12943935/KS-FT-20-002-EN- (accessed on 22 April 2021).
8. Pugh, T.A.M.; Arneth, A.; Olin, S.; Ahlström, A.; Bayer, A.D.; Goldewijk, K.K.; Lindeskog, M.; Schurgers, G. Simulated carbon emissions from land-use change are substantially enhanced by accounting for agricultural management. *Environ. Res. Lett.* **2015**, *10*, 124008. [CrossRef]
9. Rodrigo-Comino, J.; Martinez-Hernandez, C.; Iserloh, T.; Cerda, A. Contrasted impact of land abandonment on soil erosion in Mediterranean agriculture fields. *Pedosphere* **2018**, *28*, 617–631. [CrossRef]
10. Westhoek, H.; Lesschen, J.P.; Rood, T.; Wagner, S.; De Marco, A.; Murphy-Bokern, D.; Leip, A.; van Grinsven, H.; Sutton, M.A.; Oenema, O.; et al. Food choices, health and environment: Effects of cutting Europe's meat and dairy intake. *Glob. Environ. Chang.* **2014**, *26*, 196–205. Available online: http://www.sciencedirect.com/science/article/pii/S0959378014400338 (accessed on 12 April 2021). [CrossRef]

engineering proceedings

Proceeding Paper

Strategic Research Agenda for Utilisation of Earth Observation in Agriculture [†]

Karel Charvat [1], Vaclav Safar [2], Hana Kubickova [3,*], Sarka Horakova [1] and Tomas Mildorf [3]

[1] WIRELESSINFO, Cholinska 1048/19, 784 01 Litovel, Czech Republic; charvat@wirelessinfo.cz (K.C.); horakova@wirelessinfo.cz (S.H.)
[2] VUGTK, Ústecká 98, 250 66 Zdiby, Czech Republic; vaclav.safar@vugtk.cz
[3] Plan4all z.s., K Rybníčku 557, 330 12 Horní Bříza, Czech Republic; mildorf@plan4all.eu
* Correspondence: hana.kubickova@plan4all.eu
† Presented at the 13th EFITA International Conference, online, 25–26 May 2021.

Abstract: The EO4Agri Strategic Research Agenda (SRA) is a set of recommendations for future research activities in the area of Earth observation for agriculture. The EO4AGRI project provides support to all agri-food sectors based on new uses of COPERNICUS data. At first, part of the deliverable collected user needs from previous work are summarised including gaps in data, delivery platforms and knowledge management. Another input was an analysis of the current political framework and its influence on future agriculture. The implementation of the European Green Deal and the UN Sustainable Development Goals will require future collaboration of the public and private sectors. The main part of the SRA is a list of recommendations for future activities in the Group on Earth Observations (GEO), Horizon Europe (Annex 4 and Annex 6) and the Digital Europe programmes. It is not a revision of these programmes, but additional recommendations or tasks which are important to consider in updating the future programmes.

Keywords: earth observation; agriculture; strategic research agenda; green deal; destination earth

Citation: Charvat, K.; Safar, V.; Kubickova, H.; Horakova, S.; Mildorf, T. Strategic Research Agenda for Utilisation of Earth Observation in Agriculture. *Eng. Proc.* **2021**, *9*, 35. https://doi.org/10.3390/engproc2021009035

Academic Editors: Dimitrios Kateris and Maria Lampridi

Published: 14 December 2021

Publisher's Note: MDPI stays neutral with regard to jurisdictional claims in published maps and institutional affiliations.

1. Introduction

Agriculture comprises vital economic sectors producing food, agro-industrial feedstock and energy and provides environmental services through managing soil, water, air and biodiversity holistically [1]. Agriculture including forestry also contributes to managing and reducing risks from natural disasters such as floods, droughts, landslides and avalanches [2]. Farming with its close contact to nature provides the socio-economic infrastructure to maintain cultural heritage. Farmers are also conservers of forests, pastures, fallow lands and their natural resources and, in turn, of the environment [3]. Agriculture today is a composite activity involving many actors and stakeholders in agri-food chains that produce and provide food and agricultural commodities to consumers. In addition to farmers, there are farm input suppliers, processors, transporters and market intermediaries each playing their roles to make these chains efficient [4].

In FutureFarm project [5] a number of external drivers influencing the agriculture sector were detected. The following factors were recognized as the main drivers for changes in the agriculture sector: climate change, demographics (growing population, urbanization and land abandonment), energy cost, new demands on the quality of food (food quality and safety, aging population and health problems, ethical and cultural changes), innovative drivers (knowledge-based bio-economy, research and development, information and communication, education, investment), policies (subsidies, standardization and regulation, national strategies for rural development), economy and financing (economical and financial instruments, partnerships, cooperation and integration and voluntary agreements), sustainability and environmental issues (valuation of ecological performances, development of sustainable agriculture) and public opinion (press, international organization, politicians).

The common and future position of each driver can be different in reality. In many cases, two drivers can stand against each other and their future influence on the agri-production and food market depends on regulations and common policy. For example, focusing on food quality and safety could be contradictory with requests for increasing production due to the growing population. Similarly, it is with increasing production and requests on the production of bioenergy. These drivers will lead to decreasing agricultural land, but request to increase production. We need increased quality and safety and it will be connected with a decreasing number of chemical inputs. In addition, these drivers will bring requirements for decreasing the consumption of energy and water.

2. Strategic Research Agenda

Future agriculture production needs to be globally increased, with higher quality and using less land and fewer inputs at the same time. Earth observation can give us relevant information to help solve these problems on local, national and global levels. Satellite data are an important source of information for future agriculture. There is a clear need for new data, better spatial resolution, new bands and more dense data. However, the willingness and possibilities of farmers to pay for this are limited. The need for in situ data is another important issue, helping users to use remote sensing data optimally. The process of deriving useful information from satellite data that can help farmers to make precise decisions must be supported. Integration with aerial data and in-situ data is necessary. There is a need to increase spatial resolution (e.g., for the provision of soil water data) and temporal resolution (e.g., for the timely detection of crop disease and (near-)real-time monitoring). We need incorporation of more bands (e.g., thermal sensors to derive information about evapotranspiration or potentially improve often criticized Sentinel-2 cloud masks or L-Band and X-Band data for complementing information from the Sentinel-1 C-band constellation).

Satellite data, such as Copernicus and Landsat data, are available for free via many delivery platforms. The availability of other than satellite data is rather limited. A lot of data is delivered on a commercial basis. The idea is that European Data and Information Access Services (DIAS) will be self-financed, which could lead to the fact that some of them will not be operational after the end of their contracts. On the one hand, there are large investments from the public to private, to build new solutions and delivery platforms. On the other hand, agriculture is highly fragmented with enormous amounts of players in different sectors (e.g., machinery, insurance, fertiliser producers). Access to knowledge is limited and the current investments are not efficiently utilised. There is an urgent need to verify the investments for all public and private partners and get a deep understanding of the return of investment for all participants, as well as verification of climate change and/or environmental positive or negative effects.

A very innovative method that was proposed is to make use of social media and web data sources in future. As an example, in 2016, farmers tweeted about their low wheat yields; however, this situation was not picked up by early warning systems. Social media data therefore represent an interesting source of ground-based information that can inform early warning systems. The use of refined satellites with higher resolution and lower data latency will overcome the challenge and limitations of low spatial and temporal properties of current satellites. Remotely sensed data can be coupled with crop growth models and input data like climate information, soil type, plant varieties and management practices, not only to enable precise diagnosis of crop production, but also to foster crop forecasting such as crop performance and yield. These forecasts support policy makers and insurers with information about food security, trade, market access and avoidable crop damages due to weather extremes.

For the future perspective, a user-friendly dataset should be achieved that addresses the customer needs and their concrete requirements. Therefore, these data have to be characterized by long-term support through workshops, constant exchange and updates. Because satellite remote sensing data can foster decision-making activities, it is essential to use this kind of data in areas where a lack of in situ observation occurs, such as countries with ongoing conflicts and political insecurities. Eventually, it is important to share and collect land surface information in order to develop robust policies and strategies for food management to build up a dense network to assess food security worldwide.

The availability of ancillary information/data (LPIS availability, weather data) is important for the development of any agricultural service. While the weather and meteorological data are commonly available through various services, LPIS data in the EU are fragmented and not available in every part of the EU as open data. In this sense, there is sometimes a privacy problem that can be solved by the use of modern Differential Privacy and Privacy Preserving Machine Learning (e.g., Federated Learning or anonymization). As an example, even without the full information from LPIS (e.g., polygons with real coordinates), it is possible to generate valuable training data made of pixels associated with crops and a generic set of features (e.g., lat/lon intervals, slope, height, climate area) to stimulate the development of new machine learning techniques. In this sense, the release of open datasets from EU institutions can be an important step to increase the quality of the agricultural EO services. The availability of models to downscale information for small parcels is also an important aspect as models are not harmonized and this should be a direction for the new research agenda.

Adoption of the European Green Deal will require changes to the Common Agriculture Policy and this will require the cooperation of all stakeholders' groups. There are two important aspects for this to find common trust for data sharing and implementation of the FAIR Data Principles (findable, accessible, interoperable, reusable). This could be supported by a new European strategy for data, but also by international initiatives like GEO/GEOSS. There is a need for discussion between the private and public sectors about the effective sharing of data and expenses. Since some services will be in the public interest, it seems natural that the public sector will cover part of the expenses.

How to build a solution, which will connect public and private interest and which will offer possibilities for sharing data and knowledge about different stakeholders' groups? In addition, how can we do it without destroying the market? On the one side, it is a question of an agreement with data producers or service providers and the public sector. This agreement can be complicated. On the other hand, such agreements can close markets for other players and can lead to monopolies and can exclude new investments and also the entrance of new players to the market [6].

3. Conclusions

Based on the previous analysis, a set of recommendations is proposed to:

1. Organise regular workshops and conferences of all interested stakeholders. These workshops and conferences have to lead to the exchange of information, but they also need to educate all stakeholders about new methods. 2. Support cooperation of all players from the public and private sectors to fulfil the European Green Deal, Destination Earths and SD goals. It will also invite the food industry, machinery, chemical industry, IT industry and financing organizations to build a common environment. 3. Support new common multi-actor research involving both EO and agriculture/agronomy experts to develop new methods that guarantee food security and agriculture sustainability. 4. Support the farming sector with open data, including Copernicus and other EO data. This will require additional investments. Put into the practice FAIR principles. 5. Develop new metadata models and strategy for sharing all data across agriculture. 6. Reuse previous solutions. On the one side, continue with the development of new technologies and EO methods to build future Digital Twins. On the other side, there exists a large potential of existing technologies recently developed, that their potential is not fully exploited. It is necessary to prepare an overview of existing technologies and discussion among the teams on how to make solutions interoperable and how to reuse existing solutions. 7. Finance a

large number of smaller independent projects for technical development. This can bring new ideas in the short term. SmartAgriHubs can be used as an example. 8. Support standardization efforts and use of existing standards. This needs to be done in cooperation with existing standardization bodies including OGC, ISO and W3C. 9. Support large scale coordination actions, which will improve cooperation among different projects, initiatives and standardisation organizations. This needs to support both standardisation and FAIR principles. 10. There exist several technical problems, but the biggest problem will be at the level of legislation and financing. It will require a reform of the Common Agriculture Policy and also build effective strategies. This cannot be done only on a political level, but it will require communication of politicians with technical experts and researchers to define a successful strategy. For this purpose, it is necessary to establish a forum, where all these players will meet. A new strategy has to be prepared based on expert opinions and scientific results.

Author Contributions: Experiments, processing primary data and the SRA vision, K.C.; paper supervision and compilation, V.S.; writing, editing and translation, H.K. and S.H.; data pre-processing, tables and graphs, T.M. All authors have read and agreed to the published version of the manuscript.

Funding: This research was funded from the European Union's Horizon 2020 research and innovation programme under grant agreement No 821940.

Institutional Review Board Statement: Not applicable.

Informed Consent Statement: Not applicable.

Data Availability Statement: Open access to EO4Agri User Requirements is provided at https://zenodo.org/record/4596548#.YEotgmhKhnI (accessed 10 November 2021) under DOI 10.5281/zenodo.4596548 and open access to EO4Agri—The statistical summary and response to the main user requirements is provided at https://zenodo.org/record/4597166#.YEosmmhKhnI (accessed 10 November 2021) under 10.5281/zenodo.4597166.

Acknowledgments: The paper was prepared on the base of materials collected by all members of the EO4Agri consortium.

Conflicts of Interest: The authors declare no conflict of interest. The funders had no role in the design of the study; in the collection, analyses, or interpretation of data; in the writing of the manuscript, or in the decision to publish the results.

References

1. Lampridi, M.G.; Sørensen, C.G.; Bochtis, D. Agricultural sustainability: A review of concepts and methods. *Sustainability* **2019**, *11*, 5120. [CrossRef]
2. Sivakumar, M. Impacts of Natural Disasters in Agriculture, Rangeland and Forestry: An Overview. In *Natural Disasters and Extreme Events in Agriculture*; Sivakumar, M.V., Motha, R.P., Das, H.P., Eds.; Springer: Berlin/Heidelberg, Germany, 2005; pp. 1–22. [CrossRef]
3. Campbell, B.M.; Hansen, J.; Rioux, J.; Stirling, C.M.; Twomlow, S. Urgent action to combat climate change and its impacts (SDG 13): Transforming agriculture and food systems. *Curr. Opin. Environ. Sustain.* **2018**, *34*, 13–20. [CrossRef]
4. Charvat, K.; Horakova, S.; Druml, S.; Mayer, W.; Safar, V.; Kubickova, H.; Kolitzus, D.; Lopez, J.; Esbri, M.A.; Catucci, A.; et al. *D5.6 White Paper on Earth Observation Data in Agriculture*; Deliverable of the EO4Agri Project, Grant Agreement No 821940; EO4Agri: Maastricht, The Netherlands, 2020. [CrossRef]
5. Charvat, K.; Gnip, P. Analysis of external drivers for farm management and their influences on farm management information systems. *Precis. Agric.* **2009**, *9*, 915.
6. Safar, V.; Kubickova, H.; Krivanek, Z.; Krivankova, K.; Kvapil, J.; Kozhukh, D.; Kepka Vichrova, M.; Cerbova, K.; Charvat, K.; Horakova, S.; et al. *D6.6 Strategic Research Agenda Report*; Deliverable of the EO4Agri Project, Grant Agreement No 821940; EO4Agri: Maastricht, The Netherlands, 2020. [CrossRef]

Proceeding Paper

NEVONEX—The Importance of Middleware and Interfaces for the Digital Transformation of Agriculture †

Maximilian Treiber * and Heinz Bernhardt

TUM School of Life Sciences, Agricultural Systems Engineering, Technical University of Munich, Duernast 10, 85354 Freising, Germany; heinz.bernhardt@tum.de
* Correspondence: maximilian.treiber@tum.de; Tel.: +49-8161-715320
† Presented at the 13th EFITA International Conference, online, 25–26 May 2021.

Abstract: For a "system of systems" approach in agriculture, agricultural machinery is facing the challenge of bringing tractor-implement combinations and harvesting machinery to the IoT (Internet of Things). While standards such as the ISO 11783 (ISOBUS) have enabled seamless tractor-implement communications, the manufacturer-independent communication of tractors, implements and harvesting machinery with external sensors and different Cloud systems/services are still lacking. IoT ecosystems for agricultural machinery are gaining ground in the market to solve these problems. In this regard, an overview is given about the role of interfaces and middleware and how they can help to improve data flow, improve connectivity, compatibility and interoperability of digital products that farmers use.

Keywords: IoT; cloud; ISOBUS; connectivity; interoperability

Citation: Treiber, M.; Bernhardt, H. NEVONEX—The Importance of Middleware and Interfaces for the Digital Transformation of Agriculture. *Eng. Proc.* **2021**, *9*, 3. https://doi.org/10.3390/engproc2021009003

Academic Editors: Dimitrios Kateris and Maria Lampridi

Published: 19 November 2021

Publisher's Note: MDPI stays neutral with regard to jurisdictional claims in published maps and institutional affiliations.

1. Introduction

The adoption of digital agriculture technology is rising [1] and is estimated to double in the next 5 years for German farmers. With this rise, farmers start to collect bigger and more diverse datasets from their daily operations. The tools used to gather and manage this information have seen a vast improvement over the last few years with FMIS (Farm Management Information Systems), ISOBUS and government documentation systems, where farmers provide their documentation for receiving subsidies and standalone decision support systems for growers focused on finding the right time, amount and products to do applications on their crops. However, this leads to many different complex workflows that cost a farmer's time, and the decision whether or not to adopt digital technology for a certain application on a farm often involves a tedious search of combining the right digital tools to an efficient workflow to achieve the desired solution. The challenge of agricultural engineering in this regard is to bring together all the different digital tools, increase compatibility, connectivity, and interoperability so that it is easier for farmers to create seamless workflows for the digital processes/process management on their farms. Solutions to this problem can be found in standards, interfaces, and middleware.

This paper gives an overview of the systems that farmers use, the adoption of these systems, data formats, existing middleware and interfaces and shows how IoT ecosystems can improve the situation for agricultural machinery.

2. Materials and Methods

This study first provides definitions of the terms interface, standard and middleware in an agricultural context. Further, an overview of the state-of-the-art possibilities in those categories is given for the livestock farming and crop farming sectors.

3. Results and Discussion

3.1. Definitions and Examples in Agriculture

A Standard is "something set up and established by authority, custom, or general consent as a model or example" [2]. In an agricultural context, examples for standards can be found in regulations concerning the qualities of products (e.g., Regulation (EU) 1308/2013) or production processes and are set by government authorities but can also be derived from market pressure (e.g., food retailer labels). Regarding agricultural machinery, the most important standard by far is the ISO 11783 (ISOBUS) that standardizes tractor-implement communications.

An Interface is "the place at which independent and often unrelated systems meet and act on or communicate with each other" [2]. To stick with the above-mentioned examples, interfaces are necessary to provide documentation about agricultural production processes and the quality of goods to authorities or control bodies, that document the standards. In practice, these interfaces come to farmers in the form of paper documentation that needs to be filled in and slowly transitions to providing documentation in web-based frontends (e.g., application for subsidies from the European Union CAP), where software interfaces (in this case cloud APIs) to farm management information systems (FMIS) become more and more common to save farmers time and improve on the available documentation. On the side of agricultural machinery, an interface would be the control access to certain functions, sensors, or actuators in the tractor-implement combination. Examples are controlling engine speed, engine fuel rate, sections, or variable rates of an implement, etc.

"Middleware is software that lies between an operating system and the applications running on it. Essentially functioning as hidden translation layer, middleware enables communication and data management for distributed applications" [3]. As an intermediary between systems, middleware in an agricultural context can overcome the challenges of many (often proprietary) interfaces, data formats and historically grown standards by providing a layer that generates compatibility between two or more individual systems. A middleware can therefore broker data in between different cloud systems that farmers use (e.g., FMIS to government systems) or make sure that a machine can receive different data formats from the cloud to accomplish its tasks (e.g., application maps, direct control commands or messages).

Therefore, if the "system of systems" approach for agriculture [4] is taken seriously, there is no way around employing different kinds of middleware in the agricultural sector to overcome the challenges of connectivity, compatibility, and interoperability of systems. However, middleware is an often-overlooked topic, as a good middleware is not seen by the user and is hardly marketable, and with the opportunities a good middleware offers, there is the risk of companies not being able to protect their differentiation by innovation strategies long enough to recover development costs.

3.2. Situation in Livestock Farming

The adoption of Precision Livestock Farming technologies is on the rise [3]. Automatic milking systems, slat cleaning robots, autonomous feed mixers, Herd Management Systems (compare "FMIS for animal husbandry"), and sensor systems for the surveillance of animals and feedstock help farmers to focus their interests and efforts to those areas of the farm where it matters most. However, a standard such as the ISOBUS for the communication between those different systems is lacking. There are different tools and software solutions available that offer workarounds (e.g., Middleware from the home-automation sector as seen in [5].) However, a commercial and manufacturer-independent ecosystem for the integration of all precision livestock farming information from robots, wireless sensor systems and cloud documentation is still lacking. Therefore, in practice, systems either remain manufacturer-dependent (full-liner approach), farming operations must resort to software engineering providers to cover their needs, or systems simply remain incompatible. The results for farmers entail a lot of unnecessary labor and time spent on documentation.

3.3. Situation in Crop Farming

The systems farmers use in crop farming can be categorized as follows: Government documentation systems, which help farmers to apply for subsidies and meet basic documentation required by law. These are usually provided by government agencies to farmers. An FMIS helps the farmer to document all agricultural processes on the farm and the modular services also offer a lot of additional functions such as buying application maps, managing stock, tracking machine hours and fleet management and basic cost/benefit analyses. Next to the FMIS, crop management systems arise that are based on scientific models and help make decisions on application times—rates and suitable products. As those systems are not complete FMIS, they fall into the category of decision support systems, of which different ones are often offered to farmers from numerous extension services of universities, government agencies and federations. In the last few years, wireless sensor networks, drones and autonomous robots were added to the game to further complicate investment decisions for farmers. While wireless sensor networks and Wi-Fi cameras have been a huge success for individual use cases, and drones find their way to agriculture as sensor carrier platforms to create near-real-time decision map layers, robots are still in the early stages of their technology readiness level. Further, some online marketplaces for agricultural goods (and, e.g., carbon farming) have found their way to the farms and sometimes are even integrated into FMIS. Some of those services grow into FMIS as modular services by different providers. However, the transfer of machinery data and control commands to tractor-implement combinations is still in the hands of proprietary telemetry solutions. In crop farming environments, there is one basic workflow that is important to farmers during the early stages of adoption: Creating or buying application maps, transferring them to the machine, executing the task and uploading the "as-applied" map back to their documentation system, to close the loop. Experience from practice shows that the transfer to the machine often still happens via flash drives and not wirelessly, posing a threat for error on large operations regarding the availability, compatibility of flash drives and user error (loss, breakage, long distances to travel if wrong application map is used). The application map is basically the gold standard of transferring information into machine-readable formats, where an ISOBUS task-controller translates the information to machinery control commands. Limitations arise when live sensor information or live information from cloud systems should be considered to create ground truth and adjust the map value to a sensor-validated environment in real-time. Further challenges arise when next to an application map, further optimization software needs to be run on the machine. Examples for this are data-logging in the background, live optimization of a fertilizer spreader pattern or the automated control of balancing and tire pressure control systems.

This is where complex middleware such as, e.g., the NEVONEX ecosystem [6] comes into play, offering these kinds of functionalities and enabling the machine to run those in parallel in the form of digital services that are enabled by the IoT ecosystem architecture of this technology.

4. Conclusions

Middleware is a powerful set of tools to overcome the problems of connectivity, compatibility and interoperability with the different digital products that farmers use. For a "system of systems" approach, this invisible component of digital farming systems is therefore of utmost importance for the future. In precision livestock farming, commercial middleware that integrates systems to manufacturer independently is still lacking. In crop farming, the first steps towards the use of middleware can be seen in the connection of different cloud services that farmers use (e.g., agrirouter [7] and data connect [8], which are supported by the ISOBUS on the machinery side). However, to reach the next level, full interoperability of agricultural machinery and cloud systems is necessary and the different service providers need better access to the individual interfaces and functionalities of the machines. IoT ecosystems for agricultural machinery such as, e.g., the NEVONEX

Ecosystem, can help to provide these functionalities. As multi-sided platforms, their market success and operating mechanisms have yet to be better understood.

Author Contributions: Conceptualization, M.T.; methodology, M.T.; formal analysis, M.T.; investigation, M.T.; writing—original draft preparation, M.T.; writing—review and editing, M.T. and H.B.; supervision, H.B. All authors have read and agreed to the published version of the manuscript.

Funding: This research was funded by Robert Bosch GmbH.

Institutional Review Board Statement: Not applicable.

Informed Consent Statement: Not applicable.

Data Availability Statement: Data sharing is not applicable. No new data were created or analyzed in this study.

Conflicts of Interest: The authors declare no conflict of interest. The funders had no role in the design of the study; in the collection, analyses, or interpretation of data; in the writing of the manuscript, or in the decision to publish the results.

References

1. Gabriel, A.; Gandorfer, M. *Landwirte-Befragung 2020 Digitale Landwirtschaft Bayern*; Landesanstalt für Landwirtschaft: Ruhstorf, Germany, 2020.
2. Merriam-Webster. "Interface" and "Standard" Merriam-Webster.com Dictionary. Available online: https://www.merriam-webster.com/dictionary/interface (accessed on 25 July 2021).
3. Microsoft Azure—What Is Middleware? Available online: https://azure.microsoft.com/en-us/overview/what-is-middleware/ (accessed on 29 July 2021).
4. Porter, M.E.; Heppelmann, J.E. How Smart Connected Products are Transforming Competition. *Harv. Bus. Rev.* **2014**, *92*, 64–88.
5. Treiber, M.; Höhendinger, M.; Rupp, H.; Schlereth, N.; Bauerdick, J.; Bernhardt, H. Connectivity for IoT Solutions in Integrated Dairy Farming in Germany. In *2019 ASABE Annual International Meeting*; American Society of Agricultural and Biological Engineers: St. Joseph, MI, USA, 2019. [CrossRef]
6. Robert Bosch GmbH. NEVONEX Powered by Bosch: The Ecosystem for Smart, Digital Agriculture. Available online: https://www.bosch-presse.de/pressportal/de/en/nevonex-powered-by-bosch-the-ecosystem-for-smart-digital-agriculture-199552.html (accessed on 15 September 2019).
7. DKE-Data GmbH. Strong Basis—Strong Developer. 2017. Available online: https://my-agrirouter.com/en/news/archive/read/news/detail/News/strong-basis-strong-developer/ (accessed on 13 May 2019).
8. John Deere, CLAAS, CNH Industrial and 365FarmNet Form DataConnect. 2019. Available online: https://www.deere.com/en/our-company/news-and-announcements/news-releases/2019/agriculture/2019nov05-dataconnect/ (accessed on 5 July 2021).

Proceeding Paper

Molecular and Phenotypic Diversity of Indigenous Oenological Strains of *Saccharomyces cerevisiae* Isolated in Greece [†]

Aikaterini Karampatea [1,2,*], **Argirios Tsakiris** [1], **Yiannis Kourkoutas** [3] **and Georgios Skavdis** [3]

1   Department of Wine, Vine and Beverage Sciences, University of West Attica, Aigaleo, 12243 Athens, Greece; atsakiris@uniwa.gr
2   Department of Agricultural Biotechnology and Oenology, International Hellenic University, 66100 Drama, Greece
3   Department of Molecular Biology and Genetics, Democritus University of Thrace, 68100 Alexandroupolis, Greece; ikourkou@mbg.duth.gr (Y.K.); gskavdis@mbg.duth.gr (G.S.)
*   Correspondence: katerina_karampatea@yahoo.gr
†   Presented at the 13th EFITA International Conference, online, 25–26 May 2021.

Abstract: During 3 years, we explored the biodiversity of the indigenous yeast flora in five Greek wine regions by collecting five varietal grape samples, conventionally and biologically cultured. Spontaneous wine fermentations were carried out by the native microbiota of the grape juice, without the inoculation of selected industrially produced yeast. The indigenous yeast flora, isolated at three phases of these fermentations, was purified and characterized using different oenological and technological criteria. The pre-selected *Saccharomyces cerevisiae* strains, with the most promising oenological characteristics, were evaluated in microvinifications of Malagousia must and the quality of the produced wines was subjected to a sensorial descriptive analysis.

Keywords: indigenous *Saccharomyces cerevisiae*; yeast strain selection; alcoholic fermentation

Citation: Karampatea, A.; Tsakiris, A.; Kourkoutas, Y.; Skavdis, G. Molecular and Phenotypic Diversity of Indigenous Oenological Strains of *Saccharomyces cerevisiae* Isolated in Greece. *Eng. Proc.* **2021**, *9*, 7. https://doi.org/10.3390/engproc2021009007

Academic Editors: Dimitrios Kateris and Maria Lampridi

Published: 19 November 2021

Publisher's Note: MDPI stays neutral with regard to jurisdictional claims in published maps and institutional affiliations.

1. Introduction

Greece is the 17th wine-growing country in the world, representing 1.3% of European wine production with a small average vineyard ownership (3.95 acres) [1]. There are more than 300 Greek indigenous varieties, cultivated single or in combination with international varieties in 9 different wine-growing regions (mainland and islands), in soils that vary strongly, with altitudes from 0 to over 1000 m. Greece has 33 Protected Designation of Origin (PDO) zones and more than 100 Protected Geographical Indication (PGI) zones. In a highly competitive international environment, Greek agriculture has high production costs and competition from already established wine-producing countries. There is great interest in creating typical products in relation to geographical names. Value creation relies on the wine's ability to satisfy consumers in the long run. Sensory complexity is a critical indicator of quality [2] and the concept of wine complexity is increasingly of interest to scientists, winemakers, and wine lovers. Yeasts allow the complete and rapid conversion of sugars into ethanol and CO_2, and their intervention also contributes to the taste of wines [3]. A large number of microbial species, mainly yeasts, follow and replace each other in grape must during winemaking [4]. Only two species can complete the fermentation, metabolizing the total concentration of sugars: *S. cerevisiae* and, in to a lesser degree, *S. uvarum*. There are also other genera of yeast in winemaking, present in different stages, pre-fermentative or at the beginning of the fermentation. They are known as non-*Saccharomyces* (NS) and have long been considered undesirable. Scientific research has shown that many of them have a technological interest. In the context of the production of unique products, there is a lot of discussion about the "indigenous" yeasts, which are naturally present in grapes and must. These spontaneous fermentations could lead to problematic fermentations. For the best fermentation control, about 80% of them

Eng. Proc. **2021**, *9*, 7. https://doi.org/10.3390/engproc2021009007   https://www.mdpi.com/journal/engproc

are conducted by commercial yeast inoculations. *S. cerevisiae* remains the main commercial yeast marketed by oenological companies, in the form of LSA (Active Dry Yeast).

2. Materials and Methods

Grapes samples were collected in their technological maturity in three different growing seasons (2018–2020). Malagousia (mostly), Assyrtiko, Vidiano, Moschofilero, and Agiorgitiko were collected in five wine-producing PGI zones: Drama, Pangeon, Chalkidiki, Thessaloniki, and Atalanti Valley. The mode of culture was either conventional or biological. Directly fresh grapes or defrosted were destemmed and crushed by hand. The grape mass obtained, 25 L for each batch, was fermented spontaneously at 25 °C. Diammonium phosphate (20 gr hL^{-1}) was added to the grape must. All the spontaneous fermentations took place in 30 L stainless steel thermo regulated tanks, exact copies of professional winemaking tanks (Figure 1). Isolation, purification, and conservation of cultures: the indigenous yeast flora was isolated a) from grapes and b) during three phases of the fermentations (beginning, middle (6 Baume), and end (<1 Baume)). Successive dilutions of fermenting musts were placed in culture and incubated for 5 days at 25 °C [5–7]. Aliquots of several decimal dilutions in 0.1% peptone water were spread onto YPG Nutrient Agar that had been treated with streptomycin sulfate (250 mg L^{-1}). A series of coatings were performed by the method of linear coating on the agar culture plates (YPG + streptomycin sulfate) method to obtain clean cultures. The operation was renewed by randomly taking an isolated colony each time. Microbiological identification: Yeast isolates were identified by phenotypic criteria [8]. Macroscopic and microscopic observations of isolation yeasts were also performed in YPG and Chromagar (Figure 2) [9,10]. The identification system ID 32C was used for the carbon assimilation tests. ID 32 C is a standardized system for the identification of yeasts, which uses 32 miniaturized assimilation tests and a database. Molecular identification: DNA extraction was performed (Genomic DNA from tissue-Macherey-Nagel-01/2017, Rev.17). Quantification and testing of DNA purity was conducted [11]. Random amplification of polymorphic DNA (RAPD-PCR), PCR fingerprinting, and interdelta PCR typing [12], followed by detection of PCR products (electrophoresis). Enzymatic profile: An API ZYM system was used for enzymic profiling in order to evaluate the strains' potential, because of the involvement of certain enzymes in the vinification process [13]. Oenological criteria: The isolated yeasts were tested according to several criteria: production of hydrogen sulphide [14,15], flocculation properties [16], fermentation rate [17], ethanol tolerance, osmotolerance, high-temperature growth [18], malic and acetic acid production [19], and enzymatic activities [19,20]. Fermentations with selected yeast strains: Eight strains of *S. cerevisiae* selected for their oenological criteria were used for the inoculation (1%) of must obtained from Malagousia grapes. Fermentations were carried out at the Laboratory of Marketing and New Products Development, Department of Agricultural Biotechnology and Oenology, IHU. A total of 25 L of Malagousia grape musts were fermented in 30 L stainless steel thermo regulated tanks, having the following chemical characteristics: sugars, 209 g L^{-1}; pH 3.55; titrable acidity, 6.1 g L^{-1} tartaric acid; and assimilable nitrogen, 80 mg L^{-1}. Musts were supplemented with 30 ppm total of SO_2. Nutrient additions were performed before the inoculation (organic nitrogen, 40 g hL^{-1}) and after 150 g L^{-1} sugar consumption (organic and inorganic nitrogen, 40 g hL^{-1}). Pre-cultures of yeast strains were propagated in YPD agar at 26 °C. Cells were collected and re-suspended in Ringer's solution. Each strain was inoculated at approximately 100 cells mL^{-1}. Fermentations were conducted at 18 °C and the progress was monitored by density measurements. The quality of the produced wines was evaluated and subjected to a sensorial analysis [21]. **Typical chemical analysis:** The determination of chemical parameters on the must and wines was performed: reducing sugars, total and volatile acidity, pH, malic and lactic acid, and free and total sulfur dioxide [22]. Assimilable nitrogen was assayed using the formol method [23]. **Sensory analysis:** Sensory evaluation of the aroma (fruity and floral) and flavor (fruity, floral, sour, astringency, body, and after taste) of the wines fermented by the selected yeasts was performed by a panel of 10 judges/experts. Intensity ratings were scored on scale from 1 (not perceivable) to 5 (very strong). **Statistical analysis:** Statistical data processing was applied to the sensory analysis, performed using Minitab Statistical Software.

(a)   (b)   (c)

Figure 1. (**a**) Stainless steel fermentation tanks; (**b**) destemming of the grapes; and (**c**) initiation of spontaneous alcoholic fermentation.

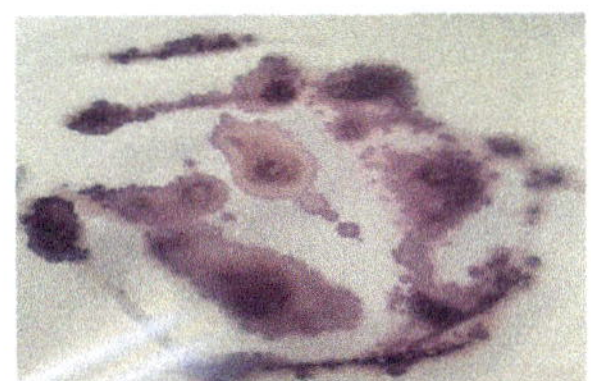

Figure 2. *Saccharomyces cerevisiae* in a Chromagar medium.

3. Results and Discussion

Phenotypic typing and characterization of *S. cerevisiae* isolates of the 76 isolates studied showed that 18 were identified as *S. cerevisiae* (Figure 3). *S. cerevisiae* were mainly derived from integrated grapes. They were found in grapes of the same variety (Malagousia) of the same sampling vineyard and repeated for 3 consecutive years, while in another PGI zone (Drama) they were isolated year after year, similarly at the same variety and vineyard each year. The same *S. cerevisiae* strain was identified in the wines produced from the grapes of the first case at the end of the fermentation. Therefore, in integrated grapes of a given area (Pangeon), the same strain was isolated in the grapes for every year of the process, before and at the end of the fermentation. In organic grapes, *S. cerevisiae* had a lower incidence. So, they were identified in a single case, in the middle of the fermentation, and in the same variety and area (Thessaloniki) for 2 consecutive years. Various species that were used to classify the genus *Saccharomyces*, but were eventually considered synonymous with *S. cerevisiae*, are named by many researchers as breeds or normal varieties [24] (Table 1). In order to distinguish various *S. cerevisiae* strains, δ1–δ2 and δ12–δ2 primers amplifying inter-delta sequences were employed [12]. **Oenological properties:** The results of the flocculation tests showed that most of the strains (89%) remained in suspension after 10 min [25,26]. The isolated strains had zero-to-medium ability of hydrogen sulphide production and zero malic and acetic acid production. All the strains had glucosidase, but low or no beta glucosidase activity. Regarding their enzymatic profile, α-fucosidase, N-acetyl β-glucosaminidase, esterase, esterase/lipase, leucine aminopeptidase, valine aminopeptidase, cystine aminopeptidase, trypsin, aminopeptidase, phosphohydrolase, and galactosidase were detectable. Although each strain had a unique enzyme pattern, the mean enzyme activity was esterase, lipase, and esterase/lipase with slightly higher activity. The action of N-acetyl β-glucosaminidase shows maximum enzymatic activity, 40 nmoles, in almost all strains except from one. Zero activity is shown by β-galactosidase, occurring only in one strain with very low activity, 5 nmoles. Alkaline phosphatase, α-chymotrypsin, and α-glucosidase were not detected in any of the strains. **Alcoholic fermentations**: Similar fermentation kinetics were observed in Sc2 and Sc6, Sc1 and Sc4, and Sc3 and Sc7. Higher fermentation rates and faster completion of the process were observed in Sc1 and Sc4. Two of the eight strains (Sc2, Sc4) did not metabolize the total initial sugars. Two strains gave wines with high volatile acidity. The duration of the fermentation varied from 9 to 17 days (Figure 4). At the end of the fermentations, a 750 mL centrifuged sample of each tank was taken for chemical analysis. In all cases, the pH values (3.56–3.89), alcoholic degree (11.75–12.95), volatile acidity (0.35–1.60 g acetic acid L^{-1}), and total acidity (5.0–6.9 g tartaric acid l^{-1}) ranged at usual levels [18,21]. **Sensory analysis:** Two months after fermentation, the lies were discarded and 3 bottles of each tank were prepared and tasted by a panel of 10 experts. Regarding the aromas, the conclusion for the overall assessment coincide with the individual assessments. From the average of the four indicators related to the aroma, some samples can be distinguished (those with the highest average values). Among these, one has an extremely low variability (standard deviation), which means that all its features are consistently high. The same sample gathered the best performance in the parameter "overall aroma rating". This is important because it demonstrates the objectivity and effectiveness of the grading method.

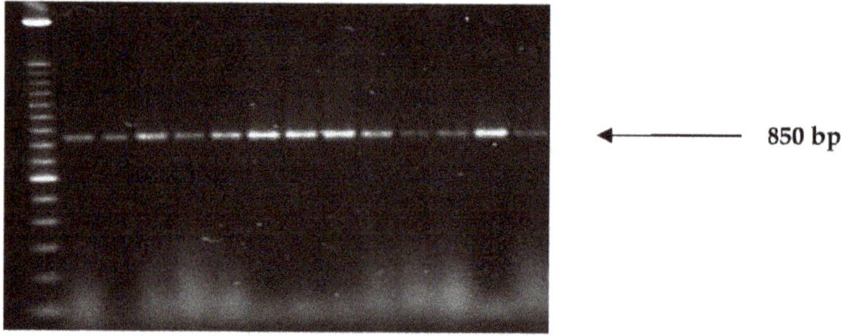

Figure 3. Agarose gel analysis of the PCR products.

Table 1. Different fermentative responses of indigenous wine *Saccharomyces cerevisiae*.

Fermentative Type	Isolate Identity *	Glu	Suc	Mal	Raf
I	MBm4	+	+	+	−
II	MCe2	+	+	+	+
III	Mci10	−	+	A	+
IV	Others	+	+	+	A

Glu, Glucose; Suc, sucrose; Mal, maltose; Raf, raffinose; +, fermentation positive; A, assimilation positive; −, fermentation and assimilation negative. * Capital letters indicate grape and type of viticulture, and small letters indicate fermentation stage: MCi, Malagousia must, Conventional culture, and initial stage. Arabic numbers represent the isolate number.

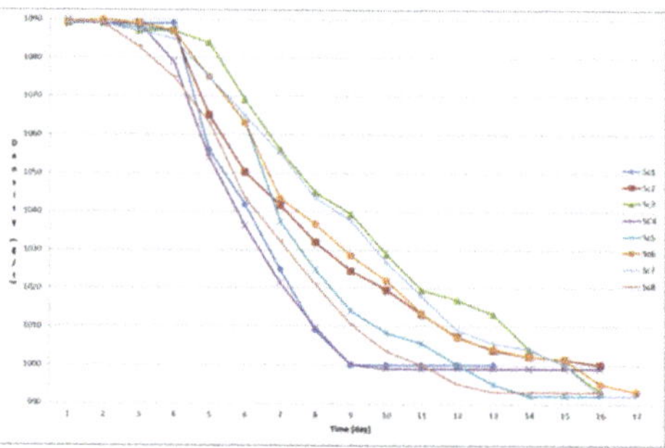

Figure 4. *Cont.*

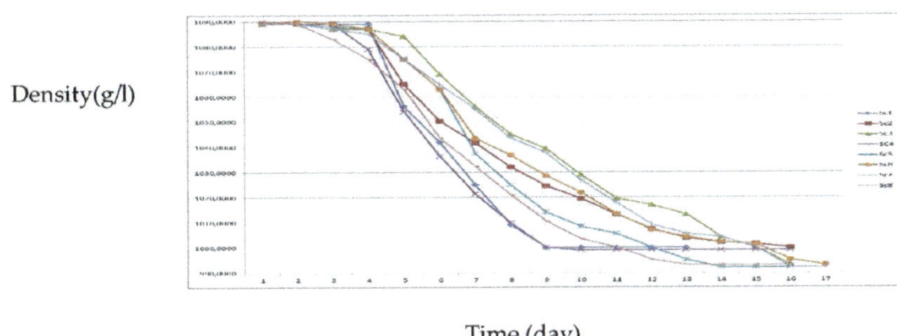

Figure 4. Kinetics of pilot-scale fermentations of inoculated grape must with indigenous *S. cerevisiae*.

Supplementary Materials: The following are available online at https://www.mdpi.com/article/10.3390/engproc2021009007/s1: Figure S1: Fermentation tanks; Figure S2: *S. cerevisiae* in Chromagar; Figure S3: Electrophoresis of the PCR products; Figure S4: Kinetics of pilot-scale fermentations; Table S1: Different fermentative responses of *S. cerevisiae* strains.

Author Contributions: Conceptualization, A.K. and Y.K.; methodology, A.K.; validation, A.K., A.T., Y.K., and G.S.; writing—original draft preparation, A.K. All authors have read and agreed to the published version of the manuscript.

Funding: This research received no external funding.

Institutional Review Board Statement: Not applicable.

Informed Consent Statement: Not applicable.

Data Availability Statement: Data are contained within the article or Supplementary Material.

Conflicts of Interest: The authors declare no conflict of interest that interfere with the results.

References

1. OIV. *2019 Statistical Report on World Vitiviniculture*; OIV: Paris, France, 2018.
2. Dubourdieu, D. La valeur du vin entre nature et culture, réflexions sur le goût mondial et la typicité des vins. In Proceedings of the ŒNO 2011, 9e Symposium International d'Œnologie de Bordeaux, Paris, France, 15 June 2012; pp. 3–6.
3. Swiegers, J.H.; Bartowsky, E.J.; Henschke, P.A.; Pretorius, I.S. Yeast and bacterial modulation of wine aroma and flavor. *Aust. J. Grape Wine Res.* **2005**, *11*, 139–173. [CrossRef]
4. Walker, G.M.; Stewart, G.G. Saccharomyces cerevisiae in the Production of Fermented Beverages. *Beverages* **2016**, *2*, 30. [CrossRef]
5. Berber, N.; Aissaoui, R.; Ali Bekada, A.M.; Coarer, M. Isolation and Molecular Identification (PCR-Delta and PCR-RFLP-ITS) of the yeast from Black muscat grape cultived in El malah (Algeria). *Adv. Environ. Biol.* **2016**, *10*, 55–61.
6. Renouf, V.; Claisse, O.; Lonvaud-Funel, A. Understanding the microbial ecosystem on the grape berry surface through numeration and identification of yeast and bacteria. *Aust. J. Grape Wine Res.* **2005**, *11*, 316–327. [CrossRef]
7. Résolution OIV-OENO 576A-2017. *Monographie Sur Les Levures Saccharomyces Sofia 2 Juin 2017*; OIV: Paris, France, 2017.
8. Barnett, J.; Payne, R.; Yarrow, D. *Yeasts: Characteristics and Identification*; Cambridge University Press: Cambridge, UK, 2000.
9. Ainscough, S.; Kibbler, C. An evaluation of the cost-effectiveness of using Chromagar for yeast identification in a routine microbiology laboratory. *J. Med. Microbiol.* **1998**, *47*, 623–628. [CrossRef]
10. Koehler, A.P.; Chu, K.C.; Houang, E.T.S.; Cheng, A.F.B. Simple, Reliable, and Cost-Effective Yeast Identification Scheme for the Clinical Laboratory. *J. Clin. Microbiol.* **1999**, *37*, 422–426. [CrossRef] [PubMed]
11. Sambrook, J.; Fritsch, E.F.; Maniatis, T. *Molecular Cloning: A Laboratory Manual*; University of Texas South Western Medical Center: Dallas, TX, USA, 1989.
12. Legras, J.-L.; Karst, F. Optimisation of interdelta analysis for Saccharomyces cerevisiae strain characterisation. *FEMS Microbiol. Lett.* **2003**, *221*, 249–255. [CrossRef]
13. Canal-Llauberes, R.M. *Effect of Pre-Fermentative Strategies on the Composition of Prieto Picudo (Vitis vinifera) Red Wines, Wine Microbiology and Biotechnology*; Harwood Academic Publishers: Reading, UK, 1993; Volume 3.
14. Giudici, P.; Kunkee, R.E. The Effect of Nitrogen Deficiency and Sulfur-Containing Amino Acids on the Reduction of Sulfate to Hydrogen Sulfide by Wine Yeasts. *Am. J. Enol. Vitic.* **1994**, *45*, 107–112.
15. Zambonelli, C. *Microbiologia e Biotecnologia dei Vini*; Edagricole-New Business Media: Bologna, Italy, 1998; p. 300.

16. Bony, M.; Bidart, F.; Camarasa, C.; Ansanay, V.; Dulau, L.; Barre, P.; Dequin, S. Metabolic analysis of *S. cerevisiae* strains engineered for malolactic fermentation. *FEBS Lett.* **1997**, *410*, 452–456. [CrossRef]
17. Pérez-Coello, M.S.; Briones Pérez, A.I.; Ubeda Iranzo, J.F.; Martin Alvarez, P.J. Characteristics of wines fermented with different Saccharomyces cerevisiae strains isolated from the La Mancha region. *Food Microbiol. Dec.* **1999**, *16*, 563–573. [CrossRef]
18. Regodón, J.A.; Pérez, F.; Valdés, M.E.; De Miguel, C.; Ramírez, M. A simple and effective approach for selection of wine yeast strains. *Food Microbiol.* **1997**, *14*, 247–254. [CrossRef]
19. Comitini, F.; Gobbi, M.; Domizio, P.; Romani, C.; Lencioni, L.; Mannazzu, I.; Ciani, M. Selected non-Saccharomyces wine yeasts in controlled multistarter fermentations with Saccharomyces cerevisiae. *Food Microbiol.* **2011**, *28*, 873–882. [CrossRef] [PubMed]
20. White, T.J.; Bruns, T.; Lee, S.; Taylor, J. Amplifcation and direct sequencing of fungal ribosomal RNA genes for phylogenetics. In *PCR Protocols: A Guide to Methods and Applications*; Innis, M.A., Gelfand, D.H., Sninsky, J.J., White, T.J., Eds.; Academic Press: San Diego, CA, USA, 1990; pp. 315–322.
21. Nikolaou, E.; Soufleros, E.H.; Bouloumpasi, E.; Tzanetakis, N. Selection of indigenous Saccharomyces cerevisiae strains according to their oenological characteristics and vinification results. *Food Microbiol.* **2006**, *23*, 205–211. [CrossRef] [PubMed]
22. *Compendium of International Methods of Wine and Must Analysis OIV, Edition 2021*; OIV: Paris, France, 2020.
23. Gump, B.H.; Zoecklein, B.W.; Fugelsang, K.C.; Spencer, J.F.T.; De Spencer, A.L.R. Prediction of Prefermentation Nutritional Status of Grape Juice: The Formol Method. *Food Microbiol. Protoc.* **2003**, *14*, 283–296.
24. Ribéreau-Gayon, P.; Glories, Y.; Maujean, A.; Dubourdieu, D. *Traite D'œnologie, 2. Chimie du vin, Stabilisation et Traitments*; Dunod: Paris, France, 2000.
25. Ortiz, M.J.; Barrajón, N.; Baffi, M.A.; Arévalo-Villena, M.; Briones, A. Spontaneous must fermentation: Identification and bio-technological properties of wine yeasts. *LWT* **2013**, *50*, 371–377. [CrossRef]
26. Suranská, H.; Vránová, D.; Omelková, J. Isolation, identification and characterization of regional indigenous Saccharomyces cerevisiae strains. *BJM* **2016**, *47*, 181–190. [CrossRef] [PubMed]

engineering
proceedings

Proceeding Paper

Susceptibility of Twenty-Three Kiwifruit Cultivars to *Pseudomonas syringae* pv. *actinidiae* †

Thomas Thomidis [1,*], **Dimitrios E. Goumas** [2], **Anastasios Zotos** [3], **Vassilios Triantafyllidis** [3] **and Efthimios Kokotos** [3]

1   Laboratory of Microbiology, Department of Human Nutrition and Diabetics, International Hellenic University, 5744 Sindos, Greece
2   Laboratory of Plant Pathology-Bacteriology, Department of Agriculture, School of Agricultural Sciences, Hellenic Mediterranean University, 71004 Heraklion, Greece; dgoumas@hmu.gr
3   Laboratory of Plant Production, University of Patras, 30100 Agrinio, Greece; tzotos@gmail.com (A.Z.); vtrianta@upatras.gr (V.T.); ekokkotos90@gmail.com (E.K.)
*   Correspondence: thomidis@cp.teithe.gr or thomi-1@otenet.gr; Tel.: +30-2310-013342
†   Presented at the 13th EFITA International Conference, online, 25–26 May 2021.

Abstract: One of the best methods to control plant disease is the use of resistant cultivars. The purpose of this study is to evaluate 23 kiwifruit genotypes and cultivars for susceptibility to four strains of Psa (biovar 3) in alaboratory setting. The results showed that all the bacterial strains were pathogenic. There was no statistical difference among the bacterial strains tested. None of the kiwifruit cultivars tested were immune to Psa. There was a statistical difference in the level of susceptibility among cultivars. The cultivars Sorelli and D495/312 were the most susceptible, while the cultivar A501/44 was the most resistant. However, the above results must be verified in field conditions.

Keywords: kiwifruit; *Pseudomonas syringae* pv. *actinidiae*; resistance

Citation: Thomidis, T.; Goumas, D.E.; Zotos, A.; Triantafyllidis, V.; Kokotos, E. Susceptibility of Twenty-Three Kiwifruit Cultivars to *Pseudomonas syringae* pv. *actinidiae*. *Eng. Proc.* **2021**, *9*, 33. https://doi.org/10.3390/engproc2021009033

Academic Editors: Dimitrios Aidonis and Aristotelis Christos Tagarakis

Published: 9 December 2021

Publisher's Note: MDPI stays neutral with regard to jurisdictional claims in published maps and institutional affiliations.

1. Introduction

Greece is one of the major kiwifruit producers in the world. Bacterial canker is a dangerous kiwifruit disease. In 1984, *Pseudomonas syringae* pv. *actinidiae* (Psa) was isolated and described in Japan, followed by Italy and South Korea on cv. Hayward (*A. deliciosa*). In 2015, the disease was identified for the first time in a Greek kiwifruit orchard, Palios Mylotopos Pella, Northern Greece [1].

Because growers do not spend money on unnecessary control methods, such as fungicides, the utilization of resistant cultivars is critical in disease control. Even when fungicides are required, cultivar resistance can be a valuable supplemental management tool. Fruit trees have been screened for resistance to bacterial diseases using a variety of approaches. Screening cultivars for bacterial canker resistance with stem inoculation is a frequent practice. In the lab, the excised twig assay is also used to determine cultivar resistance to bacterial infections. Both approaches are easy to use, convenient, and allow for a large number of replications.

A research program was carried out at the Zeus Actinidia SA company (Karitsa, Katerini Pieria, Greece) on the selection of kiwifruit genotypes suitable for Greek climate conditions. Based on their agronomic characteristics, 19 of these were identified as promising genotypes (productivity, fruit quality, etc.). However, there was no information on these genotypes' susceptibility to bacterial canker disease caused by Psa. The goal of this study was to investigate the level of susceptibility of the above genotypes to Psa in laboratory. The screening test included four commercially used kiwifruit cultivars.

2. Materials and Methods

Psa (biovar 3) strains obtained from kiwifruit with bacterial canker symptoms from Pella, Northern Greece, were used.

2.1. Excised twig Assay

Jeffers et al. [2], Krzesinska and Azarenko [3], and Thomidis et al. [4] explained the procedures used in these investigations. To obtain a layer of 10 mm, nutrient agar with 1.5 g of boric acid, 8 mg of cephalexin, and 20 mg of cycloheximide per liter was poured in sterilized Pyrex jars (9 cm diameter and 12 cm height). They were then inoculated with one of the above Psa strains and incubated for 7 days at 25 °C in the dark. From each kiwifruit genotype and cultivar tested, one-year-old dormant shoots measuring 50 cm in length and 10–15 mm in diameter were collected. Twigs measuring 10 cm in length were cut, surface disinfected for 5 min with 10% home chlorine bleach (sodium hypochlorite; 4.8%), and then rinsed three times with sterile water. Twigs were clipped to a slant at the base and put upright into the growing culture using a flamed sharp knife. Jars were resealed and placed in incubators at 20 degrees Celsius for 7 days. The length of necrosis was calculated by subtracting the depth of agar from the overall length of necrosis that formed. This experiment was carried out two times. Each jar contained ten twigs. Each genotype/cultivar had nine jars, three for each Psa strain. As a control, two non-inoculated jars were used.

2.2. Excised Shoot Assay

An excised shoot assay described by Matheron and Matejka [5] and Thomidis et al. [4] was used in a second laboratory experiment. Woody shoot segments measuring 15 cm in length and 1.5–2 cm in diameter were gathered from each kiwifruit genotype, and cultivar was evaluated. They were disinfested by soaking them in 10% household chloride bleach (sodium hypochlorite; 4.8%) for 5 min and then rinsing them three times with sterile water. The segments were inoculated with bacterial cells after removing 6 mm of epidermis from the core. To avoid desiccation, the wound was coated with petroleum jelly and sealed with adhesive tape. The length of necrosis was measured after inoculated segments were cultured for 7 days at 20 °C in moist chambers. Each cultivar had nine duplicates of five segments, and each Psa strain had three. As a control, non-inoculated portions were employed. This experiment was carried out two times.

The experimental design employed in both tests was completely randomized throughout. One-way analysis of variance was used to analyze the data. Duncan's Multiple Range Test (P0.05) was used to compare treatment means.

3. Results and Discussion

Given the importance of bacterial canker for kiwifruit agriculture, it is critical for producers to understand the level of susceptibility of cultivars to that disease. The ability of the host to thwart the pathogen's progress is referred to as resistance. Host resistance is probably the most valuable agricultural control measure, and it works best when sanitation measures are also used. In the resistant cultivar, resistance is effective throughout the colonization stage [6,7].

The major goals of this study were to conduct laboratory studies to determine the sensitivity of 23 kiwifruit genotypes and cultivars to Psa (excised twig assay and excised shoot method). These approaches can be effective for determining Psa susceptibility [4,8] since they are reliable and rapid, allow for a lot of replication, and can be used to practically any type of woody plant host. However, if relative rather than absolute levels of disease are investigated, the inoculation of stems, excised twigs, and shoots might be a valuable procedure for measuring plant sensitivity to wood diseases. The explanation for this is because changes in the physiology of excised phloem tissue caused by physical detachment from the growing plant may modify resistance to colonization by wood diseases. Furthermore, direct inoculation of the cambium and inner phloem tissues only assesses resistance mechanisms that are active once the pathogen has penetrated host tissue, obviating the need to test defense mechanisms present in the outer phloem tissue [9].

All of the bacterial strains were found to be harmful, according to the findings. There were no statistical differences between the bacterial strains that were studied (so the data were combined). All the kiwifruit genotypes and cultivars examined were infected by the bacterial strains tested (Table 1). There was a significant difference in susceptibility levels between genotypes and cultivars. Sorelli and D495/312 were the most susceptible, whereas A501/44 was the most resistant.

Table 1. The level of the susceptibility of 23 kiwifruit genotypes and cultivars on *Pseudomonas syringae* pv. *actinidiae* in laboratory.

Genotypes and Cultivars	"Excised Twig Assay"		"Excised Shoot Assay"	
Hayward	18.7 [x]	bcd [y]	14.0	abcd
Tsehelidis	17.0	abcd	14.7	abcd
Summerkiwi	19.3	bcd	15.0	abcde
Sorelli	20.7	cd	19.0	e
C496/447	17.3	abcd	15.0	abcde
C497/325	17.3	abcd	13.7	abc
A501/3	17.0	abcd	13.0	abc
B497/001	18.0	abcd	15.3	bcde
D495/312	21.3	d	18.0	de
A501/18	17.0	abcd	14.7	abcd
B501/22	18.3	bcd	15.7	cde
D4398/337	18.0	abcd	14.3	abcd
A501/44	15.0	ab	11.0	a
C497/352	16.0	abc	11.3	ab
D497/355	16.3	abc	12.0	abc
A501/103	16.7	abcd	13.3	abc
C497/76	19.0	bcd	13.7	abc
B484/56	17.3	abcd	15.7	cde
A501/28	18.0	abcd	12.7	abc
A497/25	18.0	abcd	16.0	cde
D498/325	19.7	bcd	14.0	abcd
B497/76	17.3	abcd	12.7	abc
C501/39	18.0	abcd	13.7	abc

[x] Values are the means of the two experiments; results were similar so the data were combined. The Psa strains showed statistically similar aggressiveness, so the data were combined. [y] Treatment means were separated by using Duncan's multiple range test ($p = 0.05$).

4. Conclusions

Based on the above results, the cultivars Sorelli and D495/312 should not be used when a kiwifruit orchard is established in areas where the bacterium Psa is present. In such areas, a good choice could be the cultivar A501/44. However, the above results must be verified in field conditions.

Author Contributions: Conceptualization, T.T. and D.E.G.; methodology, T.T. and D.E.G.; formal analysis, A.Z.; investigation, T.T. and D.E.G.; resources, E.K.; data curation, V.T.; writing—original draft preparation, V.T.; writing—review and editing, T.T. and D.E.G.; visualization, T.T.; supervision, T.T.; project administration, T.T.; funding acquisition, T.T. and D.E.G. All authors have read and agreed to the published version of the manuscript.

Funding: This research was funded by the Operational Programme Competitiveness, Entrepreneurship and Innovation 2014–2020 (EPAnEK) under the call "Research-Create-Innovate", project number: T1EΔK-03107, entitled "Distribution of the bacterium *Peudomonas syringae* pv *actinidiae* in Greek Kiwifruit orchards and evaluation of cultural and chemical methods to control this disease".

Data Availability Statement: The data presented in this study are available on request from the corresponding author. The data are not publicly available due to the agreement with the A.S. EPISKOPIS.

Acknowledgments: We would like to express our thanks to John Cullum for his technical support.

Conflicts of Interest: The authors declare no conflict of interest.

References

1. Holeva, M.C.; Glynos, P.E.; Karafla, C.D. First report of bacterial canker of kiwifruit caused by *Pseudomonas syringae* pv. actinidiae in Greece. *Plant Dis.* **2015**, *99*, 723. [CrossRef]
2. Jeffers, S.N.; Aldwinckle, H.S.; Burr, T.J.; Arneson, P.A. Excised twig assay for the study of apple tree crown rot pathogens in vitro. *Plant Dis.* **1981**, *65*, 823–825.
3. Krzesinska, E.Z.; Azarenko, A.N. Excised twig assay to evaluate cherry rootstocks for tolerance to *Pseudomonas syringae* pv. syringae. *HortScience* **1992**, *27*, 153–155. [CrossRef]
4. Thomidis, T.; Tsipouridis, C.; Exadaktylou, E.; Drogoudi, P. Comparison of three laboratory methods to evaluate the pathogenicity and virulence of ten *Pseudomonas syringae* pv. syringae isolates on apple, pear, cherry and peach trees. *Phytoparasitica* **2005**, *33*, 137–140.
5. Matheron, M.E.; Matejka, J.C. Differential virulence of *Phytophthora parasitica* recovered from citrus and other plants to rough lemon and tomato. *Plant Dis.* **1990**, *74*, 138–140. [CrossRef]
6. Purwantara, A.; Flett, S.P.; Keane, P.J. Colonization of roots of subterranean clover cultivars by virulent and a virulent races of *Phytophthora clandestina*. *Plant Pathol.* **1998**, *47*, 67–72. [CrossRef]
7. Widmer, T.L.; Graham, J.H.; Michell, D.J. Histological comparison of fibrus root infection of disease-tolerant and susceptible citrus hosts by *Phytophthora nicotianae* and *P. palmivora*. *Phytopathology* **1998**, *88*, 389–395. [CrossRef] [PubMed]
8. Little, E.L.; Bostock, R.M.; Kirkpatrick, B.C. Genetic characterization of *Pseudomonas syringae* pv. *syringae* strains from stone fruits in California. *Appl. Environ. Microb.* **1998**, *64*, 3818–3823.
9. Agrios, G. *Plant Pathology*, 3rd ed.; Elsevier Academic Press: San Diego, CA, USA, 1988; pp. 70–81.

MDPI

St. Alban-Anlage 66

4052 Basel

Switzerland

Tel. +41 61 683 77 34

Fax +41 61 302 89 18

www.mdpi.com

Eng. Proc. Editorial Office

E-mail: engproc@mdpi.com

www.mdpi.com/journal/engproc